Thinking Statistically

Elephants Go to School

Sarjinder Singh

St. Cloud State University

KENDALL/HUNT PUBLISHING COMPANY
4050 Westmark Drive   Dubuque, Iowa 52002

Cover image © 2006 JuipterImages Corporation

FOREWORD

Learning statistical concepts and tools has historically been one of the most difficult tasks encountered by many students during their college experience. The discipline is full of strange terms and procedures which must be mastered, yet which make little sense to the casual user and consumer of statistics. There has long been a need to redirect the approach to statistical education, to make it more friendly to those students who are not mathematically inclined. This book, entitled "Thinking Statistically: Elephants go to School," takes the discipline a giant step forward toward that goal.

The concepts presented here are done so in a light-hearted, fun-filled manner, designed to put the student at ease with the subject. Indeed, the material here is accessible to those in secondary schools, as well as university settings. The book's consistent use of animal images and pictures to represent populations and samples will warm the heart and broaden the imagination of the reader, to better enable the understanding of the wide applications of statistical methods while generating an occasional smile. One need not be well-versed in mathematical notation and theory to be able to read this book. One only needs a curiosity about the uses and tools of statistical methods.

I applaud Dr. Sarjinder Singh for his work in presenting this book to the reader. His warm, caring personality is reflected in its pages, and those beginning pursuit of statistical knowledge will surely benefit in many ways. Dear reader, enjoy what the book has to offer, and use it to enliven your imagination and brighten your day.

David H. Robinson, Professor and Chair
Department of Statistics
St. Cloud State University
St. Cloud, MN 56301

FOREWORD

Thinking Statistically: Elephants go to School. Have you ever followed your train of thought on observing the differences in behaviour in a crowd at a football match? Home team supporters (those in the red beanies) are loudest when their team is in possession; away team supporters (in the blue beanies) cheer only when their star is close to the action no matter who has the ball. This may or may not be true of the spectators taken as a whole that day. But it seems plausible – judging by the reaction of your bench mates. It is the sort of judgment made by commentators all the time. They use incidental evidence to come up with a judgement about the whole. The effort to exhaustively test that evidence against measures taken from every possible source is not justified. After all accounting for all individual data points is not the point. But how many commentators are aware that they are using some framework of probability to drive a conclusion whose main buttress in plausibility is its statistical nature? If I shifted my seat at half time – moving to the opposite side after noticing a friend in the crowd - and noticed that in fact my earlier observation no longer applied – if anything the blues seemed particular mute after their side lost possession, while the reds yelled anytime the ball was struck, I may have quietly dropped this minor insight from tomorrow's banter at work when talk got round to the post mortems. If I had had to make up a column (with a deadline of 40 minutes after full time) I may well have wanted to deploy some of this statistical thinking (perhaps turned now to tactics of the game itself); but given that my standing with readers depended on matching my inferences to their own multiplicitous observations, I may have sought some more systematic method for collecting and ordering the facts on which I would build a case, at least if I wished to retain my following among sport enthusiasts.

This book is about building frameworks for observing a population with variable characteristics. They are the core of the utility of statistics for other disciplines, as such they permeate much shared knowledge. It is directed not at those pursuing a statistical career – for whom an explicit mathematical development is unavoidable - but at those who will employ statistics in other ways. Nevertheless the mathematical apparatus of probability theory and the theory of sampling elegantly and economically set out how to correctly approach the statistical phenomena of everyday life without necessarily overwhelming practical purpose. Abstracting problems of observation to statistical frameworks is a step to an objectively grounded contestable decision system. It pays any student, mathematically inclined or not, to gain confidence in these tools over a wide variety of situations.

The statistics profession has in the past furnished underpinnings for paradigmatic shifts in social and physical science; it has given engineers vital assistance in handling error and risk; and decision makers generally capacity in dealing with uncertainty. For it to advance in the future it needs to be seen by new generations of students as grounded in the everyday, and in a clear exposition of basic, extendable, mathematical results. It is how to think through a problem that matters, not applying a formula.

The author gives a whimsical but carefully crafted guide to the core insights of modern statistics that should admirably contribute to the program set out by Wild and Pfannkuch in the International Statistical Review of 'developing a framework for thinking patterns involved in problem solving, strategies for problem solving, and the integration of statistical elements within problem solving' a tool for educators and students alike.

Stephen Horn, Statistician
Department of Family and Community Services
Box 7788, Canberra Mail Centre
ACT 2610, Australia

PREFACE

This book, entitled **Thinking Statistically: Elephants Go to School,** is for students who have little knowledge about mathematics. It is based on the author's imaginations and jokes. In short, statistics is a collection of tools used for collecting, analyzing, and interpreting data. Statisticians are like carpenters. As carpenters know how to use tools such as hammers, saws, and nails to make a chair, bed, and table etc., in the same way statisticians know how to analyze and present data using different kinds of tools such as charts, tables and formulae. You just need a little practice to know which technique can be used when, where, and how. It is my opinion that most teachers spend their lives on one or two campuses and have little knowledge about the needs of students across the world. Most of my colleagues have problems with their children and cannot understand their learning difficulties. I have always endeavored to listen to my colleagues' children's problems and teach them mathematics. My experience shows that although my colleagues are teachers, they get very easily upset while teaching their own children. This may also be true with most of the parents around the world. I listen to children because I do not have children and enjoy listening to their feelings. For example, the Venn diagram in **Figure 5.44** was corrected by one of my friend's eight-year old daughter, *Rosie*.

I have extensive teaching and research experience from seven different universities around the world in the USA, Canada, Australia, and India; including the Australian public service experience. My continuous contacts and teaching experience of about 15 years with different kinds of students led me to write a simple and easy-to-follow book in statistics. Thus, this textbook will be valuable to every student entering a university/college doing any basic statistics course. Your children or your friend's children will also like this book. It is quite common to find that most first year students having minimal background in mathematics are intimidated of statistics.

This textbook is based on class-notes that I used to teach Stat 193 entitled, *"Statistical Thinking"* at St. Cloud State University, MN. I made an effort to incorporate those notes into this book in an extended form. Learning a subject through fun has a different charm than learning from traditional books. In these days, class absenteeism is very common in schools. Thus, the last chapter has been designed to teach students the fact that missing class will have a negative effect on their GPA and that increasing study hours will have a positive effect on their GPA. Each chapter has been designed so that the readers do not require a great deal of knowledge in mathematics. In the table below, a list of various components are provided from where readers can make their judgments about the material in it.

Components	Pages		Examples	LUDIs	Pictures
Front cover	--	2	--	---	1
Copyrights	i - ii	2	--	--	1
Dedication	iii − iv	2	--	--	1
Forewords	v − viii	4	--	--	--
Preface	ix − xii	4	--	--	1
Contents	xiii − xx	8	--	--	--
1	1-84	84	06	31	85
2	85-110	26	07	13	28
3	111-172	62	17	34	78
4	173-276	104	28	30	110
5	277-362	86	25	60	109
6	363-404	42	12	25	44
7	405-450	46	11	21	62
8	451-488	38	13	15	32
9	489-586	98	16	30	121
10	587-636	50	10	21	48
Useful tables	637-642	6	--	--	1
Formulae list	643-646	4	--	--	--
Related books	647-648	2	--	--	1
Subject index	649-652	4			1
Back cover	--	2	--	--	2
Total		**676**	**145**	**280**	**726**

There are over **726** attractive pictures and graphs through **676** pages. At the end of the chapters, LUDIs are provided, which stand for "Let Us Do It". These **280** LUDIs will be useful for readers to practice and also for university, college, or school teachers to assign homework to students. In addition, there are over **145** solved numerical examples.

I would like to let you know **one truth**; that all the numerical figures (data values) and stories are imaginary. As you will read, you might feel the same way as you might while watching a movie in a theater. For example: dancing elephants, a cat following rats in a field, rats stealing lion's whiskers, bombers hitting a tank, etc. Similarly, you could make new stories and think of a situation which could happen.

Acknowledgements:

The use of the ART Explosion 600,000 has been duly acknowledged. F. Anne Zemek de Dominguez, Special Advisor to the President and Intellectual Property Officer, St. Cloud State University, help in reviewing a license agreement contained in the user manual for a clip art volume published by Nova Development entitled, "Art Explosion" has been duly acknowledged. Professor D.M. Titterington, Editor of Biometrika, and Chris Payne, Rights Executive, Oxford University Press, permissions to use the pictures of Professor Karl Pearson and Sir. R.A. Fisher have also been duly acknowledged. Professor Edmund Robertson from England, help to use some text from his web page in a couple of LUDIs has also been duly acknowledged.

The timely help from the Acquisitions Editor Edward W. Siemek, Managing Editor Ray Wood, Associate Managing Editor Jay W. Hays, Permission Editor Elizabeth Roberts, Contracts Administrator Costie Kourpias, and Project Coordinator Kimberly D. Terry, Kendall/Hunt Publishing Company has been duly acknowledged. The final acceptance letter, dated July 7, 2005, from the Senior Vice President Thomas W. Gantz, Kendall/Hunt Publishing Company has also been duly acknowledged. My sincere thanks are given to Prof. David Robinson and Mr. Stephen Horn for writing forewords explaining the need of such an elegant book. The help from Mary Shrode, Learning Resources and Technology Service, St Cloud State University in drawing a map of the USA used in **Figure 3.69** has been duly acknowledged. Special thanks are also due to a Stat 229 student Miss Lydia M Gindele, a Stat 229/332/447 student Mr. Oluseun A Odumade, a Stat 447 student Miss Thu A Le, a Stat 321/421 student Miss Cecilia A Chandra, and a Stat 193 student Mr. Russel Clark for critically checking the entire manuscript. The help from a professional English Editor Ms. Melissa Lindsey, Write Place Center, St. Cloud State University, for reading the entire manuscript twice has been duly acknowledged. Mr. Michael Scheltgen, Ph.D. scholar, University of Saskatchewan, help to correct the English of many LUDIs through e-mail is duly acknowledged.

I acknowledge to the galaxy of all friends and colleagues across the world and a few of them are listed as:

Dr. I. Grewal (India), Dr. B.R. Garg (India), Dr. S. Sidhu (India), Dr. M.L. Bansal (India), Dr. Tejwant Singh (India), Prof. L.N. Upadhyaya (India), Prof. H.P. Singh (India), Dr. P. Chandra (India), Dr. M.R. Verma (India), Dr. P.K. Mahajan (India), Dr. S.R. Puretas (Spain), Prof. M. Rueda (Spain), Prof. G. Zou (China), Prof. R. Arnab (Botswana), Dr. A. Joarder (Saudi Arabia), Er. Amarjot Singh (Australia), Mr. Qasim Shah (Australia), Mr. Stephen Horn (Australia), Prof. Maxwell L. King (Australia), Dr. Marcin Kozak (Poland), Dr. Munir Mahmood (Australia), Mr. Kuldeep Virdi (Canada), Mr. Kulwinder Channa (Canada), Prof. Balbinder Deo (Canada), Er. Mohan Jhajj (Canada), Mr. Gurbakhash Ubhi (Canada), Dr. N.S. Mangat (Canada), Prof. Patrick J. Farrell (Canada), Prof. M. Bickis (Canada), Dr. K. Khan (Canada), Prof. Sylvia R. Valdes (USA), Prof. Leonard Onyiah (USA), Dr. S.S. Osahan (USA), Dr. Jaswinder Singh (USA), Dr. Jiang Lu (USA), Mr. Gurmeet Ghatore (USA), Dr. Gurjit Sidhu (USA), Prof. Balwant Singh (USA), Prof. P. Ramalingam (USA), Prof. Jong-Min Kim (USA), Dr. Sanjay Gupta (USA), Mr. Rubi Chadda (USA), Miss Kok Yuin Ong (USA), Mr. Suman Kumar (USA), and Miss Cynthia Miller (USA). All friends cannot be listed, but none is forgotten. Thanks are also due to all the faculty and staff members in the department of statistics for their kind cooperation and help. Special thanks are given to my father Mr. Sardeep Singh Ubhi, my mother Mrs. Ranjit Kaur Ubhi for making this book possible, my brothers Jatinder, Kulwinder and late sister Sarjinder

In this book all opinions are the author's and do not reflect any institution. The names used Amy, Bob, etc. are generic and do not represent any real person or elephant in the world. Any suggestions for improvement via e-mail will be much appreciated.

Elephant is my friend

Sarjinder Singh, Ph.D.
Assistant Professor
Department of Statistics
St. Cloud State University
St. Cloud, MN 56301 U.S.A.
E-mail: sarjinder@yahoo.com

TABLE OF CONTENTS

1   BASIC CONCEPTS

2   STATISTICAL STUDIES

3   GRAPHICAL REPRESENTATION

4   NUMERICAL REPRESENTATION

5   TOUCHING PROBABILITY

6   DISCRETE DISTRIBUTIONS

7   CONTINUOUS DISTRIBUTIONS

8   SAMPLING DISTRIBUTIONS

9   THE IDEA OF HYPOTHESES TESTING

10   ANALYZING BIVARIATE DATA

USEFUL STATISTICAL TABLES

IMPORTANT FORMULAE

BIBLIOGRAPHY

HANDY SUBJECT INDEX

1. BASIC CONCEPTS

1.1 INTRODUCTION

In this chapter we introduce basic concepts such as the definition of statistics, the ideas of population, parameter, sample, statistic, the use of the random numbers table method and the lottery method. We also discuss different kinds of bias and the idea of sampling schemes: simple random sampling, stratified random sampling, systematic sampling, cluster sampling, and multi-stage sampling. The meaning of hypothesis will also be discussed.

1.2 WHAT IS STATISTICS?

It is hard to say as there are many definitions of statistics. Pick up any book and you will find a new definition of statistics. The reason may be that it is difficult to define a complete subject in two or three lines. Medical doctors, engineers, agricultural scientists, economists, space scientists, and business orientation specialists define statistics in their own way according to their needs or understanding. Statisticians define statistics with a more common trick so that it should suit every scientist in the world. Let us focus on one definition of statistics and we will end with another definition of statistics.

1.3 DEFINITION OF STATISTICS

Statisticians like to say that:

"Statistics is a science to describe or predict the behavior of a population based on a random and representative sample taken from the same population."

As we try to understand the definition of statistics, we can see that there are a few technical terms in the definition. Until we understand the meaning of the terms: population, sample, random, and representative, it is difficult to understand the above definition of statistics. Let us try to understand these terms.

1.3.1 POPULATION

We may define a population as follows:

"A population is a large body of measurements or observations or data values that are of interest to an investigator."

Let us now think of a few examples of a large body of measurements:
(a) All people living on earth
(b) All students at St. Cloud State University
(c) All employees at St. Cloud State University
(d) All machines in a factory
(e) All elephants in a zoo
(f) All rats in a house
(g) All trees in a forest

Let us ask, "What can be our interest in populations (a), (b), (c), (d), (e), (f) and (g)?"

In population (a), we may be interested in estimating the average body temperature of all people living on earth.

In population (b), we may be interested in estimating the average GPA of all students at St. Cloud State University.

In population (c), we may be interested in estimating the average salary of all employees at St. Cloud State University.

In population (d), we may be interested in estimating the average life of all machines in the factory.

In population (e), we may be interested in estimating the average diet of all elephants in the zoo.

In population (f), we may be interested in estimating the average wastage by all rats in the house.

In population (g), we may be interested in estimating the average wood per tree in the forest.

Note that in statistics, a population does not necessarily consist of only human beings. It is a collection of data values, which may come from human beings, machines, animals or trees etc.

To understand the meaning of population, let us consider an example. Suppose we have a two-liter carton of milk as shown in **Figure 1.1**, and we wish to make a cup of tea for a sick patient.

Fig. 1.1. Two liters of milk.

Note that we have only two liters of milk as shown in **Figure 1.1** and that this two liters of milk is our population. The doctor recommends that the patient should not drink milk if its fat content is more than 2%.

Fig. 1.2. The patient cannot drink milk with a fat content of more than 2%.

Now, let's say that we wish to know the fat content of the milk before making a cup of tea for the patient. To find the fat content of the milk, we have to add acid to it. Now consider the following situation of adding acid to the entire two liters of milk.

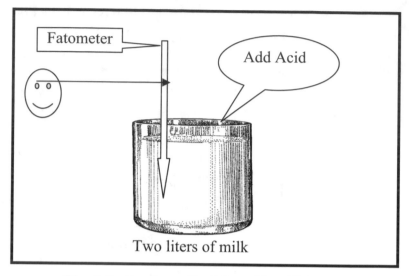

Fig. 1.3. Caution, all of the milk will be wasted!

Obviously, we can determine the fat content of the entire population of milk by adding acid to the milk as shown in **Figure 1.3**. However, the entire population will be destroyed after adding acid to it and we will be unable to make a cup of tea.

"What should we do?"

The wise step is to place one milliliter of milk in a test tube as shown in **Figure 1.4**, add acid to it, and note the fat content.

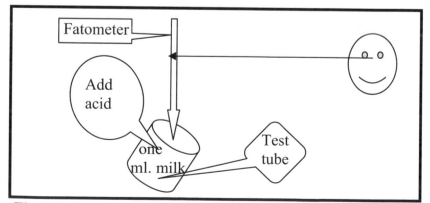

Fig. 1.4. A wise step is to only destroy one milliliter of milk.

Whatever fat content we read from the one milliliter of milk, we can assume is also the true fat content of the entire milk population. A natural question arises:

"What is one milliliter of milk?"

The obvious answer is that one milliliter of milk is called a sample. Thus we can define a sample as follows:

1.3.2 SAMPLE

A sample is a set of measurements selected from the population of interest. Thus, the obvious examples of samples of our interest from the populations (a) through (g) should be:

(a) Select a few healthy people from earth and note their body temperatures.
(b) Select a few students from the university and note their GPA.
(c) Select a few employees from the campus and note their incomes.
(d) Select a few machines from the factory and note their ages.
(e) Select a few elephants from the zoo and note their daily diets.
(f) Select a few rats from the house and watch their daily wastages.
(g) Select a few trees from the forest and weigh their wood.

Let us go back to our example of the two liters of milk. Suppose the milk has been in the pot for 5 to 6 hours and that all the fat content is saturated near the upper layer of the milk. If we took a sample of one milliliter of milk from the same place where the fat content is saturated, then naturally our sample will show a higher reading of fat. This means we are over-estimating the real fat content of the entire milk.

Fig. 1.5. Representative and random sample.

Similarly, if we took a sample of one milliliter of milk from the bottom of the pot, where the water content is saturated, then the sample will show less fat than the actual fat of the entire milk.

"What should we do now?"

It is better to stir the milk before taking the sample.

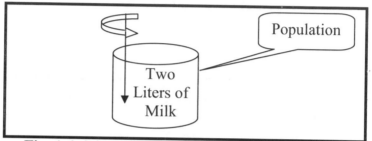

Fig. 1.6. Stir the milk before taking a sample.

Obviously, when we stir the milk before taking the sample, it will become a random and representative sample. Every drop or content of the entire milk will have a chance of selection in the sample.

1.3.2.1 RANDOM SAMPLE

A sample is said to be random if each element in the population has some chance of selection in a sample.

1.3.2.2 REPRESENTATIVE SAMPLE

A sample is said to be representative if it contains all kinds of elements or contents of the population under study.

However, a random sample may not guarantee a representative sample. For example, while estimating the average income of all persons living in a village, the sample must include rich and poor people. If we selected a random sample of only rich people then we would not be representing the village population.

In actual practice, we cannot stir a population. For example, we cannot stir all elephants in a zoo (they will be much too heavy!). There are two methods to ensure a random and representative sample: (a) Lottery Method (b) Random Numbers Table Method

1.4 POPULATION SIZE

We denote the population size by an uppercase letter N. For example, if there are five elephants in a zoo, then we say $N = 5$.

Fig. 1.7. Elephants in a zoo.

1.5 SAMPLE SIZE

We denote the sample size by a lowercase letter n. Say that the owner of the circus wishes to ship all the elephants from one zoo to another zoo. It is a cumbersome and expensive job to weigh all the elephants in the zoo, so the owner decides to weigh only two elephants. Then we say $n = 2$.

Which two elephants should the owner choose to weigh?

1.6 SIMPLE RANDOM SAMPLING

As mentioned earlier, there are two devices used to select a simple random sample (SRS) from a population. They are called:

(a) Lottery Method (b) Random Numbers Table Method.

We discuss each one of these methods as follows:

1.6.1 LOTTERY METHOD

It is clear from its name that when using this method, we have to make lottery draws. To do this, write the names of all the N units or elements in the population on identical chits (or pieces of paper) and put them in a box. Shake the box and take out n chits and record the units to be selected in the sample. There are two different ways to make lottery draws:

(a) With Replacement (WR) sampling.
(b) Without Replacement (WOR) sampling.

Let us discuss each of these methods with the help of a selection of two $(n = 2)$ elephants out of five $(N = 5)$ elephants from the zoo.

(a) With Replacement (WR) sampling: Consider the following box containing five $(N = 5)$ identical chits having the names of the elephants written on them.

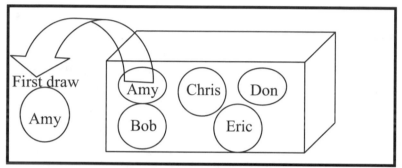

Fig. 1.8. First lottery draw.

Close your eyes, put your hand in the box and draw one chit. Assume on the first draw that **Amy** comes out. Note that we are conducting a with replacement (WR) sampling, which means we have to replace **Amy** back in the box before making the next draw.

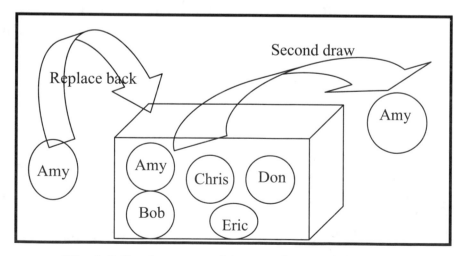

Fig. 1.9. Replacement of Amy and second draw.

Note that in the second draw, **Amy** may be selected again in the sample, but there is also a possibility of selecting Bob, Chris, Don or Eric. Further note that it is also not necessarily **Amy** who will be selected on the first draw, but anyone of these five elephants.

All possible with replacement (WR) samples are listed in the following table:

First draw	Second draw	First draw	Second draw
Amy	Amy	Chris	Don
Amy	Bob	Chris	Eric
Amy	Chris	Don	Amy
Amy	Don	Don	Bob
Amy	Eric	Don	Chris
Bob	Amy	Don	Don
Bob	Bob	Don	Eric
Bob	Chris	Eric	Amy
Bob	Don	Eric	Bob
Bob	Eric	Eric	Chris
Chris	Amy	Eric	Don
Chris	Bob	Eric	Eric
Chris	Chris		

Thus, there are total of 25 with replacement samples. Note that we have $N = 5$ and $n = 2$, so:

$$\text{Number of WR samples} = 25 = 5^2 = N^n$$

Therefore, we have the following formula: If a population consists of N units, then the possible number of Simple Random and With Replacement (SRSWR) Samples, each of n units, is given by:

$$N^n$$

To add in your memory, note that the little n is on the shoulders of big N, because a big guy can carry a little guy! Also note that with replacement sampling is not very useful, thus we will now consider without replacement sampling.

(b) Without replacement (WOR) sampling:

Consider the following box having five $(N = 5)$ identical chits with the names of the elephants written on them. Close your eyes, and put your hand in the box and draw one chit. Assume on the first draw that Amy comes out. Note that we are doing without replacement (WOR) sampling, which means we do not need to replace Amy back in the box before making the next draw. Thus, on the next draw, Amy cannot be selected if previously selected on the first draw.

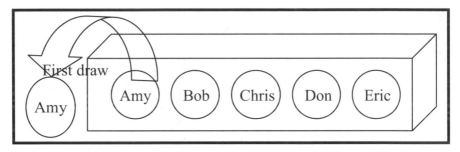

Fig. 1.10. Lottery method using WOR sampling.

Thus on the second draw any one of the remaining elephants Bob, Chris, Don, or Eric can be selected. Again, note that it is also not necessarily Amy who will be selected on the first draw. **Caution!** For a WOR sample if Bob comes on the first draw, then Amy is automatically out. Similarly, if Chris comes on the first draw, then both Amy and Bob are automatically out. Note that if we are taking a sample of two elephants, then the last elephant, Eric, cannot come out on the first draw. It is a rule, and we have to follow it.

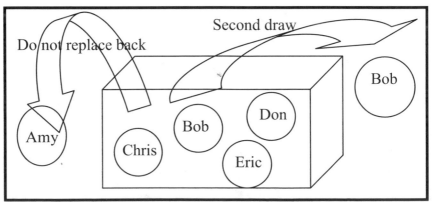

Fig. 1.11. Making a second draw.

Thus, WOR sampling seems more complicated than WR sampling when dealing with chits in a box. To construct WOR samples (sometimes called combinations), arrange the population units in a sequence and follow the forward arrows to make all possible combinations as shown in **Figure 1.12**.

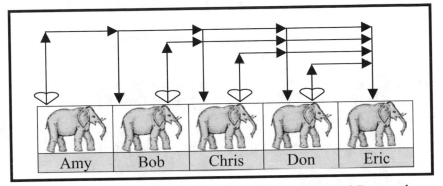

Fig. 1.12. Simple method to select all possible WOR samples.

Thus, following the directions of the forward arrows, the possible combinations (or WOR samples) are as listed in the following table:

First draw	Second draw
Amy	Bob
Amy	Chris
Amy	Don
Amy	Eric
Bob	Chris
Bob	Don
Bob	Eric
Chris	Don
Chris	Eric
Don	Eric

There are a total of 10 without replacement samples. Note that we have $N = 5$ and $n = 2$, so:

Number of WOR samples $= 10 = \begin{pmatrix} 5 \\ 2 \end{pmatrix} =$ Choose 2 elephants out of 5 elephants.

Thus we have the following formula: If a population consists of N units, then the total number of Simple Random and Without Replacement (SRSWOR) Samples, each of n units, is given by:

$$\binom{N}{n}$$

Note that we are choosing lowercase n elephants out of uppercase N elephants.

How can we understand the meaning of $\binom{N}{n}$?

Commit to memory (C.T.M.) the following procedure:

$$\binom{N}{n} = \frac{N!}{n!\,(N-n)!}$$

where ! denotes factorial, and its meaning is:

$0! = 1$ (Proof is difficult, it is a rule, so C.T.M.)
$1! = 1$
$2! = 2 \times 1$
$3! = 3 \times 2 \times 1$
.
.
$10! = 10 \times 9 \times 8 \times 7 \times 6 \times 5 \times 4 \times 3 \times 2 \times 1$
and, so on.

Now in the example of a selection of two ($n = 2$) elephants out of five ($N = 5$) elephants, we have:

$$\binom{N}{n} = \binom{5}{2} = \frac{N!}{n!\,(N-n)!} = \frac{5!}{2!\,(5-2)!} = \frac{5!}{2!\,3!} = \frac{5 \times 4 \times 3 \times 2 \times 1}{2 \times 1 \times 3 \times 2 \times 1} = 10$$

In actual practice, we have only one sample. Moreover, when a population size becomes large, say $N = 10,000$, then it is difficult to make so many identical chits and put them in a box. In this case, to draw one random and representative sample, it is better to use a Random Numbers Table.

1.6.2 RANDOM NUMBERS TABLE METHOD

In this table, the random digits from 0 to 9 are written both in columns and rows. For the purpose of illustration, we used Pseudo-Random-Numbers as given in **Table I** (This table has also been provided in the Appendix).

Table I. Pseudo Random Numbers.

Random Numbers Table

Column Number

Row	1	2	3	4	5	6	7	8	9	10	11	12	13	14	15	16	17	18	19	20
1	3	7	5	6	9	6	1	6	0	2	0	1	0	3	3	0	4	3	0	6
2	1	9	8	8	7	2	2	7	9	6	3	6	2	0	5	8	4	4	6	9
3	5	2	8	4	0	8	0	1	8	9	6	8	4	5	0	6	0	0	5	0
4	9	9	6	2	6	7	5	8	3	3	0	6	0	2	3	1	3	2	6	2
5	3	1	0	2	0	4	7	3	1	9	5	6	8	7	8	0	0	6	6	3
6	3	6	7	0	0	4	0	6	1	0	6	2	7	7	0	1	4	0	3	4
7	1	4	8	3	9	3	1	4	3	9	8	1	9	9	7	1	0	4	1	1
8	0	8	9	4	3	9	2	0	5	0	8	3	4	1	1	0	3	9	3	0
9	0	5	0	4	0	6	2	0	6	4	2	4	7	6	7	5	6	5	4	5
10	0	0	6	5	8	9	6	1	8	3	3	8	9	7	8	4	0	4	3	8
11	0	8	5	3	5	3	8	8	3	2	7	7	7	3	1	2	0	8	9	9
12	0	1	0	1	2	0	7	9	7	2	4	4	9	3	2	1	9	4	3	8
13	0	5	5	7	9	4	3	0	7	0	1	0	7	0	2	2	3	9	5	5
14	0	0	9	3	5	6	4	5	8	7	5	3	9	8	2	9	0	5	2	0
15	0	8	4	3	1	4	2	8	6	2	6	5	4	9	3	8	3	6	8	8
16	9	1	6	5	5	2	7	1	8	5	9	6	6	6	3	9	7	3	2	0
17	6	9	6	7	8	8	0	6	7	4	7	2	0	6	9	2	4	5	3	5
18	1	9	6	6	6	7	6	7	2	5	4	1	0	7	7	0	2	9	7	7
19	6	0	4	7	1	4	0	3	8	4	6	1	7	6	3	9	1	5	4	0
20	5	1	2	4	3	5	8	1	8	9	3	4	9	1	1	3	0	6	1	8
21	0	1	0	4	6	0	8	9	7	9	7	2	9	1	4	1	3	0	5	3
22	6	3	0	5	3	1	9	4	0	2	1	9	4	3	3	0	4	7	6	3
23	9	1	6	5	3	2	4	0	5	0	6	5	6	6	4	9	6	6	1	3
24	5	8	7	4	0	4	5	6	3	8	9	5	6	6	3	7	3	3	1	8
25	2	0	4	4	1	0	5	0	3	1	2	0	4	4	8	1	1	8	6	3

Source: Generated in Excel using Randbetween (0, 9).

We generally apply the following rules to select a sample:

Rule I. List all the population units (**Caution:** Do not order them). Assign them digits $1, 2, 3, ..., N$. For example, if $N = 5$, then

Units	 Amy	 Bob	 Chris	 Don	 Eric
Digits	1	2	3	4	5

Fig. 1.13. Assigning digits.

Rule II. Count the number of digits in the population size N. For example, if $N = 5$, the number of digits = 1; if $N = 25$, the number of digits = 2; and if $N = 200$, the number of digits = 3.

Rule III. Select any **random starting point, called the seed value,** from the Random Numbers Table. Then select the same number of columns (or rows) as the number of digits in the population size N.

Rule IV. From the **seed value**, move in any direction (up or down) along the columns (or rows) selected. Write all the numbers between 1 and N (both inclusive) that follow the starting point until you get n numbers. If we are using SRSWOR sampling then discard all the repeated numbers. If we are using SRSWR sampling, then retain the repeated numbers.

Rule V. Select those units that are assigned the digits in Rule I. This will constitute a required simple random sample (SRS).

Consider the problem of selecting two elephants ($n = 2$) out of all the ($N = 5$) elephants. Here the population size of five elephants consists of only one digit, so we will use only one column (or one row) from the Random Numbers Table. Suppose we are using the first row and the first column as the **random start number**. The first random number is 3, thus the third elephant, Chris, will be selected.

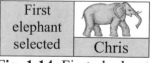

Fig. 1.14. First elephant.

Now, if we move downwards on the first column, then the next random number is 1, which means the first elephant Amy will be selected in the sample.

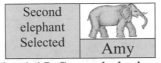

Fig. 1.15. Second elephant.

Thus the ultimate SRS of two elephants is given by:

Fig. 1.16. Both elephants.

Note that the chance of selection for one unit in SRSWR or SRSWOR sampling is equal to the inverse of the population size. So the chance of selection for any elephant in the sample is:

$$\text{Chance of selection for one elephant} = \frac{1}{N} = \frac{1}{5}$$

Why we do sampling?

Here is a story that I heard from one of my teachers. During the 19th century, a German doctor took a sample of one million healthy people and found their average body temperature was 98.6°F. Now, when we go to see a doctor, the doctor always checks our temperature and compares it with the standardized value of 98.6°F. Note that this standard result is based on only one million people selected around the world during the 19th century and is still considered valid.

What questions come to mind?

(a) How did the German doctor select one million healthy people?
(b) What is the accuracy of the estimate 98.6°F?

To get the answers to these questions, we have to read this entire book with a lot of patience. But, now we have another definition of statistics:

"Statistics is a branch of mathematics that has an application in almost every facet of our daily life."

What are the benefits of sampling?

Factors	
Cost	Less
Efforts	Less
Time consumed	Less
Accuracy of measurements	More

Note that if we are talking about a given data set, it may be a population or a sample unless it is defined or stated.

1.7 PARAMETER

Any numerical value obtained from all the units in a population is called a **parameter**. A parameter is an unknown and fixed quantity.

How can we remember it?

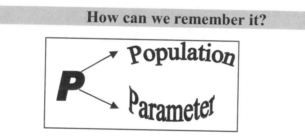

1.8 STATISTIC

Any numerical value obtained from all the units in a sample is called a statistic. Generally, a statistic is a known quantity from a given sample, and varies from sample to sample. It is used to estimate an unknown parameter.

How can we remember it?

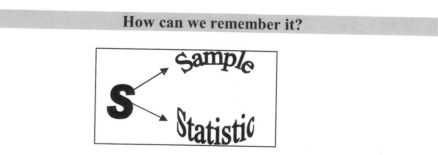

Elephants at the Hospital

Example 1.1. (BODY TEMPERATURE) Consider that all five elephants went to see a doctor, and the doctor noted their temperatures and charged each elephant $50. Their temperatures were recorded as follows:

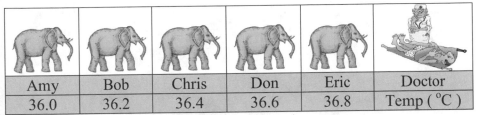

Amy	Bob	Chris	Don	Eric	Doctor
36.0	36.2	36.4	36.6	36.8	Temp ($^\circ$C)

Fig. 1.17. All elephants in a zoo.

(a) Find the average temperature.
(b) What is the unit of measurement for the average temperature?
(c) Is it a parameter or a statistic?
(d) How much does the owner of the zoo have to pay to the doctor?
Solution. (a) The average body temperature of all the elephants will be given by the sum of temperatures of all the elephants divided by the total number of elephants.
Thus:

$$\text{Average body temperature of all elephants} = \frac{\text{Amy} + \text{Bob} + \text{Chris} + \text{Don} + \text{Eric}}{5}$$

$$= \frac{36.0 + 36.2 + 36.4 + 36.6 + 36.8}{5}$$

$$= \frac{182}{5} = 36.4^\circ\text{C}.$$

(b) The unit of measurement for the average body temperature is again $^\circ$C, as for the original data.
(c) Note that it is the average temperature of all the elephants in the **population** of five elephants, so it is a **parameter.**
(d) The owner of the zoo has to pay:

$$\$50 + \$50 + \$50 + \$50 + \$50 = \$250$$

It is a very expensive job to send all of the elephants to the hospital to see the doctor, so the owner decided to do a Zoo Health Survey from time to time so that there would be some record of the health of all the animals in the zoo. The elephants also decided to participate in the Zoo Heath Survey organized by the owner. An example of such a Zoo Health Survey is as follows:

Zoo Health Survey

Example 1.2. (SAMPLED ELEPHANTS) Consider the following figure showing a sample of two elephants along with their body temperature reported by a local zoo-survey officer.

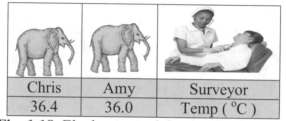

Chris	Amy	Surveyor
36.4	36.0	Temp ($^{\circ}$C)

Fig. 1.18. Elephants participating in a survey.

(a) Find the average temperature.
(b) What is the unit of measurement for the average temperature?
(c) Is it parameter or a statistic?
(d) What is the cost of the survey if the doctor charges $50 per patient?

Solution. (a) The average body temperature of the two elephants selected by the survey officer will be equal to the sum of the temperatures of the elephants in the sample divided by the total number of elephants in the sample. Thus:

$$
\text{Average body temperature of two elephants} = \frac{\text{Chris} + \text{Amy}}{2}
$$

$$
= \frac{36.4 + 36.0}{2}
$$

$$
= \frac{72.4}{2} = 36.2^{\circ}\text{C}.
$$

(b) The unit of measurement for the average body temperature will again be $^{\circ}$C, as in the original data.
(c) Note that it is the average temperature of only two elephants in a **sample** out of the population of five elephants, so it is a **statistic**.

(d) Because only two elephants are selected in the sample, the total cost of the survey is:

$50 + $50 = $100.

Why only SRS?

It is not necessary to apply SRS sampling. We can apply any probability sampling that can reduce the bias from the sample.

1.9 BIAS AND ITS SOURCES

A sampling method is biased if it gives results that differ in a systematic way from the true parameter of the population. For example, think of our example of one milliliter of milk taken from two liters of milk with more or less fat content. In general, the following are sources of bias.

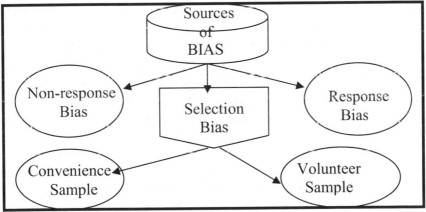

Fig. 1.19. Sources of bias.

1.9.1 SELECTION BIAS

In general, there are two sampling methods that lead to selection bias.

1.9.1.1 CONVENIENCE SAMPLE

A convenience sample is a sample consisting of units of the population that are easily accessible.

Consider a population under study that consists of both Asian elephants and African elephants as shown below:

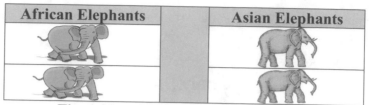

Fig. 1.20. African and Asian elephants.

Note that African elephants have bigger ears than Asian elephants. If it is more convenient to collect information from African elephants and more African elephants are deliberately included in our sample, then it is called a convenient sample. So our sample will give results in favor of African elephants.

1.9.1.2 VOLUNTEER SAMPLE

A volunteer sample is a sample consisting of the units of a population that chose to respond. For example, if we know that Asian elephants are more willing to give us a free ride, then we can chose more Asian elephants for a long tour trip. In estimating the average distance traveled per day by all elephants in the world, our estimate will favor the walking tendency of Asian elephants.

Fig. 1.21. Ride with elephants.

1.9.2 OTHER SOURCES OF BIAS

Selection bias can be reduced by any of the probability sampling methods, but the following two sources of bias are more serious.

1.9.2.1 NON-RESPONSE

Non-response bias can arise because a large number of units selected for the sample may not respond or refuse to respond. If we select 10 elephants to participate in a 10 km ride and we want to estimate the average speed of the elephants, but a few elephants refused to participate in the ride, it would create a bias. Such a bias is called non-response bias.

1.9.2.2 FALSE RESPONSE

False response can arise because of the behavior of the interviewer or the nature of the question. For example, if we ask the elephants:

"Are you happy with the behavior of the owner of the zoo?"

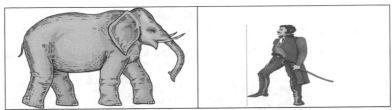

Fig. 1.22. Rude owner.

Some of the elephants may not like to disclose their answer about such a personal question. On the other hand, if you ask the elephants:

"Are you interested in playing with a football?"

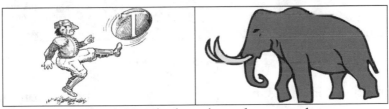

Fig. 1.23. Elephant in a playground.

Most of the elephants may be interested in responding to your second question. Most of them may respond "yes" and others may say "no" due to busy schedules, health problems, etc.

1.10 PROBABILITY SAMPLING

Probability sampling is any sampling method that gives **some** chance (not necessarily an equal chance) of selection in the sample to each unit from the population. Note that both SRSWR and SRSWOR sampling techniques are probability sampling. In addition to simple random sampling, the following are also probability samplings: stratified sampling, systematic sampling, cluster sampling, and multi-stage sampling.

1.10.1 STRATIFIED SAMPLING

In stratified random sampling, we first divide the population into mutually exclusive (disjoint) and homogeneous groups called strata. From each stratum, we take an independent random sample of the required size.

Consider the problem of estimating the average diet of all the animals in a zoo. Consider a small local zoo consisting of three types of animals: 5 Elephants, 8 Rats, and 7 Lions, as shown in **Figure 1.24**.

Fig. 1.24. Small zoo.

For stratified random sampling, first divide the population of 20 animals into three homogeneous groups: one having only elephants, the second having only rats, and the third having only lions as shown in **Figure 1.25**.

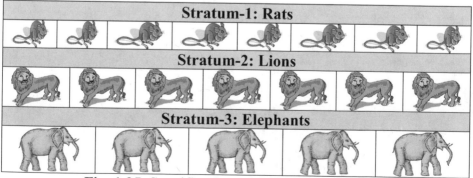

Fig. 1.25. Stratification: Homogeneous groups.

Now select a minimum of one animal from each stratum, that is, select at least one rat, one elephant and one lion from each one of the above three independent groups or strata using either the lottery method or the random numbers method. Such a scheme is called stratified random sampling.

1.10.2 SYSTEMATIC SAMPLING

Let $k = N/n$ be an integer or entire number. Arrange the units of the population into some sequence and number them from 1 to N. Select one random number between 1 and k, called a random start, and then select every k^{th} unit from the population arranged in some order.

Hungry Elephant

Elephant Bob is very hungry and wants to eat a minimum of two out of six trees grown along the fence of the zoo.

| Tree-1 | Tree-2 | Tree-3 | Tree-4 | Tree-5 | Tree-6 |

Fig. 1.26. Decorated fence of zoo.

It is true that breaking these trees destroys the beauty of the zoo, so the zoo statistician suggests using systematic sampling to select two out of the six trees.

| Zoo statistician | Hungry Bob waiting for trees |

Fig. 1.27. Use of systematic sampling.

Here $N = 6$ and $n = 2$, so $k = \dfrac{N}{n} = \dfrac{6}{2} = 3$ (which consists of only one digit). So the zoo statistician used the first column of the Random Numbers **Table I** given in the Appendix to pick a random start between 1 and $k = 3$, which is 3. Thus, the zoo statistician suggested picking the 3rd and $3+3 = 6^{th}$ tree for hungry Bob to eat. The systematic sample will consist of the two trees as follows:

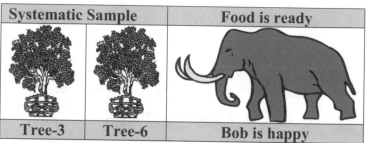

Systematic Sample		Food is ready
Tree-3	Tree-6	Bob is happy

Fig. 1.28. Applying systematic sampling.

1.10.3 CLUSTER SAMPLING

In cluster sampling, the population is first divided into heterogeneous groups, called clusters. Then a few clusters are randomly selected and information is collected from all units in the selected clusters.

If we select one cluster, then irrespective of the cluster size, the chance of selecting a particular population unit in the sample is:

$$= \frac{1}{\text{Number of Clusters}}.$$

If we select two clusters, then irrespective of the cluster size, the chance of selecting a particular population unit in the sample is:

$$= \frac{2}{\text{Number of Clusters}}.$$

In general, the chance of selecting a particular population unit in a cluster sampling is:

$$= \frac{\text{Number of clusters selected}}{\text{Total number of clusters}}.$$

Monkey Apartments

Consider the following situation of a building where a lot of monkeys are living in different apartments of one, two, and three bedrooms on the first, second and third floors as shown below:

Fig. 1.29. Situation of applying cluster sampling.

A circus owner wishes to recruit a few monkeys to play a role of highest jumper in the circus by selecting all the monkeys residing on two floors of the Monkey-Apartments.

What is the chance of selecting the dancing and singing Monkey?

Fig. 1.30. Chance in cluster sampling.

We can easily consider the first, second, and third floors (or rows) as heterogeneous groups, hence clusters, because each floor has three kinds of apartments. [Note that in contrast, if we consider vertically across all the three floors, then each column will be a good example of a homogeneous group or stratum.]

$$\text{Chance of dancing and singing Monkey} = \frac{\text{No. of clusters selected}}{\text{Total number of clusters}} = \frac{2}{3}$$

1.10.4 MULTI-STAGE SAMPLING

In cluster sampling, if a few units are selected from the selected clusters, such a scheme is called two-stage sampling.

For example, on the first-stage select a few zoos (say clusters) from the list of all zoos in a state (or country). In the second stage, select a few monkey-apartments from the selected zoos to get information about the recent activities of the monkeys. Such a scheme is called a two-stage sampling scheme. If at the third stage, a few monkeys are selected from the selected monkey-apartments at the second stage, then it is called three-stage sampling. Continuing this process is called multi-stage sampling.

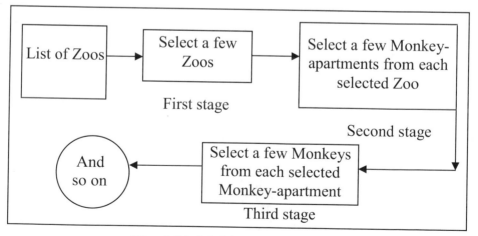

Fig. 1.31. Multi-stage sampling.

Elephants at School

Example 1.3. (TRAINING IS IMPORTANT) Consider a training class consisting of 16 elephants. Their names, scores and major (D-Dancing, R-Riding) are given in the following table:

Class	Elephants	Names	Scores	Major
		Ruth	92	D
		Ryan	97	R
		Tim	68	D
		Raul	62	R
		Marla	97	D
Instructor: Owl		Erin	68	R
		Judy	76	D
		Troy	75	D
		Tara	51	R
		Lisa	94	R
		John	70	R
		Cher	89	D
		Lona	62	R
		Gina	63	R
		Jeff	48	R
		Sara	97	R

Fig. 1.32. All elephants sitting in one queue.

(a) Select an SRSWOR sample of 4 elephants.
(b) Select a sample of 4 elephants with two majors in dancing and riding each.
(c) Select a systematic sample of $n = 4$ elephants after writing their names in alphabetic order.
(d) Consider elephants sitting in the class as shown below.

Class	Row			
	I	Cher		
	II	Erin		
	III	Gina		
	IV	Jeff	John	Judy
	V	Lisa	Lona	
	VI	Marla		
	VII	Raul	Ruth	Ryan
	VIII	Sara		
	IX	Tara	Tim	Troy

Fig. 1.33. Elephants in rows listed alphabetically.

(i) Select two rows of elephants by simple random and without replacement sampling and name the sampling scheme.

(ii) Select one elephant from each one of the two selected rows, and name the sampling scheme.

Solution. (a) Simple Random Sampling:

Here the population size is $N = 16$, so assign labels from 1 to 16 to the elephants as shown:

Fig. 1.34. Seats arranged by the instructor.

Because $N = 16$ has two digits, we have to use any two consecutive columns (or rows) of the Random Numbers **Table I** given in the Appendix.

Let us start from the Random Numbers **Table I** with the 1st row and the 1st column. The first digit selected = 3 and to make two-digits we need to take a digit from the next consecutive column which is 7. Since we are using the 1st and 2nd columns starting from the 1st row, so the number is = 37 (we also call it the **seed value** or starting point).

The first selected random number = 37. It is more than 16, so reject it.

Now move downward in the same chosen two columns, and find the next random number = 19 (which is again more than 16, so reject it).

The next random number = 52 (more than 16, so reject it).

The next random number = 99 (more than 16, so reject).

The next random number = 31 (more than 16, so reject).

The next random number = 36 (more than 16, so reject it).

The next random number = 14 (which is between 01 and 16, so the 14th elephant GINA is selected).

The next random number = 08 (which is between 01 and 16, so the 08th elephant TROY is selected).

The next random number = 05 (which is between 01 and 16, so the 05th elephant MARLA is selected).

The next random number = 00 (which is **not** between 01 and 16, so reject it).

The next random number = 08 (which is between 01 and 16, but 08 already came, so discard it, because we are doing WOR sampling).

The next random number = 01 (which is between 01 and 16, so the 01st elephant RUTH is selected).

Thus, an SRSWOR sample of 4 elephants is:

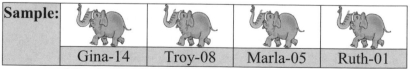

Sample:				
	Gina-14	Troy-08	Marla-05	Ruth-01

Fig. 1.35. Elephants selected in the sample.

(b) Stratified Random Sampling:

To select two elephants with a major in riding and two elephants with a major in dancing, let us first stratify the population into two strata as below:

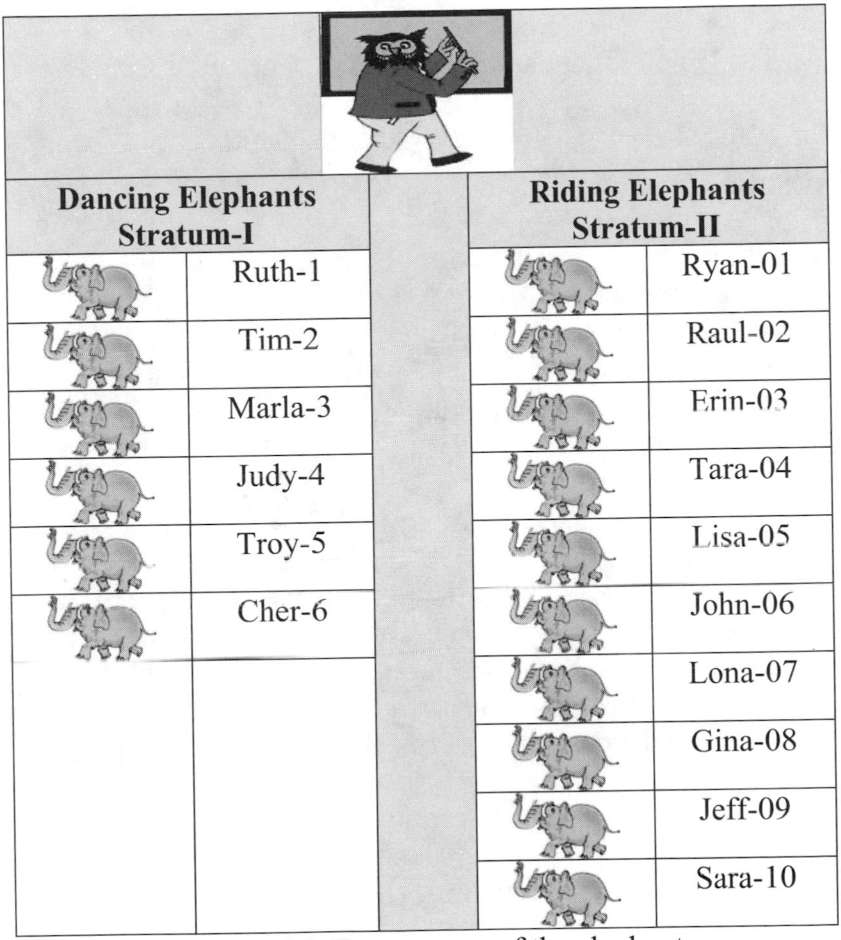

Dancing Elephants Stratum-I		Riding Elephants Stratum-II	
	Ruth-1		Ryan-01
	Tim-2		Raul-02
	Marla-3		Erin-03
	Judy-4		Tara-04
	Troy-5		Lisa-05
	Cher-6		John-06
			Lona-07
			Gina-08
			Jeff-09
			Sara-10

Fig. 1.36. Two groups of the elephants.

In the first stratum of dancing elephants, there are a total of $N_1 = 6$ elephants. We want to select two elephants $(n = 2)$ from the first stratum of dancing elephants. The first stratum population size, $N_1 = 6$, consists of only one digit, so we shall use only one column of the Random Numbers **Table I**.

Let us use the 2nd row and the 16th column, which gives us the first random number = 8 (more than 6, so reject it). Now moving downwards the next random number = 6 (which is between 1 and 6, so select the 6th elephant from stratum-I, so **Cher** is selected). The next random number = 1 (which is between 1 and 6, so select the 1st elephant from stratum-I, so **Ruth** is selected).

So the first sample from the first stratum = {Cher, Ruth}.

Note that the chance of selection for any elephant, say Troy, from the first stratum is:

$$= \frac{1}{N_1} = \frac{1}{6}.$$

In the second stratum of riding elephants, there are $N_2 = 10$ elephants, and we want to select two elephants $(n = 2)$ from the second stratum of riding elephants. The second stratum population size consists of two digits, so we shall use two columns of the Random Numbers **Table I**.

Let us start with the 1st row and the 11th and 12th columns. The first two-digit random number = 01 (which is between 01 and 10, so select the 01st elephant from stratum-II, that is **Ryan** is selected). Moving downwards, the next random number = 36 (more than 10, so reject it). The next random number = 68 (more than 10, so reject it), and the next random number = 06 (which is between 01 and 10, so select the 06th elephant from stratum-II, that is **John** is selected)

So the second sample from the second stratum = {Ryan, John}.

Note that the chance of selection for any elephant, say Gina, from the second stratum is:

$$= \frac{1}{N_2} = \frac{1}{10}.$$

Responses of the two dancing elephants selected = {Cher, Ruth}
= {89, 92}.

Responses of the two riding elephants selected = {Ryan, John}
= {97, 70}.

Stratum-I of dancing elephants:

$$\text{Estimated average} = \frac{\text{sum}}{n} = \frac{89+92}{2} = \frac{181}{2} = 90.5 \,(\text{statistic})$$

Stratum-II of riding elephants:

$$\text{Estimated average} = \frac{\text{sum}}{n} = \frac{97+70}{2} = \frac{167}{2} = 83.5 \,(\text{statistic})$$

Overall sample average from both strata:

$$\frac{\left(\text{\# of population units in Stratum - I}\right)}{N}\left(\begin{array}{l}\text{Stratum - I} \\ \text{estimated average}\end{array}\right)$$

$$+\frac{\left(\text{\# of population units in Stratum - II}\right)}{N}\left(\begin{array}{l}\text{Stratum - II} \\ \text{estimated average}\end{array}\right)$$

$$= \frac{6}{16}\times 90.5 + \frac{10}{16}\times 83.5 = 33.94 + 52.19 = 86.13 \,(\text{statistic})$$

(c) Systematic Sampling:

Let us arrange the elephants alphabetically as given below:

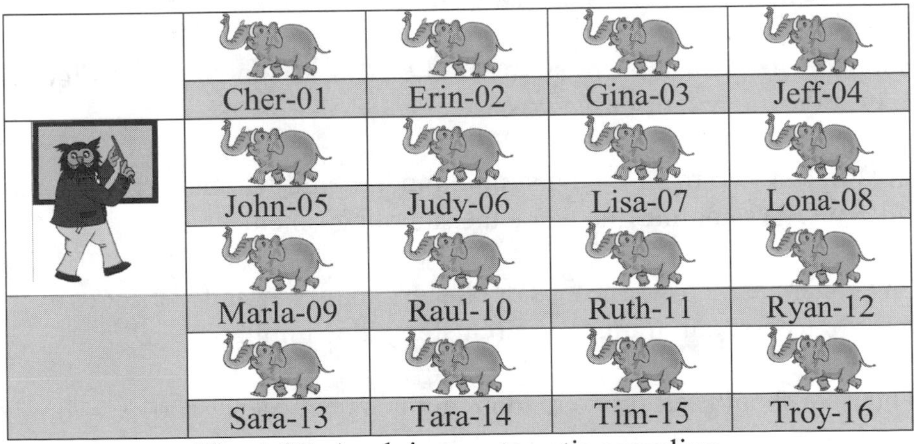

	Cher-01	Erin-02	Gina-03	Jeff-04
	John-05	Judy-06	Lisa-07	Lona-08
	Marla-09	Raul-10	Ruth-11	Ryan-12
	Sara-13	Tara-14	Tim-15	Troy-16

Fig. 1.37. Applying systematic sampling.

Here $N = 16$ and $n = 4$, so:

$$k = \frac{N}{n} = \frac{16}{4} = 4 \text{ (integer or whole number)}.$$

Select one random number between 1 and $k = 4$ and after selecting a random start select every k-th = 4th unit from the ordered population. Suppose we started from the 3rd row and 2nd column of the Random Numbers **Table I** (one digit number), the first random number = 2 (which is between 1 and $k = 4$, so select it).

Thus, the elephants selected in the sample are:

2nd	 Erin
$2 + 4 = 6^{th}$	 Judy
$6 + 4 = 10^{th}$	 Raul
$10 + 4 = 14^{th}$	 Tara

Fig. 1.38. Elephants in the systematic sample.

(d) Cluster and multi-stage sampling:

Suppose the elephants are sitting in the class as shown in the **Figure 1.39**. Here we will discuss two cases:

(i) Select two rows by the simple random and without replacement (SRSWOR) sampling and name the sampling scheme.

Since there are 9 rows, groups or clusters and we wish to select two rows (clusters) out of nine rows (clusters), this implies $N = 9$.

Thus, we should select two random numbers between 1 and $N = 9$.

Here the population size of 9 clusters consists of only one digit, so we have to select only one column from the Random Numbers Table 1 given in the Appendix.

Class	Row			
	I	Cher		
	II	Erin		
	III	Gina		
	IV	Jeff	John	Judy
	V	Lisa	Lona	
	VI	Marla		
	VII	Raul	Ruth	Ryan
	VIII	Sara		
	IX	Tara	Tim	Troy

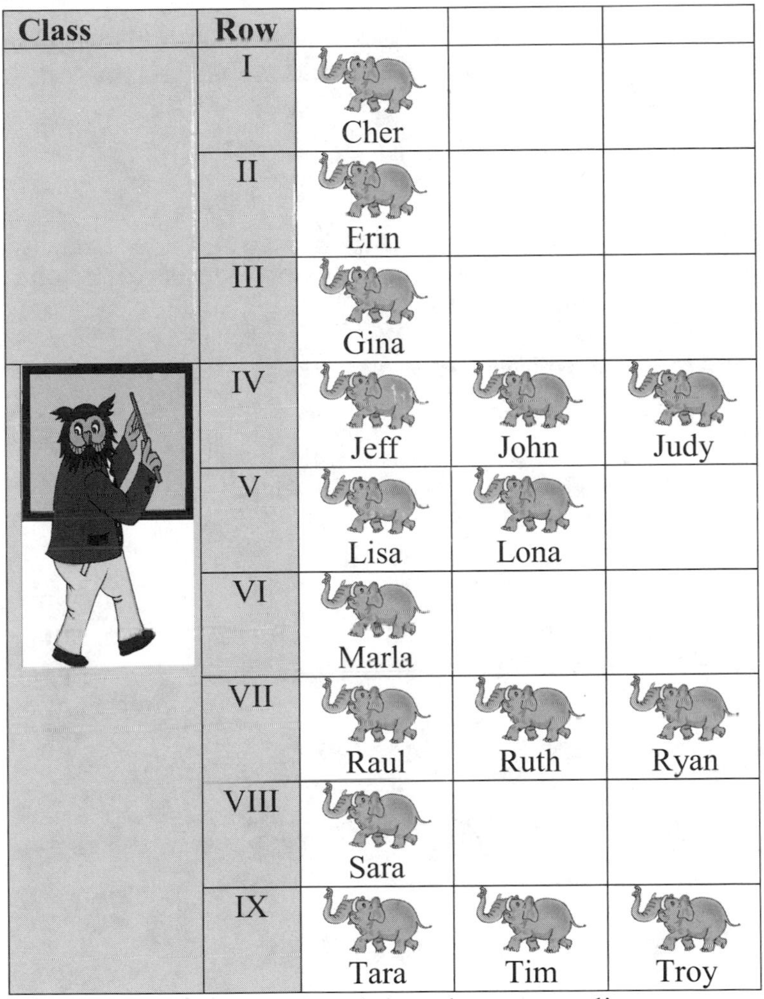

Fig. 1.39. Applying cluster sampling.

Assume that we started with the 11th row and 12th column, then the first random number = 7 (which is between 1 and 9, so the 7th row, group or cluster is selected). Moving downwards, the next random number = 4 (which is again between 1 and 9, so the 4th row, group or cluster is also selected).

Thus, the final sample of two clusters is:

Row	Cluster sampling		
VII			
	Raul	Ruth	Ryan
IV			
	Jeff	John	Judy

Fig. 1.40. Two clusters are selected.

Such a sampling scheme is called **cluster sampling**. The chance of selection for any elephant, say Jeff is:

$$= \frac{\text{\# of clusters selected}}{\text{Total \# of clusters}} = \frac{2}{9}.$$

(ii) Select one elephant from each of the two selected rows, and name the sampling scheme.

Assume we selected **Raul** from Row-VII and **John** from Row-IV at the second stage by using the Lottery Method. Such a sampling is called **two-stage sampling** or multi-stage sampling.

Two stage sampling	

Raul	**John**

Fig. 1.41. Applying two-stage sampling.

1.11 PROPORTION

Proportion is defined as the number of units that possess an attribute, say A, divided by the total number of units in the population, N. The population proportion is denoted by P. Note that the value of P is an unknown quantity and its numerical value is called a parameter. Mathematically, the population proportion is given by:

$$P = \frac{\text{COUNT}}{N}$$

where " COUNT " is the number of units possessing attribute A in the population of size N. The numerical value of the population proportion (P) always lies between 0 and 1.

The sample proportion is given by:

Read as "p-hat" ⟶ $\hat{p} = \dfrac{\text{count}}{\text{n}}$

where " count " is the number of units possessing attribute A in the sample of size n. The numerical value of $\hat{p}$ obtained from a sample is known and is called a statistic, but it varies from sample to sample. The value of sample proportion $\hat{p}$ also lies between 0 and 1.

Example 1.4. (LIVE IN THE PRESENT, THE PAST IS GONE, AND THE FUTURE IS UNKNOWN) In the above example:
(a) Find the proportion of dancing elephants in the population.
(b) Find the proportion of dancing elephants based on an SRSWOR sample of 4 elephants.
Solution. (a) Count all the elephants in the population, $N = 16$.

Count the elephants with a major in dancing, COUNT = 6.

Then the population proportion of dancing elephants is given by:

$$P = \frac{\text{COUNT}}{N} = \frac{6}{16} \text{ (Parameter)}$$

(b) Here an SRSWOR sample of 4 elephants is given by:

Elephant				
Sample	Gina-14	Troy-08	Marla-05	Ruth-01
Major	R	D	D	D

Fig. 1.42. Elephants in the SRSWOR sample.

Count all the elephants in the sample, $n = 4$.

Count the elephants with a major in dancing in the sample:

count = 3 .

Then the sample proportion of dancing elephants is given by:

$$\hat{p} = \frac{count}{n} = \frac{3}{4} \text{ (statistic)}$$

Note that we read $\hat{p}$ as p-hat.

1.12 ESTIMATE OF TOTAL IN STRATIFIED RANDOM SAMPLING

(COLLECT YOUR RECEIPT AT AN ATM) At a shopping mall consider there are three ATMs having the following denominations:

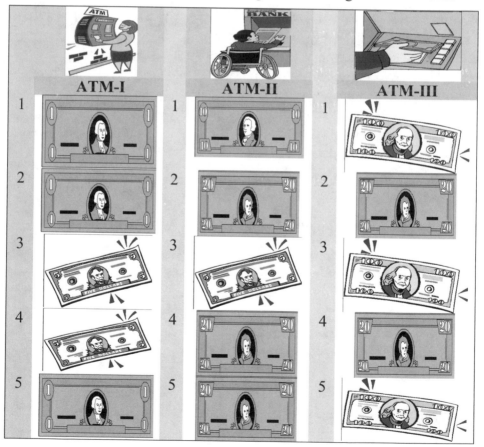

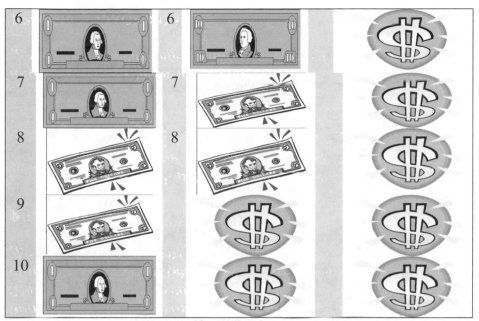

Fig. 1.43. Counting money.

Three shoppers go on a shopping spree. The first shopper withdraws an amount of $8 from ATM-I that gives a sample combination of bills at 2, 4, 5 and 7. The second shopper withdraws an amount of $35 from ATM-II that gives a sample combination of bills at 1, 4 and 7. The third shopper withdraws an amount of $120 from ATM-III that gives a sample combination of bills at 1 and 4. Assuming that ATM-I, II and III have 10, 8 and 5 notes respectively, estimate the total money available at the three ATMs. (**Remark:** The numbers of notes in each ATM are known, but not the value of each note). Also compare your estimate of total money with the true value of the total money.

Stratum-I: So from the first stratum of notes in ATM-I, we selected four notes with the value of: $1, $5, $1 and $1.

$$\textbf{average} = \frac{\text{sum}}{n_1} = \frac{1+5+1+1}{4} = \frac{8}{4} = \$2 \,\textbf{(statistic)}$$

Stratum-II: So from the second stratum of notes in ATM-II, we selected three notes with the value of: $10, $20, and $5.

$$\text{average} = \frac{\text{sum}}{n_2} = \frac{10 + 20 + 5}{3} = \frac{35}{3} = \$11.67 \text{ (statistic)}$$

Stratum-III: So from the third stratum of notes in ATM-III, we selected two notes with the value of: $100, and $20.

$$\text{average} = \frac{\text{sum}}{n_3} = \frac{100 + 20}{2} = \frac{120}{2} = \$60 \text{ (statistic)}$$

Overall sample estimate of total money in three ATMs will be:

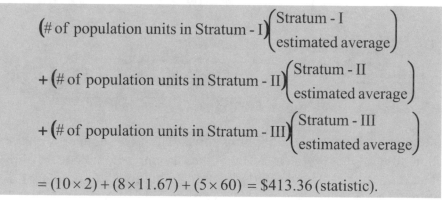

$$= (10 \times 2) + (8 \times 11.67) + (5 \times 60) = \$413.36 \text{ (statistic)}.$$

Note that the total money at ATM-I is:

$$= 1 + 1 + 5 + 5 + 1 + 1 + 1 + 5 + 5 + 1 = \$26 \text{ (Parameter)}$$

Note that the total money at ATM-II is:

$$= 10 + 20 + 5 + 20 + 20 + 10 + 5 + 5 = \$95 \text{ (Parameter)}$$

Note that the total money at ATM-III is:

$$= 100 + 20 + 100 + 20 + 100 = \$340 \text{ (Parameter)}$$

Thus, the total money at the three ATMs is:

$$= 26 + 95 + 340 = \$461 \text{ (Parameter)}.$$

Further note that this total money is an unknown quantity. The value $413.36 is an estimate of $461.

1.13 PROPORTION IN STRATIFIED RANDOM SAMPLING

(BUTTERFLY LOVERS) Consider the two gardens as shown below:

Fig. 1.44. Counting butterflies.

Count the total number of insects in both gardens: $N = 32$.
Count the number of butterflies in both gardens: $COUNT = 20$.
The population proportion of butterflies in both gardens is:

$$P = \frac{COUNT}{N} = \frac{20}{32} = 0.625 \, (\text{Parameter})$$

Count the total number of insects in the first garden: $N_1 = 16$.
Count the number of butterflies in the first garden: $COUNT_1 = 8$.
The population proportion of butterflies in the first garden is:

$$P_1 = \frac{COUNT_1}{N_1} = \frac{8}{16} = 0.50 \, (\text{Parameter})$$

Count the total number of insects in the second garden: $N_2 = 16$.
Count the number of butterflies in the second garden: $COUNT_2 = 12$.
The population proportion of butterflies in the second garden is:

$$P_2 = \frac{COUNT_2}{N_2} = \frac{12}{16} = 0.75 \, (\text{Parameter})$$

Note that:

$$\left(\frac{N_1}{N}\right)P_1 + \left(\frac{N_2}{N}\right)P_2 = \left(\frac{16}{32}\right) \times 0.50 + \left(\frac{16}{32}\right) \times 0.75 = 0.625 = P$$

Consider that we selected an SRSWOR sample of $n_1 = 5$ insects from the first garden starting from the first two columns of the Pseudo Random Numbers **Table I** given in the Appendix as follows:

SRSWOR sample from the first stratum (garden)				
14 	08 	05 	01 	02 

Fig. 1.45. Estimating butterflies from the first garden.

Count the number of units (insects) in the first sample: $n_1 = 5$.

Count the number of butterflies in the first sample: $count_1 = 3$.

Estimate of the proportion of butterflies in the first stratum:

$$\hat{p}_1 = \frac{count_1}{n_1} = \frac{3}{5} = 0.60 \text{ (Statistic)}$$

Let us say that we selected an SRSWOR sample of $n_2 = 6$ insects from the second garden starting from the first row and the 5th and 6th columns of the Pseudo Random Numbers **Table I** given in the Appendix as follows:

SRSWOR sample from the second stratum (garden)					
08	04	06	14	10	16

Fig. 1.46. Estimating butterflies from the second garden.

Count the number of units (insects) in the second sample: $n_2 = 6$.

Count the number of butterflies in the second sample: $count_2 = 4$.

Estimate of the proportion of butterflies in the second stratum:

$$\hat{p}_2 = \frac{count_2}{n_2} = \frac{4}{6} = 0.667 \text{ (Statistic)}$$

Estimate of the proportion based on the overall sample in stratified random sampling is then given by:

$$\hat{p}_{st} = \left(\frac{N_1}{N}\right)\hat{p}_1 + \left(\frac{N_2}{N}\right)\hat{p}_2$$

$$= \left(\frac{16}{32}\right) \times 0.60 + \left(\frac{16}{32}\right) \times 0.667$$

$$= 0.634 \text{ (statistic)}.$$

Note that the statistic 0.634 is an estimate of the parameter 0.625.

1.14 HYPOTHESES

Let us learn a few more statistical terms related to the testing of hypotheses.

Fig. 1.47. Demand of ropes.

Consider a rope seller and a rope buyer in a market. Assume that the rope buyer is a circus owner and is looking to buy good ropes to control his elephants in the circus or to control his ship.

The rope seller says, "The proportion of defective ropes is 0.20."

Fig. 1.48. Null hypothesis.

If the buyer believes the seller then there is nothing to test. However, a good buyer will test before making any decision, because the buyer does not trust the seller.

The buyer says, "The proportion of defective ropes is **not equal** to 0.20."

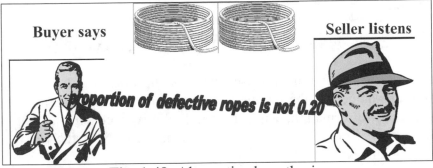

Fig. 1.49. Alternative hypothesis.

Thus there are two statements:

(a) One statement is given by the seller about all the ropes in his shop (population - parameter).

(b) The second statement is given by the buyer about all of the ropes in the shop (again, population - parameter).

"Any statement about the value of a parameter is called a **null hypothesis**." It is denoted by H_0.

Thus the null hypothesis is a statement given by the seller as follows:

$$H_0 : \text{The proportion of defective ropes is equal to } 0.20$$

"Any statement which contradicts the null hypothesis is called an **alternative hypothesis**." It is denoted by H_a or H_1.

Thus the alternative hypothesis is a statement given by the buyer:

$$H_1 : \text{The proportion of defective ropes is not equal to } 0.20$$

Now the buyer wants to know the true proportion of defective ropes. Let us say that there are 10,000 ropes in the shop, so the population size is $N = 10,000$. If the buyer breaks all 10,000 ropes to find the proportion of defective ropes, then the buyer will have nothing to buy and the seller will have nothing to sell. Thus, a good buyer will select a few ropes, say a sample of $n = 10$, to test before making any decision.

Can a buyer make an error in making his decision?

Assume the buyer randomly selected $n = 10$ ropes and found that 3 ropes are defective, that is, the proportion of defective ropes is 0.30. From this information, assume that the buyer concludes that the proportion of defective ropes is more than 0.20. Now the following question arises:

"Are you thinking that the buyer is 100% correct?"

The buyer may or may not be right. The reason is that the buyer made his decision only on the basis of information obtained from one sample of 10 ropes. Thus there are two cases:

Case I:

Suppose the true proportion of the defective ropes is 0.20, but due to some error, the buyer concluded that it is **not** equal to 0.20. Note that the null hypothesis H_0 was true, but by some error the buyer rejected it. Such an error is called a **Type I error** and it is denoted by α (alpha). Thus:

Type I error $= \alpha =$ Chance of rejecting H_0 when H_0 is true

The chance of rejecting the null hypothesis (H_0) when it is true is called a Type I error or the level of significance.

Case II:

Assume the true proportion of the defective ropes is **not** equal to 0.20, but due to some error, the buyer concludes that it is equal to 0.20.

Note that the null hypothesis H_0 was **not** true, but the buyer accepted it by some error. Such an error is called a **Type II error** and is denoted by β. Thus:

Type II error = β = Chance of accepting H_0 when H_0 is not true

The chance of accepting the null hypothesis when it is not true is called a Type II error.

Thus we can summarize the above results in a table as follows:

Decision	Hypothesis (H_0)	
	True	**False (or not true)**
Reject H_0	Type I error (α)	Correct decision
Accept H_0	Correct decision	Type II error (β)

1.14.1 SIGNIFICANCE LEVEL

The significance level is the number α and is the chance of having a Type I error, that is, the chance of rejecting the null hypothesis H_0 when H_0 is true.

1.14.2 DECISION RULE

The decision rule is any rule that helps us to reject or accept a null hypothesis (H_0).

Example 1.5. (YOUR ANSWER MAY DIFFER) For each one of the following hypotheses, decide whether a Type I error or Type II error would be more serious.

(a) Null hypothesis (H_0): The pen is not working
    Alt. hypothesis (H_1): The pen is working.

Fig. 1.50. Pen.

Solution. Type I. If you are sitting in the exam, and your pen is not working, then you may receive a poor grade.

(b) Null hypo. (H$_0$): The lion cage is open.
    Alt. hypo. (H$_1$): The lion cage is locked.

Fig. 1.51. Lion.

Solution. Type I. If you are visiting the zoo and the lion cage is open, the lion may hurt you.

(c) Null hypo. (H$_0$): The final exam is difficult.
    Alt. hypo. (H$_1$): The final exam is easy.

Fig. 1.52. Exam.

Solution. Type I. If you did not prepare for the final exam and fail it, then you may have to repeat the course!

(d) Null hypo. (H$_0$): The food-court is open.
    Alt. hypo. (H$_1$): The food-court is closed.

Fig. 1.53. Food.

Solution. Type II. If you are very hungry, you are out of luck.

(e) Null hypo. (H$_0$): The elephant owner is there.
    Alt. hypo. (H$_1$): The elephant owner is not there.

Fig. 1.54. Elephant.

Solution. Type II. If you were very near to the elephant and tried to take a ride from the elephant, it may harm you.

(f) Null hypo. (H$_0$): Student absenteeism is fun.
    Alt. hypo. (H$_1$): Student absenteeism effects GPA.

Fig. 1.55. Grade.

Solution. Type II. You paid for the class, and it also affects your GPA, so you should not miss any class.

(g) Null hypo. (H$_0$): Friends are good influences.
    Alt. hypo. (H$_1$): Friends are bad influences.

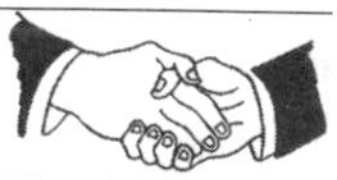

Fig. 1.56. Hands.

Solution. Type II. For example, your friends could try to get you to use drugs that may seem fun but can have very destructive long term effects. If your friends are good influences, they will allow you to think about things and make your own decisions.

Example 1.6. (UNIVERSAL TRUTH) For each one of the following hypotheses, which one doesn't require hypothesis testing?

(a) Null hypo. (H_0): The sun will rise tomorrow.
Alt. hypo. (H_1): The sun will not rise tomorrow.

Fig. 1.57. Sun.

Solution. Here H_0 is always true, so no test is needed.

(b) Null hypo. (H_0): Elephants are bigger than rats.
Alt. hypo. (H_1): Elephants are smaller than rats.

Fig. 1.58. Rat and elephant.

Solution. Here H_0 is always true, so no test is needed.

(c) Null hypo. (H_0): Public servants are lazy.
Alt. hypo. (H_1): Public servants are not lazy.

Fig. 1.59. Public servant.

Solution. Here H_0 is not always true, because a few public servants work only between 9:00am and 5:00pm, while others work during the night. Thus, testing is required based on an opinion of a large sample of public servants.

(d) Null hypo.
(H_0): Elephants can communicate over 100 miles with each other using their ears and feet.
Alt. hypo.
(H_1): Elephants cannot communicate over 100 miles with each other using their ears and feet.

Fig. 1.60. Phone.

Solution. Here H_0 is a research statement that elephants can communicate over 100 miles using their ears and feet, so we may need testing using new experiments.

1.15 SOME SECRETS OF STATISTICS

The following are some secrets of statistics:

1.15.1 SCOPE OF STATISTICS

Statistics has a scope in almost every area of life. For example: Trade, Industry, Commerce, Economics, Biology, Botany, Astronomy, Physics, Chemistry, Education, Medicine, Sociology, Psychology, Religious studies, Meteorology, National defence, and Business (Production, Sales, Purchasing, Finance, Accounting, Quality control, etc.) all involve statistics.

1.15.2 LIMITATIONS OF STATISTICS

A few limitations of statistics are:

(a) Statistics does not deal with individual measurements. This is the reason we need police to investigate individuals.
(b) Statistics deals only with quantitative characters or variables, and we have to assign codes to qualitative variables before analysis.
(c) Statistics results are true only on an average.
(d) Statistics can be misused or misinterpreted. For example, last year 90% of the pedestrians who died in road accidents were walking on paths, so some people would say it is safer to walk in the middle of the road.

1.15.3 LACK OF CONFIDENCE IN STATISTICS

A few people have the following types of views in their mind about statistics:

(a) Statistics can prove anything.
(b) There are three types of lies: lies, damned lies, and statistics.
(c) There are three types of truth: truth, damned truth, and statistics.
(d) Statistics are like clay-one can make a God or devil as he/she pleases.
(e) It is only a tool and cannot prove or disprove anything.

LUDI 1.1. (READ CAREFULLY) Fill in the blank(s):

(i) Any kind of pattern or trend in a sampling method to include or exclude certain types of subjects/units is called ----- bias
(a) Response (b) Selection (c) Nonresponse

(ii) The sampling method that first divides the population into groups and then an SRS of groups is taken is called ----- sampling.
(a) Systematic (b) Cluster (c) Multistage sampling

(iii) The sampling method that first divides the population into homogeneous groups and then an SRS is taken from each group is called ----- sampling.
(a) Systematic (b) Cluster (c) Stratified

(iv) The cluster sampling will be more effective if the units of the population were divided into ----- groups
(a) Homogeneous (b) Heterogeneous (c) Equal groups

(v) The stratified sampling will be more effective if the units of the population were divided into ----- groups
(a) Homogeneous (b) Heterogeneous, or (c) Equal groups

(vi) Statistics is a science used to describe or predict the behavior of a --- based on a random and representative --- taken from the same ---.
(a) Sample (b) Population (c) Variable (d) Unit

(vii) If you do not reject the H_0, you may be making a -----
(a) Type I error or correct decision (b) Type II error or correct decision (c) Type II error (d) None of these

LUDI 1.2. (GOOD HANDWRITING IS AN ASSET) Write a short paragraph on the Random Numbers Method and Lottery Method of sample selection. Explain using a simple example.

LUDI 1.3. (IMPROVE YOUR HANDWRITING) What are convenient and volunteer samples? What can their consequences be if we prefer to adopt one of them?

LUDI 1.4. (LEARN BASIC TERMS) Define statistics, population and sample. Explain with a few examples.

LUDI 1.5. (GROW MORE TREES) The following picture shows eight trees located at the entrance of a zoo with their heights:

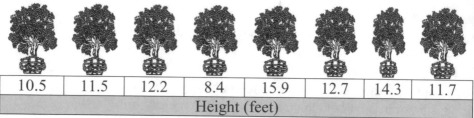

10.5	11.5	12.2	8.4	15.9	12.7	14.3	11.7
Height (feet)							

Fig. 1.61. Decorated entrance of a zoo.

Select a sample of two trees using systematic sampling, and estimate the average height. (**Rule:** Use the first column of the Random Numbers **Table I** given in the Appendix to find a random start).

LUDI 1.6. (CAT AND DOG LOVERS) At a zoo in New York, 40% of the dogs and cats are spayed. In a sample of 150 dogs and cats taken from the zoo, 36 were spayed, which is 24%. Using this information:
(i) the value 24% is a -----
(ii) the spayed status of the cats and dogs is a --
(iii) the value 40% is a -----

Hint: (a) parameter (b) statistic (c) variable

Fig. 1.62. Cat and Dog.

LUDI 1.7. (TRY IT IN YOUR CLASS) A class has 100 students with ID numbers 1 to 100. Select a 1-in-10 **systematic** sample from all the students in the class with labels given by the ID numbers.
(a) Use the first two columns of the Random Numbers **Table I** given in the Appendix to find a random start.
(b) List the student IDs selected in the sample.
(c) How many students are selected in the sample?

Fig. 1.63. A class of 100 students.

LUDI 1.8. (SINGLE WORD DECISION MAY BE DIFFICULT)
Circle T (true) or F (false) for each statement:

	Statement	Result
(i)	Simple random sampling and stratified random sampling are the same if all strata contain the same number of units.	T F
(ii)	A cluster sampling is a simple random sampling.	T F
(iii)	If the population is divided into homogeneous groups, then it is preferable to use stratified random sampling rather than cluster sampling.	T F
(iv)	In cluster sampling, the chance of selection for any unit depends on the cluster size.	T F
(v)	In probability sampling, all units in the population have an equal chance of selection in the sample.	T F
(vi)	In an SRSWR and an SRSWOR sample, each unit in the population has the same chance of selection as the others.	T F
(vii)	Mutually exclusive groups means that each unit of the population belongs to one and only one group.	
(viii)	For a 1-in-k systematic sample, we first order the units of the population in some way and randomly select one unit from the population as the first unit in the sample, and then continue by taking the next k^{th} unit in the sample until the population ends.	T F
(ix)	A parameter is a fixed value of a variable in a population.	T F
(x)	Strata are mutually exclusive groups.	T F

LUDI 1.9. (PLAYING WITH NUMBERS) Compute the following:

(I) (a) $\binom{5}{0}$ (b) $\binom{5}{1}$ (c) $\binom{5}{2}$ (d) $\binom{5}{3}$ (e) $\binom{5}{4}$ (f) $\binom{5}{5}$

(II) (a) $\binom{25}{0}$ (b) $\binom{25}{5}$ (c) $\binom{25}{10}$ (d) $\binom{25}{15}$ (e) $\binom{25}{20}$ (f) $\binom{25}{25}$

LUDI 1.10. (FEEDING MONKEYS IS FUN) Consider a small population of five Monkeys living in circus-apartments. Their various ages (in months) are given in the following table:

Monkey	M-1	M-2	M-3	M-4	M-5

Age (months)	20	25	27	24	26

Fig. 1.64. All monkeys in a circus.

(a) Compute the average or mean age of the five monkeys in the population. Is it a parameter or statistic?

(b) One possible random sample of size $n = 2$ is Monkey-1 and Monkey-2 with ages 20 and 25 respectively. Compute the average or mean age of the two Monkeys in this sample. Does it equal the population mean in (a)?

(c) Select all 10 possible without replacement random samples for each of the two Monkeys from the population of five Monkeys. Compute the average age from each sample. Are these values statistics or parameters?

LUDI 1.11. (LOST AND FOUND) Fill in the blank(s):

(i) A sample consisting of those units of the population that could be easily approached is called a ----- sample.
(a) Volunteer (b) Convenience (c) Random

(ii) A sample consisting of those units of the population that are more likely to respond is called a ----- sample.
(a) Volunteer (b) Convenience (c) Random

(iii) The sensitive nature of a question in an interview survey may lead to ----- bias.
(a) Response (b) Non-response (c) Selection

(iv) If a population can be divided into heterogeneous groups, then the most appropriate sampling method will be ----- sampling.
(a) Simple random (b) Cluster (c) Stratified random

(v) If a population can be divided into homogeneous groups, then the most appropriate sampling method will be ----- sampling.
(a) Systematic random (b) Cluster (c) Stratified random

LUDI 1.12. (A VISIT TO A ZOO) (I) The following list shows 20 units (either animal or bird) available at a special zoo in New York:

1 Antelope	8 Goose	15 Peafowl
2 Camel	9 Horse	16 Pigeon
3 Cheetah	10 Kangaroo	17 Turkcy
4 Deer	11 Lion	18 Woodpecker
5 Dove	12 Monkey	19 Yak
6 Elephant	13 Owl	20 Zebra
7 Fox	14 Parrot	ZOO

Fig. 1.65. A small zoo.

(a) Select an SRSWR sample of 6 animals and birds starting from the first two columns of the Random Numbers **Table I** given in the Appendix. What is the chance of selection for each unit (either animal or bird) on any draw in the sample?   How many WR samples are possible?

(b) Select an SRSWOR sample of 6 animals and birds starting with the first row and the 6th and 7th columns of the Random Numbers **Table I**. What is the chance of selection for each unit (either animal or bird) on any draw in the sample? How many WOR samples are possible? Are the SRSWOR and SRSWR in (a) the same?

(c) Select 3 animals and 3 birds using stratified random sampling. What is the chance of selection for an animal, say a zebra, on any draw from the first stratum of all animals? What is the chance of selecting a bird, say a dove, on any draw from the second stratum of all birds?

(d) Arrange all the animals and birds in alphabetic order on the basis of their names, and select a systematic sample of 5 birds and animals. (Start with the first column of the Random Numbers **Table I**)

(e) The following are the 5 cages having different animals and birds:

Cage No.	Animal/Bird Number
1	1, 3, 5, 7
2	2, 4, 6, 8, 10, 19
3	11, 13, 15
4	12, 14, 16, 18
5	9, 17, 20

(i) Select two cages using SRSWOR sampling, and list the names of each animal or bird in the selected cages. (Use the first column of the Random Numbers **Table I**).

(ii) Name the sampling scheme used in part (i).

(iii) What is the chance of selecting the Monkey in scheme (i) from the zoo?

(iv) From each one of the two cages selected in (i), further select one animal or bird from each cage using SRSWOR sampling. Name the animal or bird selected. (Always start from the first row and first column of the Random Numbers **Table I**). Name the sampling scheme used in (iv).

(**II**) (a) What is the proportion of birds in the whole zoo? Is it a parameter or a statistic?
(b) What is the proportion of birds in your SRSWOR sample? Is it a parameter or a statistic?
(c) What is the proportion of birds in your SRSWR sample? Is it a parameter or a statistic?

LUDI 1.13. (CLEAN YOUR WASHROOM) A company, XYZ, sells five brands of toilet paper at the same price but with different lengths depending upon the quality of the paper as given below:

Toilet Paper	A	B	C	D	E
Length (meters)	200	250	270	204	206

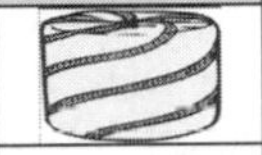

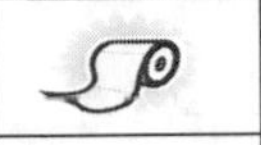

Fig. 1.66. A population of five kinds of toilet paper.

(a) Compute the average or mean length of the five brands of toilet paper in the population consisting of only five rolls. Is it a parameter or a statistic?
(b) One possible random sample of size $n = 2$ is Roll-A and Roll-D with the lengths 200 meters and 204 meters respectively. Compute the average or mean length of the two rolls in this sample. Does it equal the population mean in (a)?
(c) Select all 10 possible without replacement random samples for each of two rolls from the population of five rolls. Find the average length from each sample. Are these values statistics or parameters?

LUDI 1.14. (CARE FOR THE SENIORS) A zoo record shows that 35% of all elephants are retired. Assume that a Random Numbers **Table I** given in the Appendix is used to select 300 elephants. Out of the 300 elephants selected, 31% were found as retired. Using this information, circle the correct answer:

(a) the 35% is a (parameter, statistic).
(b) the 31% is a (parameter, statistic).

Fig. 1.67. Elephant.

LUDI 1.15. (SELECTING ELEPHANTS) A population consists of 8 elephants and 8 lions as listed below:

Clusters	Population			
	Stratum-I		**Stratum-II**	
Cluster-1	Sara	Tara	Matthew	Jeff
Cluster-2	Jennifer	Stephanie	Royal	Paul
Cluster-3	Andy	Libby	Rebecca	Emily
Cluster-4	Chris	Michael	John	Cara

Fig. 1.68. Example of elephants and lions living in different areas.

The population is divided into two strata and four clusters.

(I) A sample of four units (elephants and lions) was obtained from the population. The following table lists a few samples selected with different sampling schemes:

(a) Simple random sampling   (b) Cluster sampling
(c) Stratified sampling with equal sample size from each stratum

Choose the appropriate sampling scheme for each one of the samples listed (more than one answer is possible).

	Samples	Schemes					
		SRS		**Cluster**		**Stratified**	
1	Sara, Libby, Rebecca, Cara	Yes	No	Yes	No	Yes	No
2	Andy, Libby, Rebecca, Emily	Yes	No	Yes	No	Yes	No
3	Sara, Jennifer, Andy, Chris	Yes	No	Yes	No	Yes	No
4	Michael, Matthew, Cara, Sara	Yes	No	Yes	No	Yes	No
5	Chris, Michael, John, Cara	Yes	No	Yes	No	Yes	No
6	Stephanie, Tara, Cara, Royal	Yes	No	Yes	No	Yes	No

(II) Arrange the elephants and lions on the basis of their names from 1 to 16. Take a systematic 1-in-4 sample from the population with a random start = 3, and list the units (elephants and lions) names below:

Systematic Sample			

LUDI 1.16. (KEEP NEAT AND CLEAN APARTMENTS) The following picture shows 12 apartments in a building along with the number of members living in each apartment.

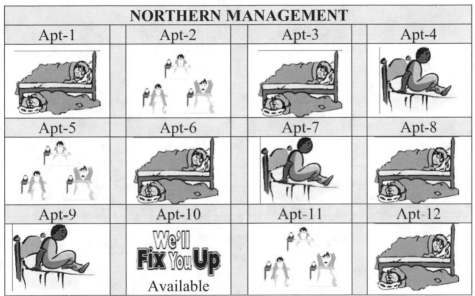

Fig. 1.69. Example of stratification and clustering.

(a) Select an SRSWOR sample of 4 apartments using the first two columns of the Random Numbers **Table I** given in the Appendix. Estimate the average number of people per apartment. Is it a statistic or a parameter?

(b) Select a 1-in-3 systematic sample of four apartments after arranging on the basis of apartment numbers and using 1 as a random start. Estimate the average number of people per apartment. Is it a statistic or a parameter?

(c) Construct three strata of your choice and comment on them.

(d) Construct three clusters of your choice and comment on them.

LUDI 1.17. (HORSE RIDING IS FUN) A boy picks 4 out of 8 toys available at a shop using the first column of the Random Numbers **Table I**.

Sr. No.	Toys	Sr. No.	Toys
1	Horse Ride	5	Bike Ride
2	Bike Ride	6	Horse Ride
3	Airplane Ride	7	Bike ride
4	Bike Ride	8	Horse Ride

Fig. 1.70. A toy shop.

(a) List the serial numbers of the toys selected by the boy.

(b) How many toys show a kid taking a horse ride in the sample?

(c) What is the estimate of the proportion of toys showing a horse ride? Is it a statistic or a parameter?

LUDI 1.18. (I) **(A WRONG DECISION MAY BE HARMFUL)**
Complete the following table.

Decision	Hypothesis (H_0)	
	True	False (or Not True)
Reject H_0		
Accept H_0		

(II) **(SOMETIMES OUR THINKING MAY NOT MATCH)** For each one of the following hypotheses, decide whether a **Type I** error or **Type II** error would be less serious.

(a) H_0 : The lion owner did not lock the cage.
       H_1 : The lion owner locked the cage.

(b) H_0 : The monkey is very friendly with visitors.
       H_1: The monkey is not very friendly with visitors.

(c) H_0 : A minimum score of 60% on the midterm is mandatory for the final exam.
       H_1: A midterm score is not mandatory for the final exam.

(d) H_0 : The book store is open.
       H_1 : The book store is closed.

(e) H_0 : The elephant owner (*mahout*) is there.
       H_1 : The elephant owner (*mahout*) is not there.

(f) H_0 : Student absenteeism is fun.
       H_1 : Student absenteeism is not fun.

(g) H_0 : Bookworms have health problems.
       H_1 : Bookworms do not have health problems.

(h) H_0 : Crossing the road is safe.
       H_1 : Crossing the road is not safe.

(i) H_0 : The scissors is sharp.
       H_1 : The scissors is not sharp.

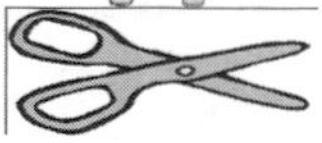

Fig. 1.71. Two types of errors.

LUDI 1.19. (JUDGEMENT MAY DIFFER FROM COURT TO COURT) For each set of hypotheses, decide whether a **Type I** error or **Type II** error would be less serious. Justify your choice.

(a) H_0 : The power is turned on.
        H_1 : The power is not turned on. 
(b) H_0 : The parachute is defective.
        H_1 : The parachute is not defective. 
(c) H_0 : The dog bites.
        H_1 : The dog barks, but does not bite. 
(d) H_0 : The snake is poisonous.
        H_1: The snake is not poisonous. 
(e) H_0: Late students cannot understand the lecture.
        H_1: Late students can understand the lecture.

Fig. 1.72. Types of errors.

LUDI 1.20. (REMEMBER THE NAMES OF ANIMALS AND BIRDS) A population consists of 20 animals and birds as shown in the following table:

Population			
Clusters	Stratum-I		Stratum-II	
I				
II				
III				
IV				
V				

Fig. 1.73. Care for our animals and birds.

The population is divided into two strata (homogeneous groups) and five clusters (heterogeneous groups).

(I) A sample of four animals and birds was obtained from the population. The following table lists a few samples selected with different sampling schemes:

(a) Simple random sampling (b) Cluster sampling
(c) Stratified sampling with equal sample size from each stratum

Choose the appropriate sampling scheme for each one of the sample listed (more than one answer is possible).

Samples				SRS		Cluster		Stratified	
1				Yes	No	Yes	No	Yes	No
2				Yes	No	Yes	No	Yes	No
3				Yes	No	Yes	No	Yes	No
4				Yes	No	Yes	No	Yes	No
5				Yes	No	Yes	No	Yes	No
6				Yes	No	Yes	No	Yes	No

Fig. 1.74. Different types of sampling techniques.

(II) Arrange the animals and birds on the basis of their names (Alphabetically: A to Z) from 1 to 20 in the following table.

1	5	9	13	17
2	6	10	14	18
3	7	11	15	19
4	8	12	16	20

Take a systematic 1-in-5 sample from the population with random start = 3, and list the animals' and birds' names below:

Systematic Sample			

LUDI 1.21. (COMPARE MOM'S HEIGHT WITH YOURS) In a circus, 61% of the working females are taller than 172 cms. In a random sample of 300 females taken from the same circus, 210 females were taller than 172 cms, which is 70%.

Fig. 1.75. Learn to use a measuring tape.

In this situation:

Q#	Question	(a)	(b)	(c)
A	The value 70% is a --	Parameter	Statistic	Variable
B	The female's height is a --	Parameter	Statistic	Variable
C	The value 61% is a --	Parameter	Statistic	Variable
D	The sample size (n) is --	300	210	510
E	The population size (N) is --	510	61%	Unknown
F	The experimental unit is --	Circus	Female	Meters
G	The value 172 cms is --	Variable	Fixed	Sample
H	The variable of interest is --	Female	Percent	Height
I	The value 0.61 is --	Proportion	Percent	Number
J	The population of interest is --	Females	Cows	Foxes
K	The child plays the role of a-	Tailor	Carpenter	Statistician

LUDI 1.22. (REVIEWING OF CLASS NOTES IS IMPORTANT)
(a) List at least three names of sampling schemes which can be used to reduce selection bias.
(b) List at least three sources of bias in survey sampling.
(c) Explain the Lottery Method and the Random Numbers Method.

LUDI 1.23. (PARROT LOVERS) Consider two cages as shown below:

Fig. 1.76. Counting birds.

Count the total number of units (birds) in both cages: $N = - - - -$
Count the number of parrots in both cages: COUNT $= - - - -$
Population proportion of parrots in both cages:

$$P = \frac{COUNT}{N} = \frac{---}{---} = ---- \text{(Is this a statistic or a parameter?)}$$

Count the total number of birds in the first cage $N_1 = ----$
Count the number of parrots in the first cage $COUNT_1 = ----$
Population proportion of parrots in the first cage:

$$P_1 = \frac{COUNT_1}{N_1} = \frac{---}{---} = ---- \text{(Is this a statistic or a parameter?)}$$

Count the total number of birds in the second cage $N_2 = ----$
Count the number of parrots in the second cage $COUNT_2 = ----$
Population proportion of parrots in the second cage:

$$P_2 = \frac{COUNT_2}{N_2} = \frac{---}{---} = ---- \text{(Is this a statistic or a parameter?)}$$

Find:

$$P = \left(\frac{N_1}{N}\right)P_1 + \left(\frac{N_2}{N}\right)P_2 = ---- \text{(Is this a statistic or a parameter?)}$$

Select an SRSWOR sample of $n_1 = 5$ birds from the first cage starting from the first two columns of the Pseudo Random Numbers **Table I** given in the Appendix and list in the following table.

SRSWOR sample from the first stratum (cage)				

Count the number of units (birds) in the first sample: $n_1 = ----$
Count the number of parrots in the first sample: $count_1 = ----$
Estimate of proportion of parrots in the first stratum:

$$\hat{p}_1 = \frac{count_1}{n_1} = \frac{---}{---} = ---- \text{(Is this a statistic or a parameter?)}$$

Select an SRSWOR sample of $n_2 = 6$ birds from the second cage starting from the first row and the 3rd and 4th columns of the Pseudo Random Numbers **Table I** and list their names in the following table:

SRSWOR sample from the second stratum (cage)					

Count the number of units (birds) in the second sample: $n_2 = - - -$
Count the number of parrots in the second sample: $\text{count}_2 = - - -$
Estimate of proportion of parrots in the second stratum:

$$\hat{p}_2 = \frac{\text{count}_2}{n_2} = \frac{- - -}{- - -} = - - - \text{ (Is this a statistic or a parameter?)}$$

Estimate of proportion of parrots based on overall sample in stratified random sampling given by:

$$\hat{p}_{st} = \left(\frac{N_1}{N}\right)\hat{p}_1 + \left(\frac{N_2}{N}\right)\hat{p}_2$$

$= - - -$ (Is this a statistic or a parameter?)

LUDI 1.24. (MISCELLANEOUS) The following list shows 63 birds, insects, and animals available on a special farm.

No.	Birds	No.	Insects	No.	Animals
1	Bunting	22	Ant	43	Boar
2	Crow	23	Bee	44	Bull
3	Duck	24	Butterfly	45	Calf
4	Eagle	25	Crawling	46	Cow
5	Finch	26	Dragonfly	47	Ewe

6	Gull	27	Fly	48	Goat
7	Hooper	28	Grasshopper	49	Lamb
8	Ibis	29	Ladybug	50	Pig
9	Kingfisher	30	Mosquito	51	Sheep
10	Loon	31	Praying Mantis	52	Cat
11	Ostrich	32	Roach	53	Dog
12	Owl	33	Scorpion	54	Horse
13	Parrot	34	Snail	55	Rabbit
14	Peacock	35	Spider	56	Deer

15	Pigeon	36	Tarantula	57	Fox
16	Quail	37	Centipede	58	Moose
17	Robin	38	Moth	59	Elk
18	Sparrow	39	Millipede	60	Frog
19	Swan	40	Locust	61	Gorilla
20	Turkey	41	Wasp	62	Llama
21	Woodpecker	42	Pincher Bugs	63	Ox

Fig. 1.77. Differentiating animals, birds, and insects.

(I) SIMPLE RANDOM SAMPLING

(a) Select an SRSWR sample of 9 units (animals, birds, and insects) starting from the first two columns of the Random Numbers **Table I** given in the Appendix:

No.	1	2	3	4	5	6	7	8	9
Name									

What is the chance of selection for each unit (either animal, bird, or insect) on any draw in the sample? How many WR samples are possible?

(b) Select an SRSWOR sample of 9 animals and birds starting with the first row and 2nd and 3rd columns of the Random Numbers **Table I**

No.	1	2	3	4	5	6	7	8	9
Name									

What is the chance of selection for each unit (either animal, bird, or insect) on any draw in the sample? How many WOR samples are possible?

(II) STRATIFIED RANDOM SAMPLING

Select 3 animals, 3 birds, and 3 insects using stratified random sampling. (**Rule:** Always start from the first row and first column of the Random Numbers **Table I**).

Stratified Random Sample		
Stratum-I (Animals)	Stratum-II (Birds)	Stratum-III (Insects)

What is the chance of selection for an animal, say an ox, on any draw from the first stratum of all animals?
What is the chance of selection for a bird, say a parrot, on any draw from the second stratum of all birds?
What is the chance of selection for an insect, say a butterfly, on any draw from the third stratum of all insects?

(III) SYSTEMATIC SAMPLING

Arrange all the animals and birds in alphabetic order on the basis of their names, and select a systematic sample of 9 units (either animal, bird or insect). (Start with the first column of the Random Numbers **Table I** to select a **random start** for the systematic sample).

1	2	3	4	5	6
7	8	9	10	11	12
13	14	15	16	17	18
19	20	21	22	23	24
25	26	27	28	29	30
31	32	33	34	35	36
37	38	39	40	41	42
43	44	45	46	47	48
49	50	51	52	53	54
55	56	57	58	59	60
61	62	63			

Systematic Sample							

(IV) CLUSTER SAMPLING

Assume the 21 rows, each having three units (animal, bird or insect), are clusters. Select three rows (or clusters) using **systematic sampling**, and list the names of each unit (either animal, or bird, or insect) in the selected row. (Use the first column of the Random Numbers **Table I**).

Selected three cages	
Cage No.	**Animal/Bird Names**

What is the chance of selecting the butterfly in cluster sampling?

(V) PROPORTION

(a) Find the proportion of insects in the whole list. Is it a parameter or a statistic?

(b) Find the proportion of insects using your SRSWOR sample. Is it a parameter or a statistic?

(c) Find the proportion of insects using your SRSWR sample. Is it a parameter or a statistic?

(d) Estimate the total number of insects on the farm using the SRSWOR sample. Is it a parameter or a statistic?

(e) Estimate the total number of insects on the farm using the SRSWR sample. Is it a parameter or a statistic?

LUDI 1.25. (SOMETIMES LITTLE DONATIONS HELP) At a shopping mall consider there are three cash registers having the following denominations:

	Cash Register-I		Cash Register-II		Cash Register-III
1	2 PFENNIG	1	10 PFENNIG	1	50 BUNDESREPUBLIK DEUTSCHLAND PFENNIG
2	1 PFENNIG	2	10 PFENNIG	2	50 BUNDESREPUBLIK DEUTSCHLAND PFENNIG
3	5 PFENNIG	3	5 PFENNIG	3	10 PFENNIG
4	5 PFENNIG	4	10 PFENNIG	4	10 PFENNIG

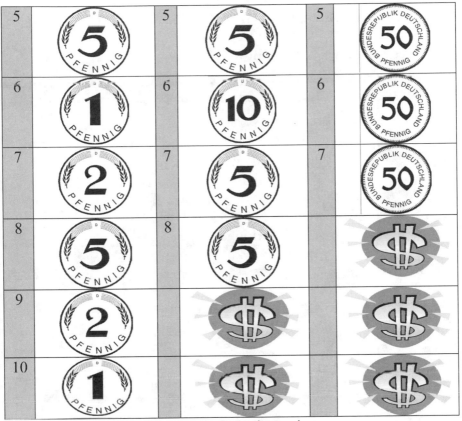

Fig. 1.78. Counting coins.

Three shoppers go on a shopping spree. The first shopper withdraws an amount of 13 *pfennig* from the first cash register that gives a sample combination of coins at the 2nd, 4th, 5th and 7th positions. The second shopper withdraws an amount of 25 *pfennig* from the second cash register that gives a sample combination of coins at the 1st, 4th and 7th positions. The third shopper withdraws an amount of 60 *pfennig* from the third cash register that gives a sample combination of coins at the 1st and 4th positions. Assume that the first, second and third cash registers have 10, 8 and 7 coins respectively.

(a) Estimate the total money available at the three cash registers. (**Remark:** Assume the numbers of coins in each cash register are known, but not the value of each coin).

(b) Compare your estimate in (a) with the true total amount (in *pfennig*) available in all the three cash registers.

LUDI 1.26. (REMEMBER THE NAMES OF ANIMALS, BIRDS AND INSECTS) A population consists of 24 units (animals, birds and insects) as shown in the following table:

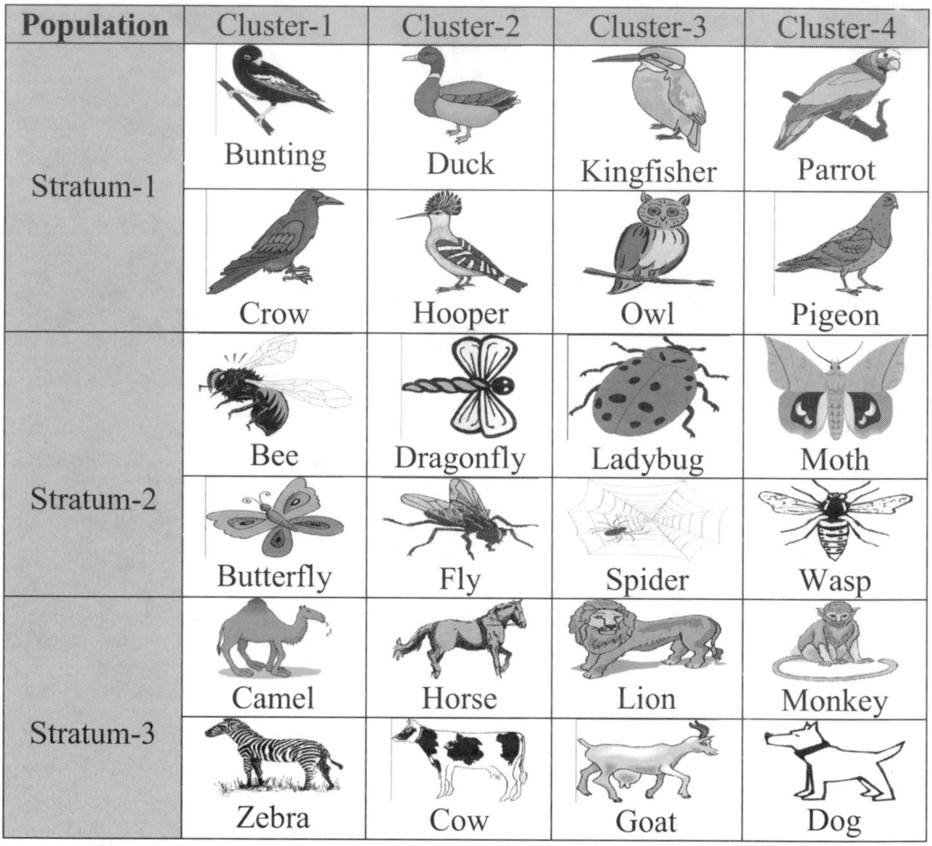

Population	Cluster-1	Cluster-2	Cluster-3	Cluster-4
Stratum-1	Bunting	Duck	Kingfisher	Parrot
	Crow	Hooper	Owl	Pigeon
Stratum-2	Bee	Dragonfly	Ladybug	Moth
	Butterfly	Fly	Spider	Wasp
Stratum-3	Camel	Horse	Lion	Monkey
	Zebra	Cow	Goat	Dog

Fig. 1.79. Learning the names of animals, birds and insects.

The population is divided into three strata (homogeneous groups) and four clusters (heterogeneous groups).

(I) A sample of **six** units (animals, birds, and insects) was obtained from the population. The following table lists a few samples selected with different sampling schemes:

(a) Simple random sampling with replacement (SRSWR)
(b) Simple random sampling without replacement (SRSWOR)
(c) Cluster sampling
(d) Stratified sampling with **equal sample size** from each stratum

Circle the appropriate sampling scheme for each one of the following samples (more than one answer is possible).

Samples	1	2	3	4
Circle one choice either "Yes" or "No" for each case				
SRSWR	Yes	Yes	Yes	Yes
	No	No	No	No
SRSWOR	Yes	Yes	Yes	Yes
	No	No	No	No
Stratified	Yes	Yes	Yes	Yes
	No	No	No	No
Cluster	Yes	Yes	Yes	Yes
	No	No	No	No

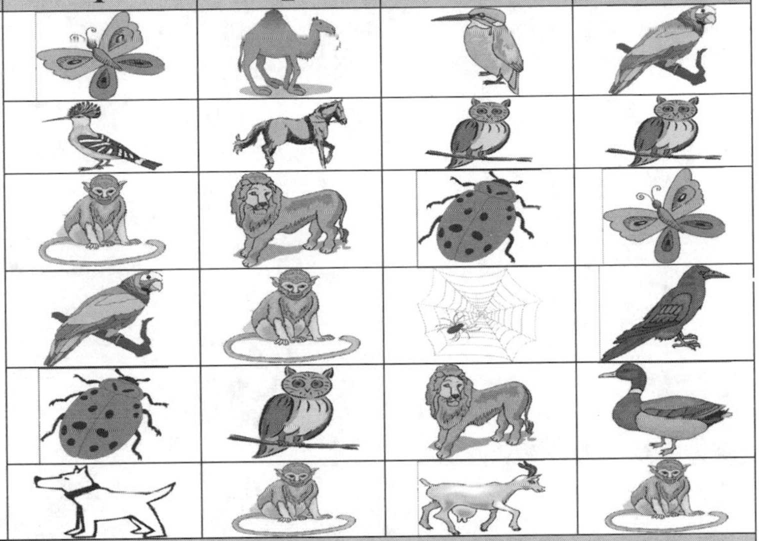

Fig. 1.80. Rewriting the names of animals, birds and insects.

(II) Arrange all the units (animals, birds and insects) on the basis of their names (Alphabetically: A to Z) from 1 to 24 in the table.

1	7	13	19
2	8	14	20
3	9	15	21
4	10	16	22
5	11	17	23
6	12	18	24

Take a systematic 1-in-4 sample from the population with a random start = 2, and list the selected animals, birds and insects below:

Systematic Sample				

LUDI 1.27. (FLAGS OF THE 50 STATES OF THE USA) The following table lists the 50 states of the United States of America (USA).

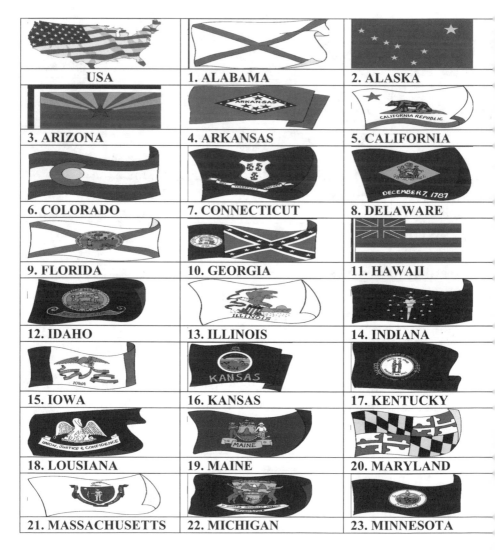

USA	**1. ALABAMA**	**2. ALASKA**
3. ARIZONA	**4. ARKANSAS**	**5. CALIFORNIA**
6. COLORADO	**7. CONNECTICUT**	**8. DELAWARE**
9. FLORIDA	**10. GEORGIA**	**11. HAWAII**
12. IDAHO	**13. ILLINOIS**	**14. INDIANA**
15. IOWA	**16. KANSAS**	**17. KENTUCKY**
18. LOUSIANA	**19. MAINE**	**20. MARYLAND**
21. MASSACHUSETTS	**22. MICHIGAN**	**23. MINNESOTA**

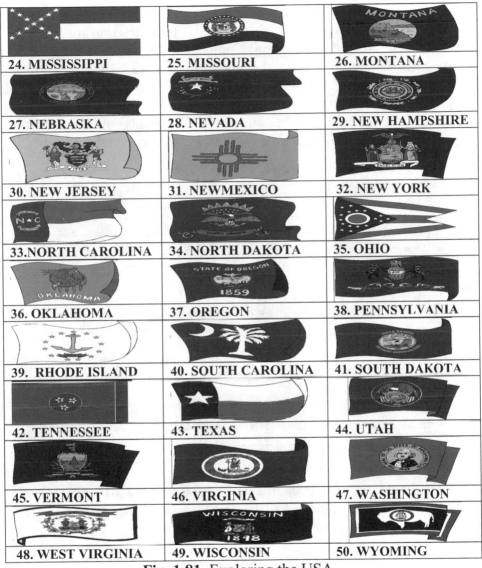

24. MISSISSIPPI	25. MISSOURI	26. MONTANA
27. NEBRASKA	28. NEVADA	29. NEW HAMPSHIRE
30. NEW JERSEY	31. NEWMEXICO	32. NEW YORK
33.NORTH CAROLINA	34. NORTH DAKOTA	35. OHIO
36. OKLAHOMA	37. OREGON	38. PENNSYLVANIA
39. RHODE ISLAND	40. SOUTH CAROLINA	41. SOUTH DAKOTA
42. TENNESSEE	43. TEXAS	44. UTAH
45. VERMONT	46. VIRGINIA	47. WASHINGTON
48. WEST VIRGINIA	49. WISCONSIN	50. WYOMING

Fig. 1.81. Exploring the USA.

(I) (a) Find the proportion of states whose names start with the letter, M (e.g., M for Minnesota).

(b) Is it a parameter or a statistic?

(II) (a) Select an SRSWR sample of 10 states starting from the first two columns of the Random Numbers **Table I**.

(b) Estimate the proportion of states whose names start with M from the above SRSWR sample of 10 states.

(c) Is it a parameter or a statistic?

(III) (a) Select an SRSWOR sample of 10 states starting from the first two columns of the Random Numbers **Table I**.

(b) Estimate the proportion of states whose names start with M from the above SRSWOR sample of 10 states.

(c) Is it a parameter or a statistic?

(IV) (a) Arrange all the states on the basis of their names (Alphabetically: A to Z) from 1 to 50 in the following table.

1	2	3	4	5
6	7	8	9	10
11	12	13	14	15
16	17	18	19	20
21	22	23	24	25
26	27	28	29	30
31	32	33	34	35
36	37	38	39	40
41	42	43	44	45
46	47	48	49	50

(b) Take a systematic 1-in-5 sample from the population with random start = 3, and list the selected states below:

Systematic Sample				
1	2	3	4	5
6	7	8	9	10

(c) Estimate the proportion of states whose names start with M from the above systematic sample of 10 states.

(d) Is it a parameter or a statistic?

LUDI 1.28. (NOTHING IS TRUE FOREVER)

(a) What is the scope of statistics?

(b) What are the limitations of statistics?

(c) Why do people have a lack of confidence in statistics?

(d) What is the purpose of sampling?

(e) Do you think statistics is a useful subject? Justify your answer.

(f) Which do you like more: statistics or mathematics? Why?

LUDI 1.29. (FRUITS AND VEGETABLES) A farmer has two farms he is using to grow 12 varieties of fruits on one farm and 12 varieties of vegetables on the second farm as shown in the following table:

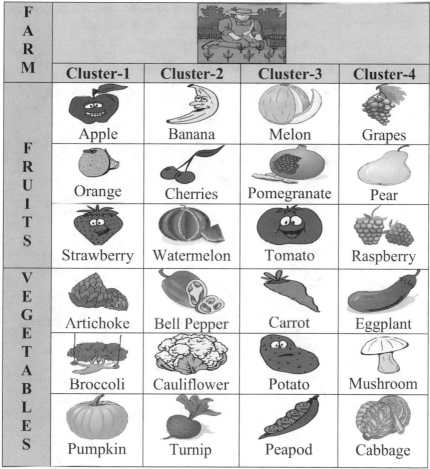

Fig. 1.82. Learning the names of fruits and vegetables.

The population of fruits and vegetables is divided into two strata (homogeneous groups) and four clusters (heterogeneous groups).
(I) One day four animals: **elephant, fox, monkey, and ox** entered both farms and each picked up **six** units (fruits and vegetables) from both fields. The following table lists a few sampling schemes used by the four animals to select fruits and vegetables:
(a) Simple random sampling with replacement (SRSWR)

(b) Simple random sampling without replacement (SRSWOR)
(c) Cluster sampling
(d) Stratified sampling with **equal sample size** from each stratum
Choose the appropriate sampling scheme used by the animals for each
one of the samples listed below (more than one answer is possible).

	Monkey	Elephant	Fox	Ox
Samples	**1**	**2**	**3**	**4**
Circle one choice either "Yes" or "No" for each case				
SRSWR	Yes	Yes	Yes	Yes
	No	No	No	No
SRSWOR	Yes	Yes	Yes	Yes
	No	No	No	No
Stratified	Yes	Yes	Yes	Yes
	No	No	No	No
Cluster	Yes	Yes	Yes	Yes
	No	No	No	No

Fig. 1.83. Rewriting the names of fruits and vegetables.

(II) After arriving at home, the four animals found that they were missing a few varieties of fruits and vegetables available on the farms.

Next time they arranged all the units (fruits and vegetables) on the basis of their names (**Alphabetically**: A to Z) from 1 to 24 in the following table.

List the names of the fruits and vegetables alphabetically			
1	7	13	19
2	8	14	20
3	9	15	21
4	10	16	22
5	11	17	23
6	12	18	24

(a) The elephant took a systematic 1-in-4 sample from the farms with a random start = 1. List the fruits and vegetables selected by the elephant:

Systematic sample by the elephant					

(b) The fox took a systematic 1-in-4 sample from the farms with a random start = 2. List the fruits and vegetables selected by the fox:

Systematic sample by the fox					

(c) The monkey took a systematic 1-in-4 sample from the farms with a random start = 3. List the fruits and vegetables selected by the monkey:

Systematic sample by the monkey					

(d) The ox took a systematic 1-in-4 sample from the farms with a random start = 4. List the fruits and vegetables selected by the ox:

Systematic sample by the ox					

(e) How many units (fruits or vegetables) are left on both farms?

LUDI 1.30. (VEGETABLES LOVERS) A lady entered a market and found that **seven** baskets were lying on a clearance table.

Contents in the baskets	A lady at the shopping center 			
I Peppers				
II Corn				
III Eggplants				
IV Mushrooms				
V Onions				
VI Pumpkin				
VII Carrots				

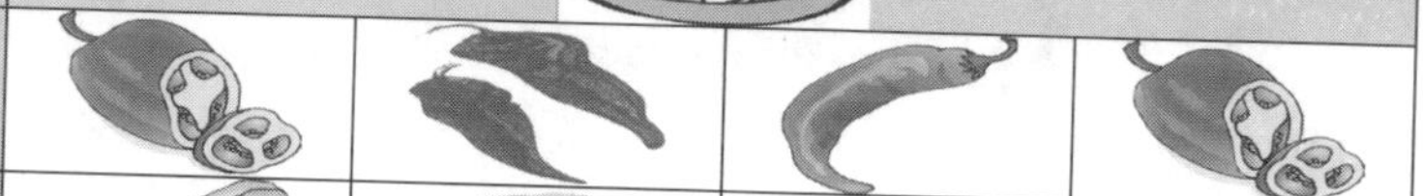

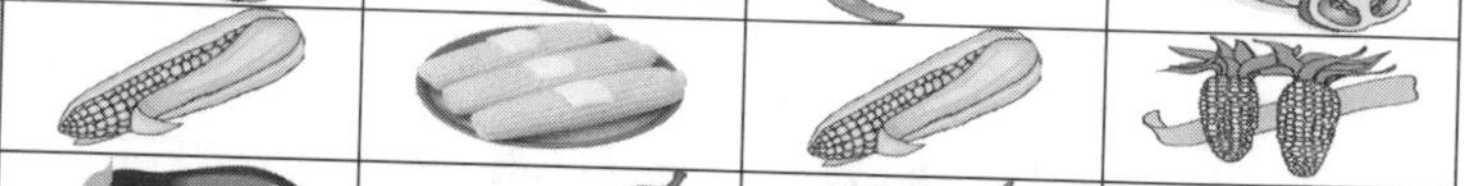

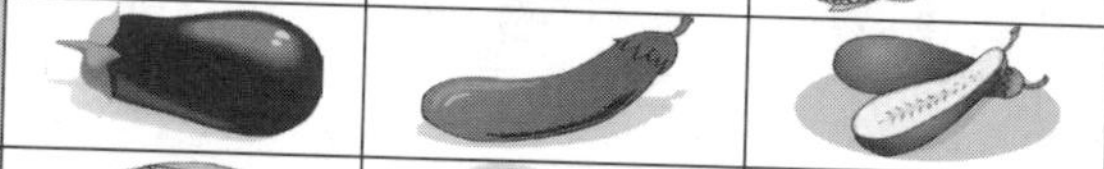

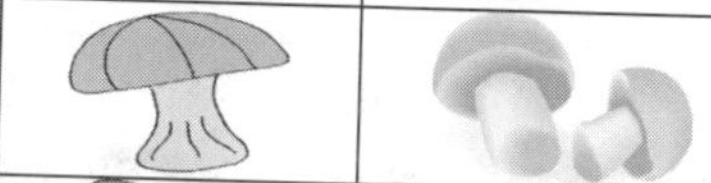

Fig. 1.84. Vegetables are good for your health.

In a hurry, she randomly picked up three out of seven baskets. What is the chance that she picked up the basket with the pumpkin?

LUDI 1.31. (A VISIT TO A FARM) (I) The following list shows 36 fruits and vegetables available at a special farm in New York:

 1 Apple	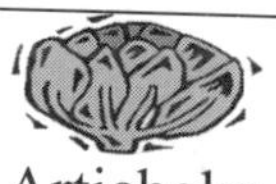 2 Artichoke	 3 Beet
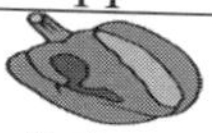 4 Bell Pepper	 5 Tomato	 6 Raspberry
 7 Lemon	 8 Bananas	 9 Eggplant

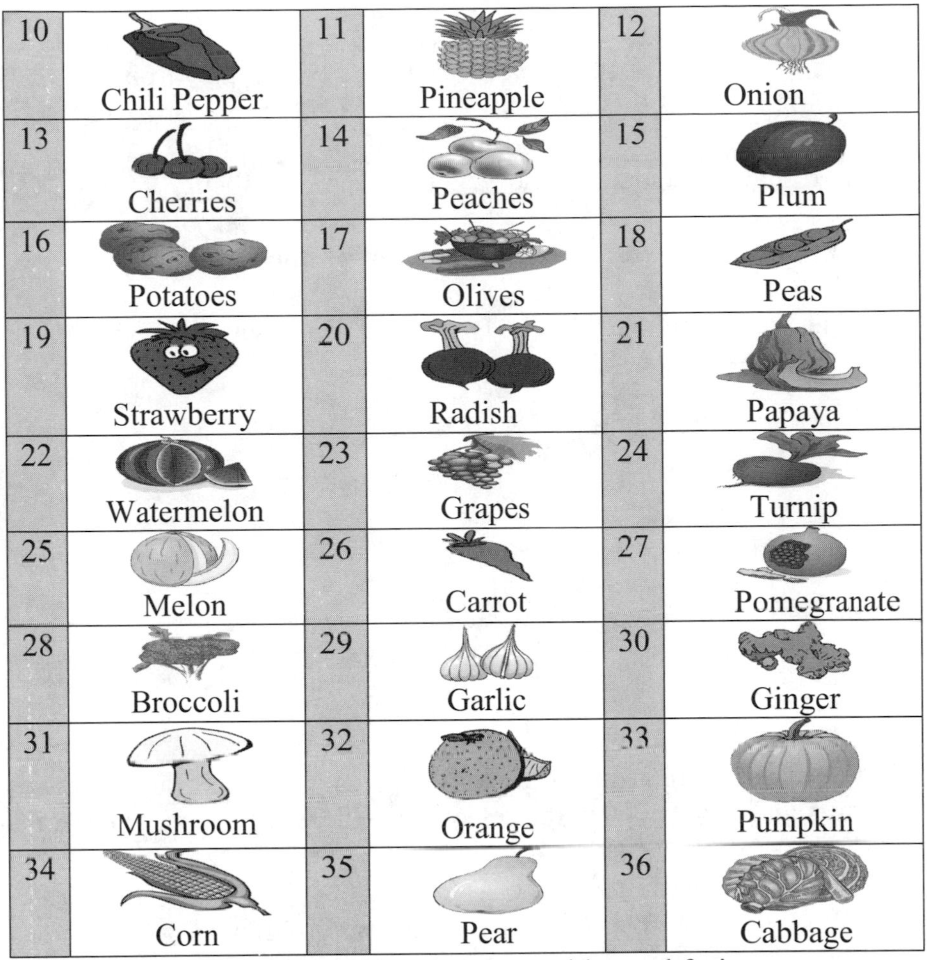

10	Chili Pepper	11	Pineapple	12	Onion
13	Cherries	14	Peaches	15	Plum
16	Potatoes	17	Olives	18	Peas
19	Strawberry	20	Radish	21	Papaya
22	Watermelon	23	Grapes	24	Turnip
25	Melon	26	Carrot	27	Pomegranate
28	Broccoli	29	Garlic	30	Ginger
31	Mushroom	32	Orange	33	Pumpkin
34	Corn	35	Pear	36	Cabbage

Fig. 1.85. Differentiate vegetables and fruits.

(a) Select an SRSWR sample of 10 fruits and vegetables starting from the first two columns of the Random Numbers **Table I**. What is the chance of selecting each unit (either fruit or vegetable), say pumpkin, on any draw in the sample? How many SRSWR samples are possible?

(b) Select an SRSWOR sample of 10 fruits and vegetables starting with the first row and the 2th and 3rd columns of the Random Numbers **Table I**. What is the chance of selecting each unit (either fruit or vegetable), say pumpkin, on any draw in the sample? How many SRSWOR samples are possible? Are the SRSWOR and SRSWR in (a) the same?

(c) Select 5 fruits and 5 vegetables using stratified random sampling. What is the chance of selection for a fruit (say, tomato) from the first stratum of all fruits? What is the chance of selection for a vegetable (say, pumpkin) from the second stratum of all vegetables?

(d) Arrange all the fruits and vegetables in alphabetic order on the basis of their names, and select a systematic sample of 6 fruits and vegetables. (Start with the first column of the Random Numbers **Table I** to select a random start).

(e) The following are the 8 beds on the farm having different fruits and vegetables:

Bed No.	Fruit /Vegetable No.
1	1, 3, 5, 7
2	2, 4, 6, 8, 10, 19, 35
3	11, 13, 15, 32
4	12, 14, 16, 18
5	9, 17, 20, 33
6	21, 23, 25, 27, 29
7	22, 24, 26, 28, 30
8	31, 34, 36

(i) Select two beds using SRSWOR sampling, and list the names of the fruits and vegetables in the selected beds. (Use the first column of the Random Numbers **Table I**).

(ii) Name the sampling scheme used in part (i).

(iii) What is the chance of selecting the pumpkin in (i)?

(iv) From each one of the two beds selected in (i), further select two units from each bed using SRSWOR sampling. Name the units (fruits and vegetables) selected. (Always start from the first row and first column of the Random Numbers **Table I**). Name the sampling scheme used in (iv).

(II) (a) What is the proportion of vegetables on the farm? Is it a parameter or a statistic?

(b) Find the proportion of vegetables using your SRSWOR sample. Is it a parameter or a statistic?

(c) Find the proportion of vegetables using your SRSWR sample. Is it a parameter or a statistic?

2. STATISTICAL STUDIES

2.1 INTRODUCTION

In this chapter we discuss two types of statistical studies: (i) experimental studies and (ii) observational studies. In experimental studies, we do experiments in fields, factories, and institutions; whereas in the observational studies, we do surveys related to human beings.

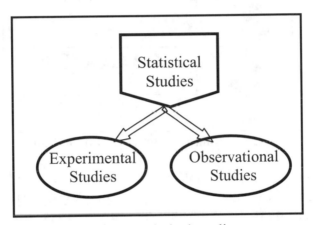

Fig. 2.1. Statistical studies.

For example, growing any crop in a field is an experimental study, whereas asking customers in a shopping mall to fill out a questionnaire is an observational study. Experimental or observational studies can be made on human beings, animals, crops, machines or rocks, etc. These are called **units** in statistical language.

2.2 UNIT

Any **objects** on which observations are made are called **units** or experimental units. Note that if the objects are human beings then these units are called **subjects**. For example, consider the gender of a student. Then the student is a subject, and gender is a variable as it varies from student to student. Consider the age of a machine. Then the machine is a unit, and age is a variable as it varies from machine to machine.

Now let us consider the following information collected on five monkeys:

Monkey's ID No.	Photo ID	Tail size (feet)	Special diet (gms)	Jump (feet)
1		2.5	200	10.0
2		2.7	225	10.3
3		3.8	250	10.4
4		4.2	200	10.6
5		4.5	250	9.6

Fig. 2.2. Monkey as an experimental unit.

In the above table, each monkey is a unit, experimental unit or a statistical unit. Note that there are five monkeys, so there are five experimental units. Note that the "tail size" varies from monkey to monkey. Any characteristic that varies from object to object is called a variable. For each monkey, we are observing three variables: "tail size", "special diet", and "jump".

2.3 EXPERIMENTAL AND OBSERVATIONAL STUDIES

Fig. 2.3. A natural plot.

Consider a natural plot of length l and width w. An elephant-gardener grew a few plants of grapes and observed the yield at the end of the season, say 500 kg of grapes. Note that the grapes were grown in a **natural** plot, and hence it is called an observational study, also called the **control**. Note that the control is the first principle of design of an experiment.

Fig. 2.4. Add 20 kg nitrogen to the field.

Now consider the situation where the gardener selected another plot of the same length and width (all other conditions like temperature, irrigation, soil type etc. are the same) and grew the same number of grape plants as before, but added 20 kg of nitrogen. At the end of the season, the gardener observed 530 kg of grapes. Now the gardener studied the effect of nitrogen on the yield of grapes, which is called an experimental study. A natural question arises, why did the gardener add 20 kg of nitrogen? The gardener can add any amount of nitrogen, say 25 kg, 30 kg etc. Now note that the yield of the grapes and the use of nitrogen will vary from plot to plot. Thus, both nitrogen and yield are variables. Further note that the input of the amount of nitrogen is in the gardener's hands, but the output yield of the grapes is in God's hands. The gardener can manipulate the input usage of nitrogen, but cannot directly manipulate the output yield of grapes.

Thus, these variables can be classified into two categories as follows:

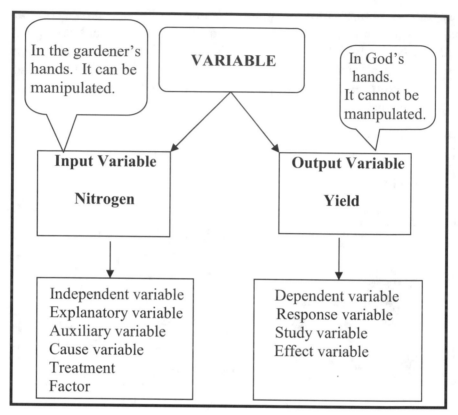

Fig. 2.5. Input and output variables.

In experimental studies, the **input variable**, nitrogen, is most often called a **treatment** or a **factor**. As said earlier, the elephant-gardener could also use 20 kg, 25 kg, or 30 kg of nitrogen. Note that each value of the input factor or treatment is called a **level**.

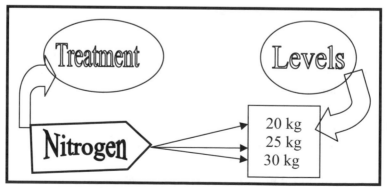

Fig. 2.6. Treatment and levels.

An old gardener suggested to the elephant-gardener that the yield of grapes also depends upon the use of potassium.

Elephant-gardener with an old gardener

Fig. 2.7. Always listen to your seniors.

Then the elephant-gardener made a plan to do a few experiments to study the yield of grapes by using different amounts of nitrogen (N) and potassium (K). The elephant-gardener plans to study nitrogen at three levels: say 20 kg, 25 kg and 30 kg; and potassium at two levels: say 50 kg and 100 kg. A **design layout** for the elephant-gardener's experiment can be seen from the following 2 x 3 contingency table:

		Nitrogen		
		20 kg	25 kg	30 kg
Potassium	50 kg	50 X 20	50 X 25	50 X 30
	100 kg	100 X 20	100 X 25	100 X 30

Thus, there are six treatments, and hence six plots are allocated to these six treatments in a random way. Random allocation of units (or plots) to the treatments under study is called **randomization**. At the end of the experiment, there will be six observations of the yield.

Counting Treatments

Total No. of treatments = (Levels of 1st treatment) × (Levels of 2nd treatment)

= 2 × 3 = 6

From each one of the six plots, the elephant-gardener noted the grapes' yield, and wants to know if there is any difference between the treatments applied to different plots. So the elephant-gardener consulted a statistician to reach a certain conclusion.

Elephant-gardener with a statistician

Fig. 2.8. Experts are on demand.

The statistician cautioned, "You should not trust a single data value from each plot and you should really repeat the experiment." The repetition of an experiment is called a **replication**.

If the elephant-gardener repeats the above experiment, then the total number of data values or observations will be 12:

$$\text{Total No. data values} = (\text{Treatments}) \times (\text{Replications})$$
$$= \quad 6 \quad \times \quad 2 \qquad = 12$$

FOX AND GRAPES

One day, a fox visited the garden and the elephant-gardener offered her a few grapes and asked about the quality of the grapes.

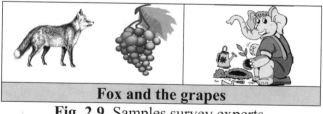

Fox and the grapes

Fig. 2.9. Samples survey experts.

The fox said that the grapes are seedless and very sweet. The response given by the fox about the quality of the grapes is an **observational study.** Note that for the elephant-gardener, it is also an **experimental study** because he knows the effect of the use of different amounts of nitrogen and potassium on the yield of grapes.

Example 2.1. (TALKING PARROTS) A mechanic wishes to make parrot cages from brass plates and conducts an investigation of the bending capacity of brass plates. The mechanic believes that the **bending capacity** depends upon the **temperature** and the **brass content** of the plates. So the mechanic conducted an investigation for temperatures: 50°C, 75°C, 100°C, and 125°C; and percentages of brass contents: 40%, 60%, and 80%. There were **two observations (replications)** for each treatment combination.

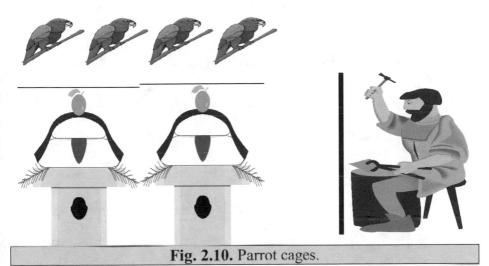

Fig. 2.10. Parrot cages.

(a) Give the response variable(s).
(b) Give the factors(s) and corresponding number of levels for each factor.
(c) How many treatment combinations are there?
(d) How many total number of observations would there be for this investigation?
(e) Is this an observational or an experimental study? Explain.

Solution. (a) There is only one response variable called the **bending capacity.**

(b) There are two variables related to the **bending capacity** called the **temperature** and the **percentage of brass content.** We know that the related variable(s) (or independent variable(s)) are called **factor(s).** Thus there are only two factors: temperature and percentage of brass content. We also know that each value of the factor (or independent variable or explanatory variable) is called a **level.**

Now **temperature** has the four values of 50°C, 75°C, 100°C and 125°C, so it has four levels, and the **percentage of brass content** has the three values: 40%, 60%, and 80%, so it has three levels.

(c) The first treatment (or factor) **temperature** has 4 levels, and the second treatment (or factor) **percentage of brass content** has 3 levels, thus the total number of treatments will be:

$$\text{Total no. of treatments} = (\text{Levels of first treatment}) \times (\text{Levels of second tretment})$$
$$= 4 \times 3 = 12.$$

We can also see this from the following (3 rows)×(4 columns) table:

Brass (%)	Temperature (°C)			
	50°C	75°C	100°C	125°C
40%	Treat-1	Treat-2	Treat-3	Treat-4
60%	Treat-5	Treat-6	Treat-7	Treat-8
80%	Treat-9	Treat-10	Treat-11	Treat-12

(d) Note that there are 12 treatments and each treatment is repeated two times, called replication, thus:

$$\text{Total number of observations} = (\text{No. of treatments}) \times (\text{No. of replications})$$
$$= 12 \times 2 = 24.$$

(e) This is an experimental study because the treatments are actively imposed on the brass plates.

Example 2.2. (PIZZA LOVERS) A local cat food company is trying to introduce a new cat-chow-pizza-crust to the market.

Cat waiting for pizza   Fig. 2.11. Pizza shop.   Cat waiting for pizza

An experiment is conducted to find the optimal baking time (15 or 20 minutes), baking temperature (200°C, 210°C, or 220°C), and number of toppings (4 or 5). Six batches of **Cat-Chow-Pizza-Experts (CCPEs)** will be assigned to each treatment. The **CCPEs** will rate the quality of pizza on a five-point scale to determine the appearance, taste, size and satisfaction of the pizza.

(a) What is the 5 point scale?
(b) How many response variable(s) are there? List the name(s).
(c) Give the factors and number of levels for each factor.
(d) How many treatment combinations are there?
(e) What are the experimental units?
(f) How many experimental units are used in this experiment?
(g) How many total number of observations are in this experiment?
(h) Give a design layout table for this experiment.

Solution. (a) A scale used by CCPEs for rating the appearance, taste, size and satisfaction of Cat-Chow-Pizzas. For example:

1	2	3	4	5
No reply	Poor	Good	Better	Best

(b) There are four response variables: appearance, taste, size and satisfaction.

(c) There are three factors: baking time, baking temperature, and number of toppings. The baking time has two levels (15 or 20 minutes), the baking temperature has three levels (200°C, 210°C, or 220°C) and the number of toppings has two levels (4 or 5).

(d) Note that:

Total number of treatments =
$$= (\text{no. of levels of 1st factor}) \times (\text{no. of levels of 2nd factor}) \times (\text{no. of levels of 3rd factor})$$
$$= (\text{no. of levels of time}) \times (\text{no. of levels of temp.}) \times (\text{no. of levels of toppings})$$
$$= \quad 2 \quad \times \quad 3 \quad \times \quad 2$$
$$= 12 \ (\text{treatements})$$

(e) The experimental units are **Cat-Chow-Pizza-Experts (CCPEs)**.

(f) The number of experimental units will be same as the number of observations. Note that there are six CCPEs attached to each treatment, which means the number of replications is six.

We know that the total number of observations is shown by:

$$= (\text{Total number of treatments}) \times (\text{Replications}) = 12 \times 6 = 72$$

Thus, there will be a total of 72 observations of CCPEs involved in the experiment.

(g) Same question as (f)

(h) Design layout in tables:

This design can be presented with the help of two tables as below:

Table 1. Pizza with 4 toppings.

Baking Time (Min.)	Baking Temperature (^{o} C)		
	200°C	210°C	220°C
15 minutes	Treat-1	Treat-2	Treat-3
20 minutes	Treat-4	Treat-5	Treat-6

Table 2. Pizza with 5 toppings.

Baking Time (Min.)	Baking Temperature (^{o} C)		
	200°C	210°C	220°C
15 minutes	Treat-7	Treat-8	Treat-9
20 minutes	Treat-10	Treat-11	Treat-12

Experimental Design

2.4 RANDOMIZED BLOCK DESIGN

A block is a collection of experimental units or subjects that are known before the experiment to be similar in some way. These units or subjects are expected to affect the response to the treatments. In a Randomized Block Design (RBD), the random assignment of units to treatments is carried out separately within each block.

2.5 COMPLETELY RANDOMIZED DESIGN

In the above randomized block designs, if one ignores the blocks but does randomization of treatments among the groups, then it is called Completely Randomized Design (CRD). In CRD, we can only compare groups or treatments with each other. Sometimes CRD is also called **one ways classification**, and RBD is called **two ways classification**. In other words, in CRD we can compare either only columns (only treatments) or only rows (again only treatments) at a time. In RBD, we can compare rows (blocks) as well as columns (groups) with each other at the same time.

Example 2.3. (UNDERSTANDING BLOCKING) Suppose there are 18 animals (units) in an experiment. We wish to apply Randomized Block Design (RBD) to compare the performance of different types of animals while each animal is provided with different types of training.

Fig. 2.12. Making blocks.

Solution. The above 18 animals can be classified into six types: deer, elephant, kangaroo, koala, lion, and zebra. Then each type of animal within the blocks can be given different types of training such as dancing, riding, or playing.

Note that there are six types of animals, thus six different blocks (homogeneous collections) can be formed. For example, all three deer are in the first block, all three elephants are in the second block, all three kangaroos are in the third block, all three koalas are in the fourth block, all three lions are in the fifth block, and all three zebras are in the sixth block.

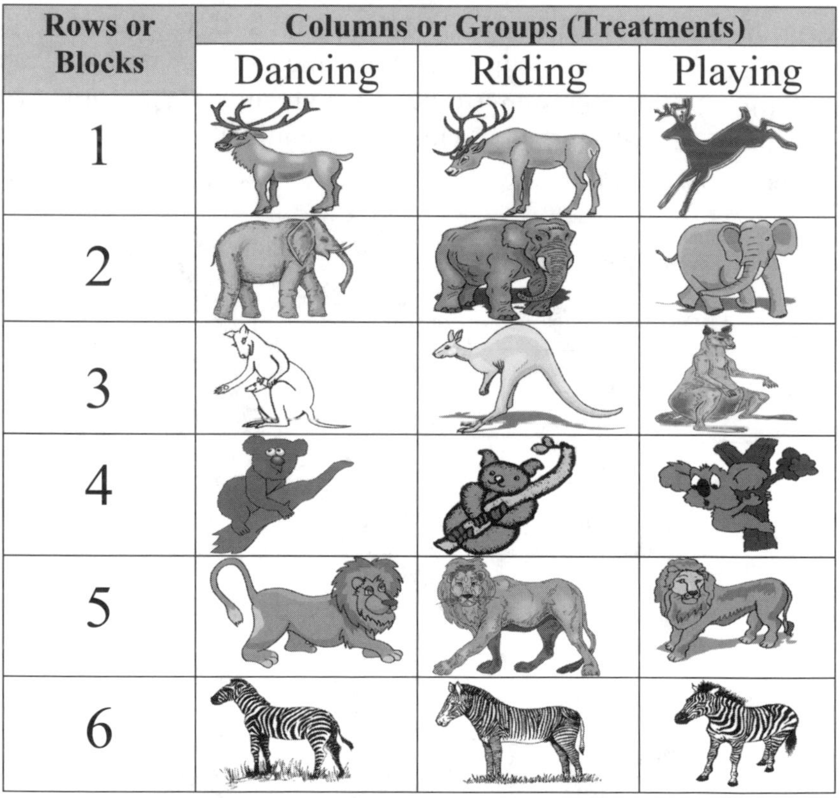

Fig. 2.13. Blocks and groups.

Thus, each block (row) consists of particular types of animals (units), but each group (column) consists of different types of animals (units). In other words the rows (blocks) are homogeneous like strata and the columns (groups) are heterogeneous like clusters. Within each block we have three animals, thus three different random treatments can be applied within each block. This means that in the first column (group) all six animals are under the same kind of training, maybe dancing. The six animals in the second group are again under the same training, say riding, and the other six animals in the last column are under the same training, say playing. Now if we compare six rows with one another (different types of animals) and three columns (different types of trainings) with each other, it is called two ways classification or randomized block design (RBD). If we consider the comparison of only three columns, ignoring the comparison among rows (types of animals), it is called one ways classification or completely randomized design (CRD). Note that if equal numbers of units (animals) are assigned to each treatment (training), it is called a **balanced design**.

Blinding Experiments

When we use such experiments on human beings (subjects), we sometimes do blind experiments. If a patient taking the medicine knows about it, it is called an experiment. For example, consider the following situation of a patient having headache. The doctor gives a tablet of Tylenol to the patient and tells them about it. This is called an experiment.

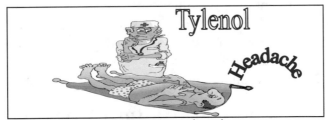
Fig. 2.14. Doctor and patient.

If the patient taking the medicine does not know about it, but the doctor does, it is called a **single blind experiment**.

If both the doctor and the patient do not know about the exact name of the medicine (say Tylenol), but a third party knows, it is called a **double blind experiment**.

A **placebo** is a dummy drug – it looks like the real drug, but has no active ingredients. Note that if a patient is recovering due to a dummy drug it is called a **placebo effect**.

A **confounding variable** is a variable whose effect on the output variable cannot be separated from the effect of the input variable on the output variable. The confounding variables are not of primary interest and may not even be measured, but are associated with the response variable.

For example, a hungry person may have a headache and may be taking Tylenol due to the feeling of a headache. Thus, hunger is a confounding variable.

Confounding variables more often occur in observational studies. In experimental studies, it is possible to control confounding variables.

Basic principles of a design of an experiment

The basic principles of the design of an experiment are:

(a) **Control:** Control confounding variables. Do not apply any treatment to the control unit.
(b) **Randomization:** This ensures that the experiment does not favor one treatment over the other.
(c) **Treatments:** Give treatments of interest to some units for study.
(d) **Replication:** Do not trust only a single observation, assign at least two units to each treatment.
(e) **Blinding:** If possible, give a placebo to the control group.

Example 2.4. (PRESCRIPTIONS ARE IMPORTANT) A study of the anti-depression drug REMERON–30 mg (Mirtazapine) was tested with two groups, each consisting of 10 patients. The treatment group was given the drug while the control group was given a placebo. The doctor also informed the patients whether he/she received the actual drug or just the placebo. Which one of the following basic principles of experiment design was violated?
(a) Randomization (b) Control (c) Blinding (d) Replication
(e) No principle violated.
Solution. (c) Blinding.

2.6 SURVIVAL ANALYSIS

Survival analysis is a sequence of statistical procedures used to analyze statistical data whose study variable is the amount of **time** until an **event** of interest occurs.

We can understand the above two lines if we know the meaning of **time** and **event** in survival analysis.

2.6.1 EVENT

In survival analysis an event means death, disease incidence, exam failure, return to school (or recovery) or any other event of interest. Note that recovery is a positive event whereas others such as death, disease incidence, failure etc. are negative events. We denote event with uppercase E or X.

2.6.2 TIME

In survival analysis the time corresponds to the age of an individual under study when the event occurs. We generally measure time in years, months, weeks, days, hours, minutes, or seconds. The time variable is generally called survival time, and we denote it by uppercase T. Note that any particular value of survival time (T) is denoted by lowercase t. In other words, $0 \leq t \leq T$.

2.6.3 CENSORED DATA

In survival analysis it is most important to learn about the censored data. If for an individual we have some information about the survival time, but we do not know the exact survival time, then such a record is called a censored record.

To understand it in a better way, consider the situation of seven elephants under study for a period of 11 months who are leukemia patients.

Amy enters the study at the beginning and goes 'out of remission' after five months, thus her survival time is five months. Note that 'out of remission' means the event of interest has happened.

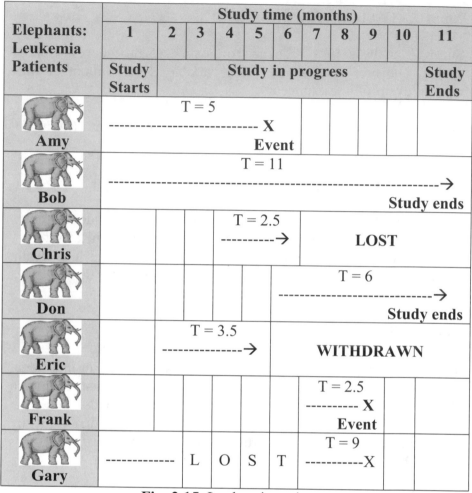

Fig. 2.15. Leukemia patients.

Bob remains in remission until the study ends. Thus the minimum survival time of **Bob** is eleven months, but his exact survival time is unknown. In other words, Bob's survival time is censored.

Chris enters the study at the end of the third month and remains in remission for a period of two and a half months and then gets lost from the study. Again, the exact survival time of Chris is not known, and is censored.

Don enters the study at the end of five months and remains in remission until the study ends. Thus Don's exact survival time is unknown and is censored.

Eric enters the study at the start of the second month and remains in remission for a period of three and a half months, but later withdraws himself (maybe due to adverse effect of medicine etc.). This means his time is also censored.

Frank enters the study at the end of the sixth month and goes 'out of remission' after a period of two and a half months. Thus, the survival time of Frank is two and a half months.

Gary enters at the beginning of the study and gets lost for a few weeks and then re-enters the study. He goes out of remission at the end of the nine months. In this case, Gary's survival may be recorded as nine months although he was lost for a couple of weeks.

2.6.4 SURVIVAL DATA STORAGE

Note that the survival time is of two types: one is exactly known and the other is censored. To differentiate between exact survival time and censored survival time, we define an indicator variable *psi* as:

$$\psi_i = \begin{cases} 1 & \text{if event occured} \\ 0 & \text{if censoring occured} \end{cases}$$

Thus the following table is generally used to present survival data.

Patient Elephant	Survival Time (months)	Event (1) Censored (0) (ψ_i)
Amy	5	1
Bob	11	0
Chris	2.5	0
Don	6	0
Eric	3.5	0
Frank	2.5	1
Gary	9	1

The first column generally represents the patient's name or ID number, while the second column represents survival time (may or may not be censored). The third column is generally used to distinguish if the survival time is censored or not.

In the previous example, the patients Chris and Frank have the same survival time of two and a half months. But, note that the survival time of Frank is not censored whereas that of Chris is censored. This is distinguished in the third column by inserting an indicator variable with values 0 or 1.

Sometimes we also have information about the units (say patients) such as their age, sex, and smoking status etc. These can be tabulated in the consecutive columns of the above table. The discussion of survival analysis is beyond the scope of this book, but two well known mathematical techniques are (a) Survivor function and (b) Hazard function, which can be taken from advanced books.

Similarly, you may think of some new situations where the data may or may not be considered as censored.

2.7 HUMAN BLOOD TYPES

All humans have ABO blood type. Every human being has two blood type alleles, one inherited from their mother and another from their father. Thus, the parents can have alleles A, B, O or a combination of any of these two inherits (or called Mendelian Inherits) results in the blood type of their child. The following table lists such possibilities:

Parent's Alleles		Child's blood type
O	O	O
A	A	A
B	B	B
A	O	A
B	O	B
A	B	AB

Remark: Note that animals and birds have many blood types, which you can find on the internet.

Example 2.5. (DREAMS MAY COME TRUE) Angie and **Bob** each have alleles **A** and **B**, respectively. What blood type can their child have?

Fig. 2.16. Angie and Bob's expectations.

Solution. From the above table, if **Angie** has **A** and **Bob** has **B** alleles, then their child should have **A** $\oplus$ **B = AB** blood type.

Example 2.6. (HAPPY FAMILY) Amy and **Adam** both have alleles **A** and **A.** What blood type can their child have?

Fig. 2.17. Amy and Adam's family.

Solution. From the above table, if **Amy** has **A** and **Adam** has **A** alleles, then their child should have **A** $\oplus$ **A = A** blood type.

Example 2.7. (ONE GOOD KID IS ENOUGH) Anna and **Olson** each have alleles **A** and **O**, respectively. What blood type can their child have?

Fig. 2.18. Anna and Olson's family.

Solution. From the above table, if **Anna** has **A** and **Olson** has **O** alleles, then their child should have **A** $\oplus$ **O = A** blood type.

LUDI 2.1. (IMPROVE YOUR HANDWRITING) Write a short note (not more than one page) on the basic principles of designing an experiment.

LUDI 2.2. (EXERCISE TO REDUCE YOUR WEIGHT) A study was conducted on elephants to find out what kind of diet was best for weight loss. Four levels of fiber content were chosen: low, medium, moderate, and high; while three levels of carbohydrate content were chosen: low, medium and high. Twenty-five elephants were randomly assigned to each treatment combination of fiber and carbohydrate content. Their weights were measured two months after administering the diets.

Fig. 2.19. Elephant.

(a) The study variable is …..
(b) The auxiliary variables are ……
(c) The total number of elephants required for this experiment are …
(d) What are the levels for each one of the auxiliary variables?

LUDI 2.3. (BIRD WATCHING) Consider the situation where a bird-watcher captured seven parrots. The bird-watcher attached a ring to the legs of the parrots and let them go in the forest. Then the bird-watcher tried for a period of 12 months to capture birds from the same forest. The bird-watcher's interest is to know if anyone among these parolee parrots got recaptured or not. Thus, the bird-hunter's event of interest is 'parolee parrot getting recaptured' over a period of 12 months. Explain situations where censorship of data may occur.

Fig. 2.20. Parrot lovers.

LUDI 2.4. (EGG CONSUMPTION) In a study, 5000 females were asked about their frequency of egg consumption and then their blood cholesterol level was recorded. At the end of the study, it was observed that females who ate more eggs had higher cholesterol levels.

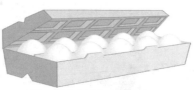

Fig. 2.21. Egg carton.

(a) Is this an observational study or an experimental study?
(b) What is the response variable?
(c) List at least one confounding variable.
(d) What is the explanatory variable?

LUDI 2.5. (LAB EXPERIMENTS) A mechanical engineering group plans to conduct an investigation of the curvature of iron plates. It is believed that the amount of curvature depends upon the temperature and the iron content of the plates. The investigation was conducted for the temperatures 50°C, 75°C, 100°C, 125°C and 150°C; and percentage of iron content 40%, 60%, 80% and 90%. There were two observations (replications) for each treatment combination.

Fig. 2.22. Mechanic.

(a) Give the response variable(s).
(b) Give the factors(s) and the corresponding number of levels for each response variable.
(c) How many treatment combinations are there?
(d) How many total number of observations would there be for this investigation?
(e) Is this an observational or an experimental study? Explain.

LUDI 2.6. (PIZZA LOVERS) A local pizza company is trying a new pizza crust. An experiment is conducted to find the best cooking time (20, 25, 30, 35 minutes), cooking temperature (400°F, 425°F, 450°F), and amount of cheese (2 cups, 2.5 cups). Five batches of pizza crust dough will be assigned to each treatment. Expert taste testers will rate the crusts on a five point scale with respect to flavor, outlook and size.

Fig. 2.23. A pizza shop.

(a) What is the 5 point scale?
(b) How many response variable(s) are there? List the name(s).
(c) Give the factors and number of levels for each.
(d) How many treatment combinations are there?
(e) What are the experimental units?
(f) How many experimental units are used in this experiment?
(g) Give a design layout table for this experiment

LUDI 2.7. (DOES EXPERIENCE MATTER?) A survey was conducted to learn about the effect of work experience on the hiring decision of experts. Assume that there are 110 city blocks and the interviewer decides to take a 1-in-10 systematic sample of these blocks. All of the eligible candidates living in these selected blocks will be used in the survey.
(a) Is this an observational or an experimental study? Explain.
(b) What is the response variable?
(c) Is there any explanatory variable?
(d) How many blocks will be selected in the systematic sample?
(e) Use the first two columns of the Random Numbers **Table I** given in the Appendix to decide a random start between 1 and 10, and later select a systematic sample. Circle the blocks selected.

BLOCKS									
1	2	3	4	5	6	7	8	9	10
11	12	13	14	15	16	17	18	19	20
21	22	23	24	25	26	27	28	29	30
31	32	33	34	35	36	37	38	39	40
41	42	43	44	45	46	47	48	49	50
51	52	53	54	55	56	57	58	59	60
61	62	63	64	65	66	67	68	69	70
71	72	73	74	75	76	77	78	79	80
81	82	83	84	85	86	87	88	89	90
91	92	93	94	95	96	97	98	99	100
101	102	103	104	105	106	107	108	109	110

(f) Note that we are collecting information from all the adults living in the selected blocks, such a sampling scheme is called …………...

LUDI 2.8. (A ONE WORD DECISION MAY BE DIFFICULT)
State if the following statements are true or false.

(a) An experiment is said to be double-blinded if a placebo is given to some of the subjects. (T or F)
(b) Assume an experiment has five treatment combinations. If the experiment uses random allocation, then the number of units assigned to each treatment combination must be the same. (T or F)
(c) If there is only one independent variable, the number of treatments equals the number of levels of that variable. (T or F)
(d) In a randomized, controlled, double blinded experiment the primary role of randomization is to allocate the experimental units (subjects) to the treatment and control groups in such a way that neither the subject nor the experimenter know whether the treatment or the placebo will be given to that group. (T or F)
(e) A **placebo** is a dummy drug. It looks like the real drug, but has no active ingredients. (T or F)

LUDI 2.9. (BEAUTIFUL HANDWRITING REFLECTS YOUR BEAUTIFUL MIND) Write a short paragraph on survival analysis. What is meant by censored data? Explain with an example.

LUDI 2.10. (BURN YOUR FAT WITH EXERCISE) A study was conducted to find out what kind of diet was best for reducing weight. Three levels of fiber contents: low, medium and high; and three levels of carbohydrates: low, medium and high, were chosen. Twenty people were randomly assigned to each treatment made of a combination of fiber and carbohydrates. Their weights were recorded after a 20 weeks period of treatment.

(a) Is this an experimental or an observational study?

(b) The response variable is -----

(c) The explanatory variables are (list all of them) -----

(d) List the corresponding levels for each explanatory variable -----

Fig. 2.24. Patients assigned to each treatment.

(e) The total number of treatments is -----

(f) The total number of patients required for this experiment is …

(g) Provide a design layout.

LUDI 2.11. (TECHNICAL TERMS ARE IMPORTANT) In the following table write two names of output variables and two names of input variables used in statistical science.

	Input variable
1	
2	

	Output variable
1	
2	

LUDI 2.12. (ALWAYS REMEMBER YOUR BLOOD TYPE)
(a) Amy and Michael each have alleles A and O, respectively. What blood type can their child have?

Fig. 2.25. Amy and Michael's family.

(b) Heather and Chris each have alleles A and B, respectively. What blood type can their child have?

Fig. 2.26. Heather and Chris's family.

(c) Janie and Kevin each have alleles B and O, respectively. What blood type can their child have?

Fig. 2.27. Janie and Kevin's family.

LUDI 2.13. (DESIGN OF EXPERIMENT EXPERTS) Consider 18 fruits (units) in an experiment. We wish to compare the survival time for each type of fruit as well as three different types of cold storages with each other.

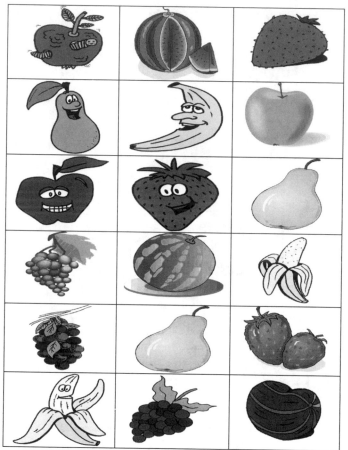

Fig. 2.28. Making blocks.

(a) Design the appropriate randomized block design (RBD) to compare the three types of cold storage as well as the six types of fruits.

(b) If we consider only the comparison of three types of cold storage with each other, ignoring the type of fruits, name the design layout you are using to compare these three types of storages.

3. GRAPHICAL REPRESENTATION

3.1 INTRODUCTION

In this chapter, we discuss different types of variables and their presentation using graphical techniques. Consider the following table showing a class of four students who have already taken Stat 100:

	Student Name	Gender	Grade in Stat 100	GPA	# of "A" grades
	Amy 	F	A	3.8	5
	Bob 	M	F	2.6	2
	Cara 	F	B	2.2	1
Instructor	Don 	M	C	3.6	4

Fig. 3.1. Qualitative and Quantitative variables.

In the above table, the experimental units or subjects are "students," and for each student we are measuring four variables: "Gender," "Grade in Stat 100," "GPA," and "# of A grades".

3.2 QUALITATIVE RANDOM VARIABLES

Note that the variable "Gender" tells us that the quality of the students with the names Amy, Bob, Cara and Don are female, male, female and male, respectively. Thus the variable "Gender" classifies the students into two categories: male and female. Such a variable is called **categorical variable or qualitative variable**. These random variables assume values that are not necessarily numerical, but can be categorized. Adding, subtracting, or averaging such qualitative variables have no meaning. Thus all numerical variables such as, "Phone numbers," "Area codes," "Zip codes," and "Visa card numbers" etc. are all qualitative variables. "Gender" has two possible categories: male and female. These two categories can be arbitrarily coded numerically as female = 0 and male = 1. Such coded variables are called **nominal** variables. Now consider "Grade in Stat 100" which can take five possible categories: A, B, C, D and F. These five categories can be arbitrarily coded numerically as: A = 4, B = 3, C = 2, D = 1, and F = 0. Note that here the magnitude of coding tells us the quality of the grade. If the code is 3 then the grade is better than if the code is 2. Such a coded variable is called an **ordinal** variable. Also note that in the case of the **nominal** variable, the code male = 1 and female = 0 does not mean that males are superior to females. **Pie** or **bar charts** are generally used to represent **qualitative** variables. Thus qualitative variables are of two types: (a) nominal variables and (b) ordinal variables.

3.2.1 NOMINAL VARIABLES

If the categories of a qualitative variable are simply renamed with numbers, such a variable is called a **nominal** variable. For example: gender (male = 0, female =1); and color (red = 1, blue = 2) etc.

3.2.2 ORDINAL VARIABLES

If the categories of a qualitative variable can be ordered based on the magnitude of the coded numbers, such a variable is called an **ordinal** variable. For example: grades (A = 4, B = 3, C = 2, D = 1, F = 0) etc.

3.3 QUANTITATIVE RANDOM VARIABLES

Quantitative random variables can take numerical values for which adding, subtracting, or averaging does have meaning. For example, the variable "GPA" and "# of A grades" are quantitative random variables. Note that the variable, "# of A grades" cannot be a fraction, but the variable "GPA" can take any fractional value. Thus, quantitative variables are also of two types: without and with fractional values, and are called: (a) discrete random variables, and (b) continuous random variables.

3.3.1 DISCRETE RANDOM VARIABLES

If a random variable takes a countable number of non-negative integer values, it is called a discrete random variable. For example, the variable "# of A grades" takes a value of whole numbers 0, 1, 2, etc. Note that the "# of A grades" cannot be a decimal like 2.4, and we have to jump from one value to another value as shown below:

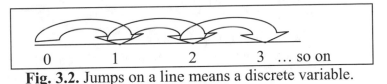

Fig. 3.2. Jumps on a line means a discrete variable.

A few examples of quantitative and discrete random variable are: "# of kids", "# of chairs", "# of apples", etc.
Note that "# of anything" is a discrete random variable.

3.3.2 CONTINUOUS RANDOM VARIABLES

A random variable is said to be continuous if it can take any possible values on a real line or interval. For example, the variable "GPA" can take any real value on an interval from 0 to 4, and we can move smoothly from one point to another point without any jumps as shown below. Such a variable is called a **continuous random** variable.

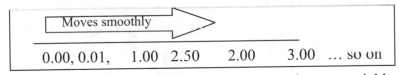

Fig. 3.3. Smooth move on a line means a continuous variable.

A pictorial representation to differentiate between **qualitative** and **quantitative** variables is given in the following figure.

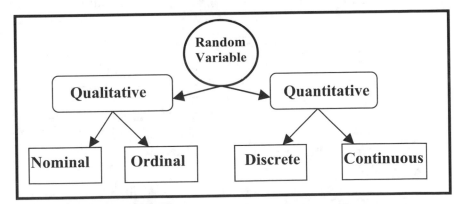

Fig. 3.4. Types of random variables or data.

A few examples of quantitative and continuous random variable are: age, height, weight, pressure, volume, length, salary, temperature, and speed etc. Note that "age" itself is a quantitative variable whereas "age groups" is a qualitative variable.

Pie charts, bar charts, dot plots, line (or time) plots, stem and leaf plots, histograms, box plots, and scatter plots are generally used to represent **quantitative** variables.

Example 3.1. (VARIABLES) Identify the following variables:

	Variable	Picture	Answer
1	Number of elephants		Quantitative and discrete
2	Length of a rat's tail		Quantitative and continuous
3	Frog jump		Quantitative and continuous
4	Color of a rat		Qualitative and nominal
5	Number of parrots		Quantitative and discrete
6	Weight of elephants		Quantitative and continuous
7	Grade in the class		Qualitative and ordinal
8	Age of a parrot		Quantitative and continuous
9	Volume of a pumpkin		Quantitative and continuous
10	Number of marbles in a bag		Quantitative and discrete

Fig. 3.5. Differentiating variables.

3.4 GRAPHS FOR QUALITATIVE DATA

If a variable of interest is qualitative (or categorical), the statistical table is a list of categories along with a measure of how often each category occurred. If the frequency, f, or the number of data values in each category is known, and n is the total number of data values in all the categories, then the following formulae are applicable:

$$(1) \quad \text{Relative frequency (RF)} = \frac{\text{Frequency}\,(f)}{n}$$

$$(2) \quad \text{Percent} = (RF) \times 100\%$$

$$(3) \quad \text{Angle} = (RF) \times 360^{o}$$

Such a statistical table can be presented with two types of charts:

(a) Pie chart (b) Bar chart

The category that occurs most frequently is called **modal category** or **mode**.

3.4.1 PIE CHART

A *pie* chart is a well-known circular graph, which shows how data values are distributed among the categories.

3.4.2 BAR CHART

A *bar* chart shows the distribution of data values in categories with the height of the bar measuring how often each category occurred.

Any data set having only one mode is called *uni-modal,* that having two modes is called *bi-modal,* and so on.

Example 3.2. (PIE AND BAR CHART) An elephant, a rat and a man were admitted into the hospital. A doctor noted their pulse rate per minute as follows:

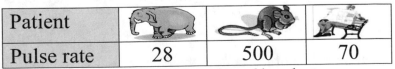

Patient			
Pulse rate	28	500	70

Fig. 3.6. Illustrating pie and bar chart.

(a) Is it uni-modal data set?
(b) Construct a pie chart (or circular chart).
(c) Construct a bar chart.
Solution. (a) Consider the following table:

Patient	Medical ID	Pulse Rate (f)	RF	Percent	Angle
Elephant		28	0.047	4.7	16.92
Rat		500	0.836	83.6	300.96
Man		70	0.117	11.7	42.12
Sum		598	1.000	100.0	360.00

Fig. 3.7. Steps for making a pie and a bar chart.

In the above table, we used:

$$n = 500 + 70 + 28 = 598.$$

Then, the relative frequencies (RF) of the pulse rates of elephant, rat, and man are as follows:

$$RF(\text{Elephant}) = \frac{f}{n} = \frac{28}{598} = 0.047$$

$$RF(\text{Rat}) = \frac{f}{n} = \frac{500}{598} = 0.836$$

and

$$RF(\text{Man}) = \frac{f}{n} = \frac{70}{598} = 0.117$$

Note that the sum of the relative frequencies (or proportions) is always one. Sometimes we may get a sum equal to 0.999 or 0.998 etc. due to an error in rounding. Further note that the terms: relative frequency, proportion, probability, and area under a density curve have same meaning.

Here *rat* has maximum frequency, the data set shows only one modal category, hence it is *uni-modal* data set.

Percent: The percentages of pulse rates are given by:

$$\text{Percent(Elephant)} = \text{RF(Elephant)} \times 100\% = 0.047 \times 100\% = 4.7\%$$

$$\text{Percent(Rat)} = \text{RF(Rat)} \times 100\% - 0.836 \times 100\% = 83.6\%$$

and

$$\text{Percent(Man)} = \text{RF(Man)} \times 100\% = 0.117 \times 100\% = 11.7\%$$

Note that the sum of the percentages is always 100%

Angle: The angle of the pulse rates in each category is given by:

$$\text{Angle(Elephant)} = \text{RF(Elephant)} \times 360° = 0.047 \times 360° = 16.92°$$

$$\text{Angle(Rat)} = \text{RF(Rat)} \times 360° = 0.836 \times 360° = 300.96°$$

and

$$\text{Angle(Man)} = \text{RF(Man)} \times 360° = 0.117 \times 360° = 42.12°$$

Note that the sum of angles around a point is always 360°.

(b) Pie chart: A pie chart displaying the pulse rates of the elephant, the rat, and the man is given by:

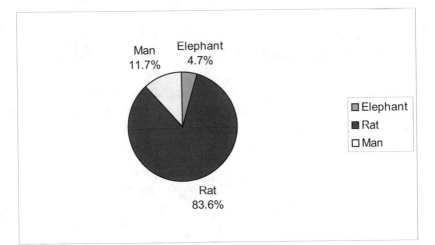

Fig. 3.8. Pie chart.

The above graph shows the pulse rates of the rat, the man, and the elephant are 83.6%, 11.7% and 4.7% per minute, respectively.

(c) Bar chart: A bar chart showing the pulse rates of the elephant, the rat, and the man is given below:

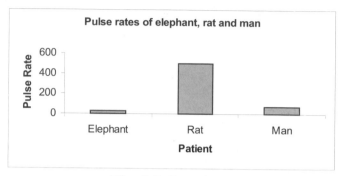

Fig. 3.9. Bar chart.

Note that there should be some space from bar to bar in a bar chart. There is no rule to determine how much space between bars or to decide the width of each bar. However, the height of the bar must be proportional to the pulse rates of the elephant, the rat, and the man.

3.5 RAW DATA

Data set recorded in the sequence in which it is collected before it is processed or ranked is called raw data.

3.6 FREQUENCY DISTRIBUTION TABLE FOR A QUALITATIVE VARIABLE

A frequency distribution table for a qualitative variable lists all the categories and number of units (or elements) that belong to each category. For example, consider the following table showing the colors of 49 marbles in a bag:

Color	Frequency
Red	14
Pink	25
Yellow	10

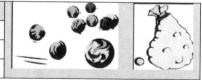

Fig. 3.10. Marbles.

Note that in the above table, color is a qualitative variable and that marbles are experimental units. This table is called a frequency distribution table for the qualitative variable "color."

3.6.1 PARETO CHART

A bar chart in which the bars are ordered from largest to smallest is called a *Pareto* chart. A reverse process may be called an *anti-Pareto* chart.

Example 3.3. (A VISIT TO THE ZOO) A class of 20 students visited a local zoo and after a period of one month the school teacher asked each student to write the name of one animal on a piece of paper and submit it to the teacher. The teacher read the names of the animals from the pieces of paper in the sequence.

Fig. 3.11. Memory of students toward animals.

(a) Construct a frequency distribution table from the above raw data.
(b) Construct a Pareto chart and define it.
Solution. (a) The frequency distribution table is given by:

Category	Tally	Frequency
	||	2
	⦚⦚⦚	5
	⦚⦚⦚ ||	7
	⦚⦚⦚ |	6

Fig. 3.12. Frequency distribution table.

(**b**) From the above table in (a), the majority of the students can remember a lion, followed by an elephant, a monkey, and then a donkey. Thus lion is a modal category. Let us now make a bar chart with the bars arranged in the order of the memory of the students from highest to lowest, called a Pareto chart, as follows:

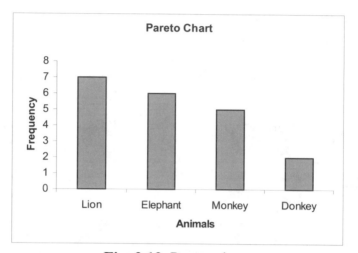

Fig. 3.13. Pareto chart.

Example 3.4. (ALWAYS REMEMBER YOUR BLOOD TYPE) Consider the following pie chart showing the distribution of blood types for 200 people in a hospital.

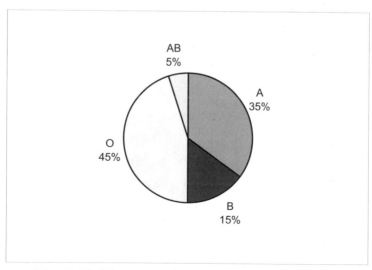

Fig. 3.14. Pie chart: distribution of blood types.

(a) How many people have A blood type?
(b) How many people have AB blood type?
(c) How many people have O blood type?
(d) How many people have B blood type?
(e) How many people have A or O blood type?
(f) How many people have A and O blood type?
(g) Which category has the maximum number of people?
(h) Which category has the minimum number of people?

Solution. Here the total number of people is $n = 200$ and out of them, 35% have A blood type, 15% have B blood type, 45% have O blood type, and 5% have AB blood type.

(a) The number of people having A blood type:

$$= 200 \times 35\% = 200 \times \frac{35}{100} = 70$$

(b) The number of people having AB blood type:

$$= 200 \times 5\% = 200 \times \frac{5}{100} = 10$$

(c) The number of people having O blood type:

$$= 200 \times 45\% = 200 \times \frac{45}{100} = 90$$

(d) The number of people having B blood type:

$$= 200 \times 15\% = 200 \times \frac{15}{100} = 30$$

(e) Number of people having either A or O blood type:
$$= 70 + 90 = 160$$

(f) Number of people having A and O blood type:
$$= 0$$

(g) Blood type - O

(h) Blood type - AB

3.7 GRAPHS FOR QUANTITATIVE DATA

The following graphs are used to present a quantitative variable:

(a) Pie chart (b) Bar chart (c) Dot plot (d) Line (or Time) plot
(e) Stem and leaf plot (f) Histogram (g) Box plot and (h) Scatter plot.

The following figure lists different types of graphs for qualitative and quantitative data sets:

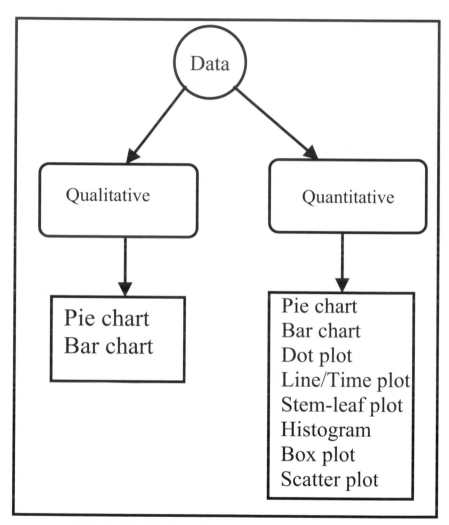

Fig. 3.15. Choice of graphs for a given data set.

The use of scatter plots will be discussed when two or more related variables are introduced, and box plots will be discussed when we understand the computation of quartiles and the median of a data set. The rest of the graphs, which generally deal with a single variable, and a line (or time) plots that deal with two variables (say, time versus a variable) will be discussed in this chapter.

3.7.1 PIE CHART

A pie chart for a quantitative variable is a well-known circular graph that shows how data values are distributed among different classes.

3.7.2 BAR CHART

A bar chart for a quantitative variable shows the distribution of data values in different classes with the height of the bar measuring how often a particular class was observed.

Let us explain each one of these graphs with an example:

Example 3.5. (MEDICAL BILLS GIVE A HARD TIME) The amount of money spent on the medical bills of five classes of animals consisting of elephants, lions, monkeys, donkeys and rats in a national zoo is listed below:

Animal		Medical Bills ($ millions)
Elephants		$70.8
Lions		$90.9
Monkeys		$55.0
Donkeys		$34.7
Rats		$6.8

Fig. 3.16. Horrible bills!

(a) Construct a pie chart.
(b) Construct a bar chart.

Solution. Note that here "medical bills" is a quantitative variable, animals are experimental units, and each type of animal forms a class. To construct a pie chart, the angle and percentage for each type of animal (or for each class) is calculated as follows:

Classes	Amount ($ millions)	$\text{Angle} = \dfrac{\text{Amount}}{\text{Sum}} \times 360°$	$\text{Percent} = \dfrac{\text{Amount}}{\text{Sum}} \times 100\%$
 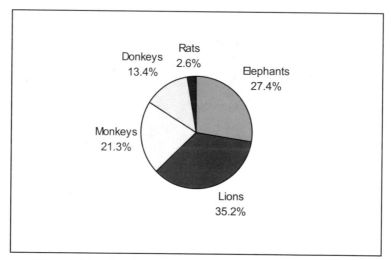	70.8	98.71°	27.4%
	90.9	126.74°	35.2%
	55.0	76.69°	21.3%
	34.7	48.38°	13.4%
	6.8	9.48°	2.6%
Sum	258.2	360°	99.9%

Fig. 3.17. Steps to find the angles and percentages.

(a)Then we have the following pie chart showing the distribution of medical bills among the five classes of animals in the zoo:

Fig. 3.18. Pie chart: distribution of bills in classes.

(b) The bar chart to present the distribution of the amount of medical bills among the five classes of the animals in the zoo is given by:

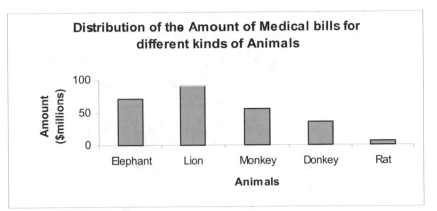

Fig. 3.19. Bar chart: distribution of bills.

3.7.3 DOT PLOT

A plot showing the "dot" corresponding to each quantitative data value is called a **dot plot**.

A dot plot could be used to know the distribution of a data set, that is, to know if the distribution is symmetric or skewed. A dot plot could be used to detect **outliers**. Note that any strangely large or small value in a data set is called an **outlier**. A dot plot could also be used to find the most frequently occurred value(s) called the **mode(s)** and mid value called the **median**.

For example, consider the following table listing elephant Bob's number of working hours at a circus during the last six days:

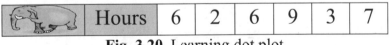

| Hours | 6 | 2 | 6 | 9 | 3 | 7 |

Fig. 3.20. Learning dot plot.

The dot plot of such a data set is given by:

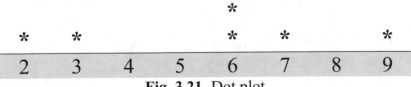

Fig. 3.21. Dot plot.

This dot plot is an example of an asymmetric and uni-modal data set with mode = 6 and median = 6.

Example 3.6. (HOCKEY MATCH) The following is the number of goals scored by a hockey team in 21 matches:

3, 5, 2, 2, 1, 4, 0, 0, 3, 3, 2, 1, 10, 2, 4, 1, 1, 3, 2, 4, 5

A dot plot was used to display the goal scores:

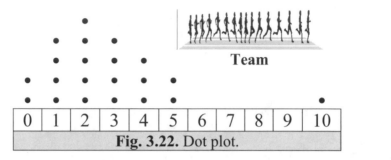

Fig. 3.22. Dot plot.

Then an instructor raised the following questions:

	 Instructor	Answers by a  talking parrot
(a)	What are the experimental units?	Hockey matches
(b)	What is the study variable?	Goals
(c)	What is the minimum number of goals?	0 goals
(d)	What is the maximum number of goals?	10 goals
(e)	Is there any outlier?	Yes, 10
(f)	How many times were 3 goals scored?	4 times
(g)	What is the median score?	11th value = 2 goals
(h)	How many times was the score more than 3?	6 times
(i)	How many times was the score 3 or more?	10 times
(j)	How many times was the score below 2?	6 times
(k)	How many times was the score 2 or less?	11 times
(l)	How many times was the score below 6?	20 times
(m)	Is the distribution symmetric?	No, skewed to the right
(n)	How many times was the score 7?	Never
(o)	How many times was the score more than 7?	1 time
(p)	How many total matches were played?	21
(q)	Which value occurred most frequently? or what is the mode value?	2 goals, it occurred times

Fig. 3.23. Reading a dot plot.

Example 3.7. (THREE DISTRIBUTIONS) (a) Explain the three types of distributions.
(b) Construct three dot plots showing the three distributions.

Solution. (a) (i) Symmetric distribution: If a vertical line can be placed in the center of a graph such that one side can be seen as a mirror image of the other, then the graph is a symmetric distribution. Thus the right and left tails are of the same length.

Left tail Right tail

Fig. 3.24. Symmetric distribution.

(ii) Skewed to the right distribution: If the right tail of a graph is longer than its left tail, then the graph is a skewed to the right distribution.

He cuts his
left tail  Right tail
is long

Fig. 3.25. Positively skewed distribution.

(iii) Skewed to the left distribution: If the left tail of a graph is longer than its right tail, then the graph is a skewed to the left distribution.

Left tail  He cuts his
is long right tail

Fig. 3.26. Negatively skewed distribution.

(b) A dot plot can be used to see if a distribution is skewed to the left, symmetric or skewed to the right as shown below:

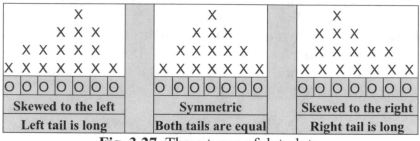

Fig. 3.27. Three types of dot plots.

3.7.4 LINE PLOT

A plot obtained by joining consecutive dots on a dot plot is called a line plot. One such example is a time plot.

(a) **Time plots:** When data values are plotted against time (at equally spaced intervals such as hourly, daily, weekly etc.) or against the order they are collected, it is called a **time plot**. For example, if we collected raw data on the speed of a car, and later plotted speed versus the order of collection of the data as follows:

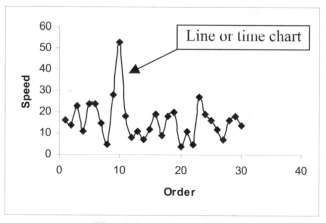

Fig. 3.28. Time plots.

Such a graph is called a line or time plot. Note that there are two variables: order and speed. The variation around the 10th attempt is greater in comparison to the rest of the data set. Such very large or small values are called outliers and one should look for its reason.

(b) **Trend:** A trend in a time series is a persistent, long-term rise or fall. If a **best-fit line** across a time series data has a positive slope, then the trend is positive. If a **best-fit line** across a time series data has a negative slope, then the trend is negative. If a **best-fit line** has a slope of zero, then there is no trend. For learning about the best-fit line, please refer to **Chapter 10.**

(c) **Seasonal variation:** A pattern in a time series that repeats itself at known regular intervals of time is called **seasonal variation.** For example, consider the price of commodity changes every month:

Month	Price ($)	Month	Price ($)
Jan	28	Sept	24
Feb	26	October	26
March	24	Nov.	28
April	20	Dec.	30
May	18	Jan.	28
June	15	Feb.	26
July	16	March	25
August	22		

Thus we have the following line/time plot:

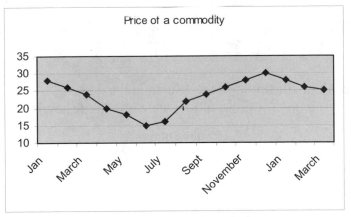

Fig. 3.29. Seasonal variation.

Note that there are two variables: month and price. If there is any seasonal variation, then we need to adjust it before analyzing and interpreting the results. This is also called seasonally adjusted data.

(d) Index number (Fixed basis): In a fixed base, a particular year is chosen as the base year and index numbers are expressed relative of that year. In other words, we compute the percentage of price relative to the price of a commodity of the base year as:

$$P_1 = \frac{\text{Price of the commodity for the current year}}{\text{Price of the commodity for the base year}} \times 100\%$$

This is also called the index number. Now consider the following table with 1982 as the base year:

Year	Price of commodity ($)	Index (base = 1982)
1982	120	$\frac{120}{120} \times 100 = 100\%$
1983	140	$\frac{140}{120} \times 100 = 116.67\%$
1984	270	$\frac{270}{120} \times 100 = 225\%$
1985	110	$\frac{110}{120} \times 100 = 91.67\%$

Example 3.8. (INSURANCE IS IMPORTANT) The following table shows the percentage of disability insurance for elephants in a circus for the years 1980 to 2000 as:

Year	1980	1985	1990	1995	2000
Percent	2.9	3.0	3.8	4.2	4.9

Construct a line (or time) plot to illustrate the data.
Solution. The line (or time) plot is shown in the following figure:

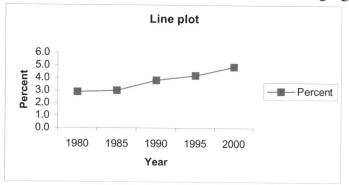

Fig. 3.30. Illustrating a line plot.

3.7.5 STEM AND LEAF PLOT

A stem and leaf plot is a simple way to represent data. It could be used to detect outliers, modal value(s), median and distributions etc. The following rules apply to construct a **stem and leaf** plot.

(i) Divide each data value into two parts: the **stem** and the **leaf**.

For example:

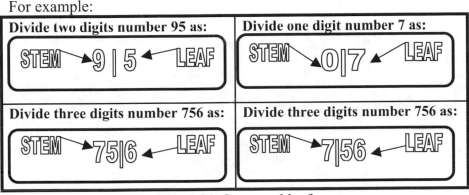

Fig. 3.31. Stem and leaf.

Caution! We prefer one digit leaf, but it also depends upon the other data values.

(ii) List the stems in a column, with a vertical line to their right.
(iii) For each data value, record the leaf portion in the same row as the corresponding stem.
(iv) Order the leaves in ascending order in each stem.
(v) Provide a leaf unit or key to read the graph, which is given by:
    (Stem Leaf) x (Key) = (Data Value).
For example if:
(a) Data value = 95, Stem = 9, Leaf = 5; thus (95)x(Key) = 95 implies: Key = 1.
(b) Data value = 9.5, Stem = 9, Leaf = 5; thus (95)x(Key) = 9.5 implies: Key = 0.1
Here, for simplicity, we will always keep: Key = 1.

Example 3.9. (STEM AND LEAF PLOT) The age (in months) of 19 elephants in a local zoo is given below:

	90	70	70	70	75	70	65	68	60	74
	70	95	75	70	68	65	40	65	70	

Fig. 3.32. An elephant.

Construct a **stem and leaf** plot.

Solution. (i) Divide each data value into two parts: the **stem** and the **leaf**.

9|0	7|0	7|0	7|0	7|5	7|0	6|5	6|8	6|0	7|4
7|0	9|5	7|5	7|0	6|8	6|5	4|0	6|5	7|0	

(ii) – (iii) List stems in a column, and leaves in each stem.

Stems	Leaves									
4	0									
5										
6	5	8	0	8	5	5				
7	0	0	0	5	0	4	0	5	0	0
8										
9	0	5								

(iv) Put the leaves in each stem in ascending order:

Stems	Leaves									
4	0									
5										
6	0	5	5	5	8	8				
7	0	0	0	0	0	0	0	4	5	5
8										
9	0	5								

(v) Leaf unit or key =1.

Example 3.10. (LEARN TO READ COMPUTER OUTPUTS) A zoo statistician used a computer package, like MINITAB, to display the age in years of animals using a stem and leaf plot as given below:

Stems	Leaves								
2	0	2	4	5					
3	3	4	6	8	9				
4	0	0	0	3	7	9			
5	5	5	5	5	6	8	8	8	9
6	2	3	4	5	6	8			
7	2	3	5	7	9				
8	0	4	5	8					

Fig. 3.33. Reading computer outputs.

Then the zoo statistician answered the following questions from the stem and leaf plot:

Questions	Answers
(a) How many animals are there?	39
(b) What are the experimental units?	Animals
(c) What is the study variable?	Age
(d) What is the minimum age?	20 years
(e) What is the maximum age?	88 years
(f) How many animals are 55 years old?	4 animals
(g) What is the median age?	20th value = 56 years
(h) How many animals are older than 68?	9 animals
(i) How many animals are older than or equal to 68?	10 animals
(j) How many animals are younger than 30?	4 animals
(k) How many animals are older than 10?	All 39 animals
(l) How many animals are younger than 90?	All 39 animals
(m) Is this distribution symmetric?	Yes
(n) How many animals are 30 years old?	None
(o) How many animals are 77 years old?	1 animal
(p) What is the key or leaf unit for this graph?	Key = 1
(q) Which value occurred most frequently or what is mode value?	55, it occurred 4 times
(r) Is there any outlier?	No

Example 3.11. (READING A STEM AND LEAF PLOT) (a) Use any computer package, like MINITAB, to construct a stem and leaf plot and provide a leaf unit for the following data set:

52	52	80	96	65	79	71	87
93	95	69	72	81	61	76	86
79	68	50	92	83	84	77	64
71	87	72	92	57	98	81	83
76	65	74	82	77	90	75	91

Fig. 3.34. Computers can make a stem and leaf plot.

(b) Find the minimum and maximum values.
(c) Find the median value.
(d) Find the range.
(e) Is it unimodal?

(f) What are the benefits of the use of a computer?
(g) What are the limitations of the use of a computer?
(h) Is there any outlier?
Solution. (a) Stem and leaf plot:

Stems	Leaves
5	0 2 2 7
6	1 4 5 5 8 9
7	1 1 2 2 4 5 6 6 7 7 9 9
8	0 1 1 2 3 3 4 6 7 7
9	0 1 2 2 3 5 6 8

Key = Unit leaf = 1.

(b) Minimum value = 50; Maximum value = 98.
(c) Median = (77+79)/2 = 78.
(d) Range = 98-50 = 48.
(e) No.
(f) Benefits: It is fast and easy to use.
(g) Limitations: No steps or details are given.
(h) No.

Example 3.12. (STEM AND LEAF PLOT)   Construct three hypothetical situations of a stem and leaf plot showing three kinds of distributions.
Solution. Let Stem = S, and Leaf = X, then stem and leaf plots showing three distributions are as given below:

S	X				S	X				S	X			
S	X				S	X X				S	X X X		
S	X X			S	X X X			S	X X X X	
S	X X X		S	X X X X		S	X X X	
S	X X X X		S	X X X		S	X X	
S	X X X		S	X X		S	X	
S	X			S	X			S	X	
Skewed to the left			**Symmetric**			**Skewed to the right**		

Fig. 3.35. Three types of stem and leaf plots.

Example 3.13. (THE QUALITY OF A PRODUCT MATTERS)
The following back-to-back stem and leaf plot shows the lifetime in hours of 15 paper lanterns and 19 torches.

Paper lantern						Torch					
					41						
					42	5					
		6	3	1	0	43	1				
9	9	8	7	6	2	44	4	6			
		9	7	3	0	45	2	3	4	5	
						46	0	2	2	5	7
						47	1	1	3	4	6	8
					2	48					
						49					
Key = 1							Key = 1				

Fig. 3.36. Reading a back-to-back stem and leaf plot.

	Question	Answer
(a)	What is the longest that a paper lantern lasted?	482
(b)	What is the longest that a torch lasted?	478
(c)	What is the median age of paper lanterns?	448
(d)	What is the median age of torches?	462
(e)	How many paper lanterns are older than torches?	1
(f)	What is the minimum age of paper lanterns?	430
(g)	What is the minimum age of torches?	425
(h)	What is the distribution of age of the paper lanterns?	Skewed to the right
(i)	What is the distribution of age of the torches?	Skewed to the left
(j)	Is there any outlier paper lantern?	Yes, one
(k)	What is the lifetime of the outlier paper lantern?	482

3.8 HISTOGRAM

A histogram is a graph in which classes are marked on the horizontal axis and either the frequencies, or relative frequencies or percentages are marked on the vertical axis. Note that in a histogram the bars are drawn adjacent to each other without any space from bar to bar.

A graph formed by joining the midpoints of the tops of successive bars in a histogram with straight lines is called a **polygon**. The polygon is sometimes also called a **line plot**.

3.8.1 TOOLS FOR MAKING HISTOGRAM

To make a histogram, we should first learn a few tools as follows:

3.8.1.1 FREQUENCY DISTRIBUTION

A frequency distribution for quantitative data lists all the classes and the number of values that belong to each class.

To understand it, let us consider the following data set showing the weekly earnings of 100 elephants working in a circus.

Weekly earnings ($)	Number of elephants
301 to 400	9
401 to 500	16
501 to 600	33
601 to 700	20
701 to 800	14
801 to 900	8

The first column of the above table lists the classes, which represent the quantitative variable: the weekly earnings. The numbers 301 and 400 denote the lower and upper limits of the first class; the numbers 401 and 500 denote the lower and upper limits of the second class, and so on. Thus there are six classes. The second column represents the frequency, that is, the number of elephants in each class. The frequency is generally denoted by the lowercase letter f. The class having maximum frequency is called a ***modal class***. Thus the third class 501 to 600 is an example of a modal class.

3.8.1.2 CLASS BOUNDARIES

The class boundary is given by the midpoint of the upper limit of one class and the lower limit of the next class.

For example, the midpoint of the upper limit of the first class and the lower limit of the second class will give us the upper boundary of the first class and the lower boundary of the second class. In this case, the upper limit of the first class = 400, and the lower limit of the second class = 401, so the midpoint = (400+401)/2 = 400.5, which is now the upper boundary point of the first class and lower boundary point of the second class.

In the same way, we have the following table:

Class limits	Class Boundaries
301 to 400	300.5 to less than 400.5
401 to 500	400.5 to less than 500.5
501 to 600	500.5 to less than 600.5
601 to 700	600.5 to less than 700.5
701 to 800	700.5 to less than 800.5
801 to 900	800.5 to less than 900.5

3.8.1.3 CLASS WIDTH

Class width is defined as the difference between the upper boundary point and the lower boundary point.

Class Boundaries	Class width
300.5 to less than 400.5	400.5-300.5 = 100
400.5 to less than 500.5	500.5-400.5 = 100
500.5 to less than 600.5	600.5-500.5 = 100
600.5 to less than 700.5	700.5-600.5 = 100
700.5 to less than 800.5	800.5-700.5 = 100
800.5 to less than 900.5	900.5-800.5 = 100

In the above table the class width is equal for all the classes, but in practice this may not be the case. Unbounded classes are not acceptable to make a histogram.

3.8.1.4 CLASS MIDPOINT

Class midpoint is defined as:

$$\text{Class midpoint} = \frac{\text{Lower limit} + \text{Upper limit}}{2}$$

Note the use of upper limit and lower limit points.

Class limits	Class Midpoints
301 to 400	(301+400)/2 = 350.5
401 to 500	(401+500)/2 = 450.5
501 to 600	(501+600)/2 = 550.5
601 to 700	(601+700)/2 = 650.5
701 to 800	(701+800)/2 = 750.5
801 to 900	(801+900)/2 = 850.5

3.8.1.5 STEPS TO MAKE A FREQUENCY DISTRIBUTION

There are three steps:

(i) Number of classes:

There is no rule, but generally we prefer to make 5 to 20 classes depending upon the number of data values.

The number of classes can be approximated by:

$$c = 1 + 3.3\log(n)$$

where c denotes the number of classes, and n denotes the number of data values. We can also approximate the number of classes as:

$$c = \sqrt{n}$$

Note that sometimes both formulae give different results.

(ii) Class width: The approximate class width is given by:

$$\text{Approximate class width} = \frac{\text{Maximum value} - \text{Minimum value}}{\text{No. of classes}}$$

(iii) Starting point or lower limit of the first class: Any convenient value that is equal to or less than the smallest value in the data set.

Example 3.14. (RAW DATA) Consider the following raw data set that gives the dancing time (in minutes) of an elephant in a circus during the last 30 weeks:

		81	84	79	76	73	74	77	82	75	81
		76	76	80	82	78	72	80	83	80	77
		78	78	79	84	73	86	83	79	83	79

Fig. 3.37. An elephant.

(a) Construct a frequency distribution table.
(b) Construct a histogram and a polygon.
Solution. (a) To construct a frequency distribution table, we have to first decide three things:
　　　　(i) Number of classes,
　　　　(ii) Class width,
and
　　　　(iii) Lower limit of the first class,
as follows:

(i) Number of classes: We approximate the number of classes as:

$$c = 1 + 3.3 \log(n) = 1 + 3.3 \log(30) = 1 + 3.3(1.477) = 5.87 \approx 5 .$$

(ii) Approximate class width:

$$\text{Approximate class width} = \frac{\text{Maximum value} - \text{Minimum value}}{\text{No. of classes}}$$

$$= \frac{86 - 72}{5} = \frac{14}{5} = 2.8 \approx 3 .$$

(iii) Lower limit of the first class: Minimum data value $= 72$.

Thus the frequency distribution table is given by:

Class limits: Time(min)	Class boundaries	Tally	Frequency (f)	Relative frequency
72 – 74	71.5 to < 74.5	||||	4	4/30 = 0.133
75 – 77	74.5 to < 77.5	ⅢⅠ	6	6/30 = 0.200
78 – 80	77.5 to < 80.5	Ⅲ Ⅲ	10	10/30 = 0.333
81 – 83	80.5 to < 83.5	Ⅲ||	7	7/30 = 0.233
84 – 86	83.5 to < 86.5	|||	3	3/30 = 0.100
		Sum	**30**	**0.999**

(**b**) On the horizontal axis take the **class boundaries**, and on the vertical axis take the frequency (or RF or %) of different dancing times during the last 30 weeks. Note that there is no gap between bars in a histogram. Thus, we have the histogram and polygon as follows:

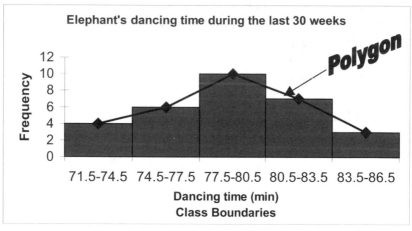

Fig. 3.38. Histogram and polygon.

Remarks: (a) If **class limits** with gaps are considered on the horizontal axis, then the histogram becomes a **bar chart** for a quantitative variable.
(b) A histogram can also be used to detect outliers, modal classes, median class and distributions of data values.

(c) As the number of data values increases, the number of classes also increases, and the polygon becomes smooth and smooth as shown in **Figure 3.39**:

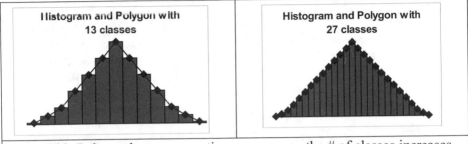

Fig. 3.39. Polygon becomes continuous curve as the # of classes increases.

As the number of data values become sufficiently large, the polygon tends to a ***continuous curve*** as shown in **Figures 4.18**, **4.19**, and **4.20** for a skewed to the left, symmetric and bell shaped, and skewed to the right distributions, respectively.

Example 3.15. (READING A HISTOGRAM) The histogram below gives the prices of 22 different pairs of monkey shoes:

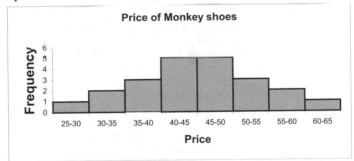

Fig. 3.40. Reading a histogram.

	Questions	Answers
(a)	How many classes are there?	8
(b)	Which class occurred most frequently?	Two: 40-45; 45-50
(c)	How many shoes' price is $40 or more?	5+5+3+2+1 = 16
(d)	How many shoes' priced is >$40?	5+3+2+1 = 11
(e)	What is the distribution?	Symmetric
(f)	How many shoes' price is $55 to $60?	2
(g)	How many shoes' price is $30 to $35?	2
(h)	Is there any outlier?	No

Example 3.16. (HEALTH IS WEALTH) The owner of a zoo is worried about the health of certain birds who become sick due to heavy rain. A zoo statistician noted the rainfall data (mm) for 12 months: 1.5, 3.5, 4.4, 5.8, 5.2, 6.7, 9.7, 4.1, 7.8, 7.8, 5.2, and 5.2. Now the statistician wishes to present data to the owner using a histogram.
(a) Create a histogram for the rainfall data. Use a lower limit of 0, an upper limit of 10, and a class width of 2. Your first class should be [0, 2), and your last class should be [8,10). Be sure to label your axis and provide some values.
(b) How many months had less than 6 mm of rain?
(c) What is the midpoint of the second class?
(d) What are the lower and upper boundaries of the third class?
(e) Describe the distribution of the rainfall.

Rainbow

Solution. (a) We have the following situation:

Class Boundaries	Tally Marking	Frequency	Relative Frequency	Percentage = RFx100
[0, 2)	I	1	1/12 = 0.083	= 8.3%
[2, 4)	I	1	1/12 = 0.083	= 8.3%
[4, 6)	IIIII I	6	6/12 = 0.500	= 50.0%
[6, 8)	III	3	3/12 = 0.250	= 25.0%
[8, 10)	I	1	1/12 = 0.083	= 8.3%
Sum		**12**	**1**	**99.9%**

Thus the histogram is given as below:

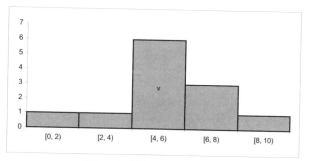

Fig. 3.41. Histogram.

(b) No. of months with less than 6 mm rain = 1 + 1 + 6 = 8 months.
(c) Midpoint of the second class = (2 + 4)/2 = 3.
(d) Third class: Lower boundary = 4 and upper boundary = 6.
(e) Distribution is slightly skewed to the left.

Example 3.17. (BOAT LOVERS) A company, XYZ, made the following number of boats during 30 months as

24	32	27	23	33	33	29	25	23	28
21	26	31	22	27	33	27	23	28	29
31	35	34	22	26	28	23	35	31	27

Fig. 3.42. A boat company's 30 months of production.

(a) Construct a frequency distribution table using classes: 21 - 23, 24 - 26, 27 - 29, 30 - 32, and 33 - 35.
(b) Calculate the relative frequencies and percentage for all classes.
(c) Construct a histogram and a polygon for the data. What is modal class? What is median class?
(d) What percentage of months are the number of boats made in the interval 27 to 29?
(e) Approximate the number of classes with the following formula
$$c = 1 + 3.3 \log(n)$$
and give your opinion about the choice of the above 5 classes.
(f) What proportion of months is the number of boats made between 24 and 29?
(g) Construct a pie chart to present the same classes.
Solution. (a) and (b)

Class limits	Class boundaries	Tally	f	RF	Percent = RFx100%
21–23	20.5 to< 23.5		7	0.233	23.3
24–26	23.5 to< 26.5		4	0.133	13.3
27–29	26.5 to< 29.5		9	0.300	30.0
30–32	29.5 to <32.5		4	0.133	13.3
33–35	32.5 to <35.5		6	0.200	20.0
		Sum	30	0.999	99.9

(c) Histogram and polygon:

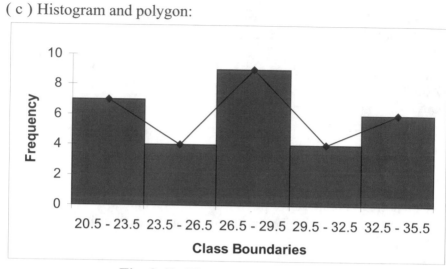

Fig. 3.43. Histogram and polygon.

Modal class: 26.5-29.5. Median class: 26.5-29.5.
(d) 30%.
(e) Number of classes:

$$c = 1 + 3.3\log(n) = 1 + 3.3\log(30) = 1 + 3.3(1.477) = 5.87 \approx 5$$

Thus, the choice of 5 classes seems to be a good one.

(f) The required proportion will be: $0.133 + 0.300 = 0.433$.

(g) The pie chart for a quantitative variable can be constructed as:

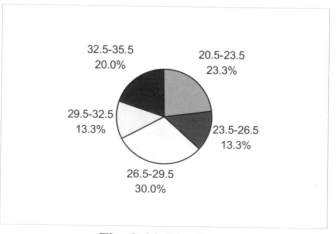

Fig. 3.44. Pie chart.

LUDI 3.1. (DOCTOR'S APPOINTMENT)

A doctor notes the variables listed in the following table from every patient before admitting them into his hospital. Which of the following variables are qualitative and which are quantitative? Which quantitative variables are discrete and which are continuous? Which qualitative variables are ordinal and which are nominal?

Fig. 3.45. Patient and doctor.

	Variable(s)	Answer
(i)	Name	
(ii)	Gender (male or female)	
(iii)	Age	
(iv)	Smoker (yes or no)	
(v)	Weight (lbs)	
(vi)	Systolic blood pressure (mm of Hg)	
(vii)	Height (cm)	
(viii)	Level of calcium in blood (mg/ml)	
(ix)	Postal code (or area code)	
(x)	Phone number	
(xi)	Marital status (Single, Married, Divorced, Widow)	
(xii)	Vegetarian (yes or no)	
(xiii)	Social Security Number	
(xiv)	Temperature ($^\circ$F)	
(xv)	Family Income	
(xvi)	Number of family members	
(xvii)	Do you have children?	
(xviii)	Are you taking any medicine? Yes or No	
(xix)	Have you eaten anything today? Yes or No	
(xx)	Who sent you here?	
(xxi)	Do you have medical insurance?	
(xxii)	Are you employed?	
(xxiii)	Eye color: Blue, Green, Brown etc.	
(xxiv)	Do you have any special needs?	
(xxv)	How do you feel today? (good or bad)	

LUDI 3.2. (COLLECTING POINTS) If you select a certain residential long distance telephone company, you can earn points for different magazines containing information on animals and birds. There are 49 magazines to choose from. The points needed to redeem each are listed below:

470	570	1030	470	570	470	470	410	470
510	570	470	730	730	600	910	470	350
570	410	470	1010	570	410	570	660	510
470	730	570	470	850	380	660	470	410
410	600	970	470	350	410	540	1010	1290
600	590	410	690	1280	540	500	850	660
470	470	1350	570	1010	630	970	690	690

Fig. 3.46. Reduce your phone bill.

(a) Make a histogram by starting from 350 with a class width of 100.
(b) How many magazines have less than 550 points?
(c) Repeat part (a) with a class width of 50. Which histogram gives you more information and why?

LUDI 3.3. (WATERFALLS) A pigeon delivered letters to Michael who was visiting the six highest waterfalls in the world at the heights (meters) given in the following table:

Michael receiving letter on the waterfalls	
Waterfalls	**Height**
Angel (Venezuela)	1000 m
Tugela (South America)	914 m
Cuquenan (Venezuela)	610 m
Sutherland (New Zealand)	580 m
Takkakaw (British Columbia)	503 m
Ribbon (California)	491 m

Fig. 3.47. Pigeon delivering mail.

(a) Construct a bar graph for this data set.
(b) Construct a pie graph for this data set.
(c) What proportion of these six waterfalls are located in California?

LUDI 3.4. (PHONE AND FAX ORDERS) The following table gives the number of phone and fax orders received by a bolt company for their special kind of bolts from 16 different countries:

BOLT ORDERS FROM DIFFERENT COUNTRIES 	 Phone Orders	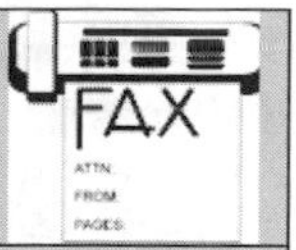 Fax orders
Australia	17	06
Austria	40	15
Canada	22	07
Denmark	40	21
France	25	10
Ireland	06	04
India	12	11
Israel	08	04
Japan	22	13
Netherlands	12	08
Norway	19	07
Poland	22	04
Sweden	28	11
Switzerland	37	15
United Kingdom	11	07
United States	19	06

Fig. 3.48. Fax and phone orders.

(a) Construct a back to back stem and leaf plot to present the phone and fax orders. Are there any outliers? What are the distributions?

(b) From how many countries did the phone orders exceed the highest number of fax orders received in the entire sample of 16 countries?

(c) Find the total number of phone and fax orders and again present the new data set with a new stem and leaf plot.

(d) Construct bar charts for presenting these three data sets.

(e) Find the percentage of the total number of orders (phone + fax) from each country. Classify the countries into four groups having orders between 0% - 25%, 26% - 50%, 51% - 75%, and 76% - 100%. Suggest an appropriate graph to present such data, and comment on your suggestion.

LUDI 3.5. (WHO IS WHO?) State if the following variables are qualitative or quantitative. If it is quantitative, then state if it is a continuous or a discrete variable. If it is qualitative, then state if it is a nominal or an ordinal variable.

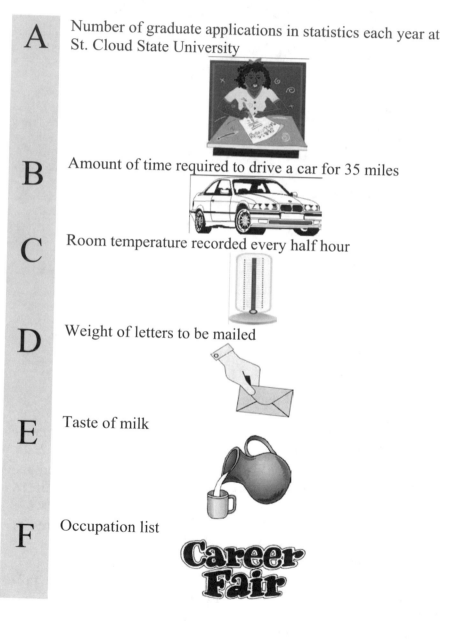

A   Number of graduate applications in statistics each year at St. Cloud State University

B   Amount of time required to drive a car for 35 miles

C   Room temperature recorded every half hour

D   Weight of letters to be mailed

E   Taste of milk

F   Occupation list

Career Fair

G   Phone number

H   Number of mistakes on an examination

I   Time to finish an examination

J   Shoe size

K   Gender of a student

L   Discipline of a student

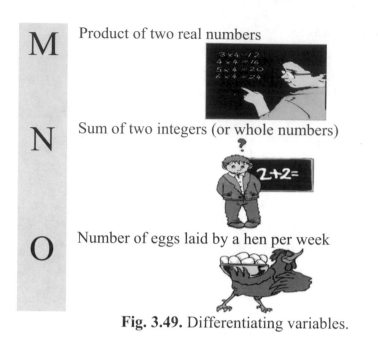

M Product of two real numbers

N Sum of two integers (or whole numbers)

O Number of eggs laid by a hen per week

Fig. 3.49. Differentiating variables.

LUDI 3.6. (a) Name the following graphs:

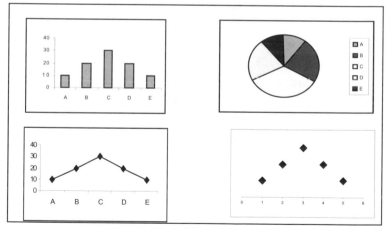

Fig. 3.50. Different kinds of graphs.

(b) In a stem and leaf plot if stem = 5, leaf = 8 and key = 0.01, then data value = ---.

(c) In a stem and leaf plot if data value = 34.2, then stem = ---, leaf = --- and key = ---.

(d) In a stem and leaf plot if data value = 4729, then stem = ---, leaf = --- and key = ---.

LUDI 3.7. (WHICH CAME FIRST- THE HEN OR EGG?) The following table shows the number of eggs laid by seven hens during a week.

| 3 | 5 | 12 | 3 | 6 | 5 | 5 |

Fig. 3.51. Hens.

Construct a dot plot. Is the distribution skewed? Is there any outlier?

LUDI 3.8. (A VISIT TO THE ZOO) A class of 30 students visited a local zoo. After a period of one month, the school teacher asked each student to write the name of one animal on a piece of paper and submit it to the teacher. The teacher read the names of the animals on the pieces of paper submitted by the students and the names came in the following sequence:

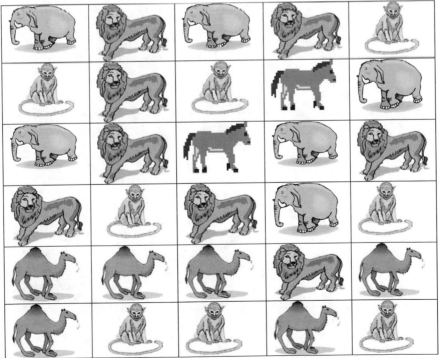

Fig. 3.52. Memorizing animals at the zoo.

(a) Construct a frequency distribution table using tally marking.
(b) Construct a Pareto chart and define it.
(c) Construct a pie chart.

LUDI 3.9. (GOOD GRADES ARE IMPORTANT) Last spring, all students in Stats 193 received a final grade of either A, B, C or F. Assume the distribution of grades for the 240 students is given in the following table:

Final Grade	A	B	C	F
Percentage of students	15%	60%	20%	5%

(a) How many students received an A in the class?
(b) How many students received an F in the class?
(c) Complete the following table:

Final Grade	Frequency	Relative Frequency	Percent (Given)	Angle
A			15%	
B			60%	
C			20%	
F			5%	
Sum	240	1	100%	360°

(d) Complete the following pie-chart:

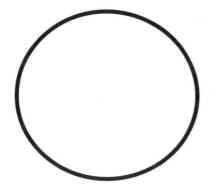

Fig. 3.53. Examination hall.

LUDI 3.10. (A VISIT TO A ZOO) A class of 30 students visited a local zoo. After a period of one month, the school teacher asked each student to write the name of one bird on a piece of paper and submit it to the teacher. The teacher read the names of the birds on the pieces of paper submitted by the students in the following sequence:

Chough	Chough	Eagle	Owl	Pigeon
Swan	Chough	Eagle	Pigeon	Pigeon
Pigeon	Eagle	Chough	Pigeon	Swan
Pigeon	Pigeon	Hen	Chough	Pigeon
Pigeon	Parrot	Eagle	Hen	Chough
Peafowl	Parrot	Woodpecker	Peafowl	Woodpecker

Fig. 3.54. Memorizing birds at the zoo.

(a) Construct a frequency distribution table using tally marking.
(b) Construct a Pareto chart, and define it.
(c) Construct a pie-chart.

LUDI 3.11. (QUALITY OF A PRODUCT MATTERS) The following back-to-back stem and leaf plot shows the lifetime in hours of 15 A-grade light bulbs and 15 B-grade light bulbs

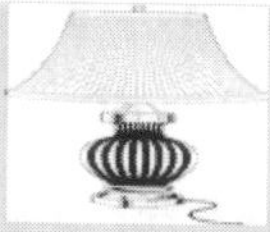

A-grade						B-grade
					51	
					52	5
7	3	1	0		53	1
9 8 8	7	6	2		54	6
8	6	3	0		55	2
					56	0 2 2 4 6
					57	1 1 3 4 6 8
			1		58	
					59	

Key = 1	Key = 1

Fig. 3.55. Back-to-back stem and leaf plot.

(a) What is the longest that any light bulb lasted?
(b) Give a good reason why someone might prefer an A-grade light bulb. Give supportive details.
(c) Give a good reason why someone might prefer a B-grade light bulb. Give supportive details.
(d) What is the longest that an A-grade light bulb lasted?
(e) What is the longest that a B-grade light bulb lasted?
(f) What is the median age of A-grade light bulbs?
(g) What is the median age of B-grade light bulbs?
(h) How many A-grade light bulbs lasted longer than B-grade bulbs?
(i) What is the minimum age of A-grade light bulbs?
(j) What is the minimum age of B-grade light bulbs?
(k) What is the distribution of the ages of the A-grade light bulbs?
(l) What is the distribution of the ages of the B-grade light bulbs?
(m) Are there any outliers?
(n) Repeat (a) to (m) with Key = 0.1, and comment.

LUDI 3.12. (GOOD MARKS NEED HARD WORK) The following table gives the marks of 300 students in Stat 193 during the 1998-1999 school year at St. Cloud State University.

39	88	88	59	39	40	38	63	60	86
97	90	47	76	99	79	99	62	72	79
63	56	76	68	46	70	34	69	91	62
88	86	69	41	69	79	96	34	75	39
71	75	37	63	86	95	40	63	78	88
55	43	97	96	58	93	69	60	76	57
46	94	38	50	35	90	34	77	88	44
84	55	33	57	62	74	89	64	37	44
40	57	56	83	51	72	88	95	37	69
98	66	43	63	81	52	61	48	86	97
59	30	64	63	45	97	82	63	89	36
77	59	42	98	98	57	81	74	36	30
84	92	88	49	70	85	97	81	42	61
79	64	87	71	33	60	53	56	85	56
59	39	69	83	36	41	80	98	42	91
69	71	87	34	59	87	45	90	47	72
79	93	92	60	93	70	99	48	99	32
76	37	30	56	61	67	50	76	81	77
32	38	72	54	45	59	61	58	94	56
48	90	39	33	71	69	88	57	58	67
76	46	32	98	32	97	73	33	52	48
59	82	90	90	37	88	97	96	78	44
34	80	93	94	37	57	69	54	61	96
98	77	78	98	58	72	31	83	60	90
35	70	76	87	92	45	65	72	62	99
81	95	30	80	70	96	74	86	70	71
71	54	36	81	86	65	68	76	69	32
98	96	89	61	96	68	76	37	98	41
82	76	57	76	39	89	34	67	75	48
59	55	95	86	34	38	37	90	48	90

(a) Find:

$$\text{No. of classes} = 1 + 3.3\log(n)$$

(b) Find the starting point = Minimum value

Fig. 3.56. Exams.

(c) Approximate the class width as:

$$\text{Class width} = \frac{\text{Maximum value} - \text{Minimum value}}{\text{No. of classes}}$$

(d) Complete the following frequency distribution table:

Class Limits	Class boundaries	Tally	Frequency
Sum			**300**

(e) Construct a histogram:

(f) Construct a polygon on the above histogram. Is it symmetric?

LUDI 3.13. (QUALITY OF A PRODUCT) The following back-to-back stem and leaf plot shows the lifetime in hours of 15 paper lanterns and 19 torches.

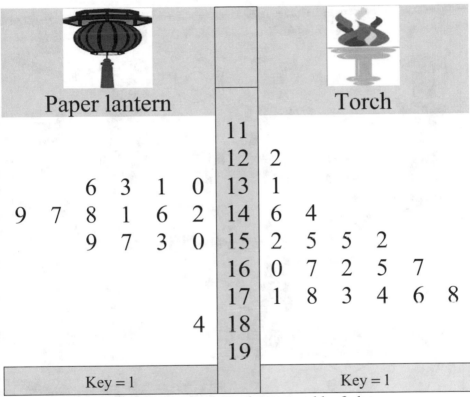

Paper lantern						Torch					
				11							
				12	2						
6	3	1	0	13	1						
9 7 8	1	6	2	14	6	4					
9	7	3	0	15	2	5	5	2			
				16	0	7	2	5	7		
				17	1	8	3	4	6	8	
			4	18							
				19							
	Key = 1							Key = 1			

Fig. 3.57. Back-to-Back stem and leaf plot.

(a) Make some amendments in the above plot if required.
(b) What is the longest that a paper lantern lasted?
(c) What is the longest that a torch lasted?
(d) What is the median age of the paper lanterns?
(e) What is the median age of the torches?
(f) How many paper lanterns lasted longer than torches?
(g) What is the minimum age of the paper lanterns?
(h) What is the minimum age of the torches?
(i) What is the distribution of the age of the paper lanterns?
(j) What is the distribution of the age of the torches?
(k) Give a good reason why someone might prefer a paper lantern.
(l) Give a good reason why someone might prefer a torch.
(m) Is there any outlier?

LUDI 3.14. (BE CAREFUL WHEN USING SHARP TOOLS) A construction company orders the following tools: axe, bolt, chainsaw, drill, hammer, nut, and nail cutter.

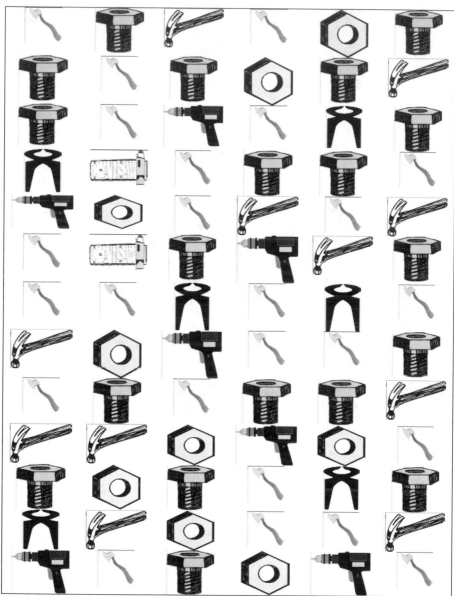

Fig. 3.58. Tools.

(a) Construct a Pareto diagram for this raw data.
(b) What is the maximum proportion of tools?
(c) Construct a pie chart for representing the same distribution.

LUDI 3.15. (REDUCE YOUR PHONE AND FAX BILLS) The following table gives the number of phone calls and faxes received by a company during the period of a year.

Month	 ☎	 FAX
January	26	08
February	60	17
March	42	09
April	59	23
May	45	12
June	08	06
July	10	06
August	42	13
September	32	10
October	39	09
November	32	06
December	48	12

Fig. 3.59. Comparing use of phone and fax.

(a) Construct a back to back stem and leaf plot. Provide a key.
(b) In how many months did the phone call rate exceed the highest fax rate during the year?
(c) Construct two bar charts to display both data sets.
(d) Construct two pie charts to display both data sets.

LUDI 3.16. (COMPARE YOUSELF TO THE CLASS) The following table gives a list of the scores of two students: Amy and Bob.

 **Amy**	72	86	82	76	78	82	84	74	80	76
Bob	79	52	86	94	76	48	92	69	79	45

Fig. 3.60. Comparing two students.

(a) Construct a back-to-back stem and leaf plot of the data.
(b) Which student, Amy or Bob, do you think has done a good job? Explain your answer in two to three lines.

LUDI 3.17. (DESIGN THE FRONT OF YOUR HOME) Michael contacts a construction company that makes readymade material to construct buildings. For easiness the company numbers the tiles to be used in each pillar for making a building. The front of the Michael's plot is 90 feet wide.

1-10, 11-20, ...81-90

Numbers on the 250 tiles									
11	27	27	26	30	36	36	46	46	27
29	41	45	32	51	19	51	42	15	42
22	37	48	52	56	21	52	44	39	55
33	43	57	66	34	47	65	34	58	58
67	18	48	10	32	19	9	55	59	15
59	43	37	57	43	49	54	64	39	42
22	1	2	69	50	29	69	22	61	61
72	62	48	31	45	60	43	60	46	24
33	24	38	71	32	5	54	66	13	69
76	63	38	26	50	48	43	23	46	72
81	75	48	53	45	26	8	77	39	31
67	18	63	67	3	65	40	55	13	78
33	41	41	53	33	32	54	30	38	54
82	79	41	68	56	20	7	41	38	42
11	25	80	57	12	45	44	23	12	61
70	70	70	36	57	49	4	73	46	28
59	17	37	53	56	29	44	34	39	58
83	84	63	31	50	20	36	49	49	31
28	42	85	68	71	47	6	23	14	24
89	86	75	25	50	53	44	64	47	16
35	59	35	78	49	35	40	34	35	62
76	77	79	60	74	21	51	45	14	28
90	17	37	74	60	65	44	66	40	58
88	62	80	25	56	47	51	64	73	16
51	87	68	52	50	21	40	30	47	55

Fig. 3.61. Learning the importance of histograms.

The company suggests to Michael that he divide the front side of the plot into 9 equal sized pillars and number as: 1-10, 11-20, ...,81-90, and then fix all the tiles in these pillars according to the numbers assigned to the 250 tiles as shown in the above table. Construct a histogram by using the steps provided, and justify that each pillar corresponds to one bar in the resultant histogram.

(a) Use the following formula to guess the number of classes:

$$\text{No. of classes} = 1 + 3.3\log(n)$$

(b) Find the starting point = Minimum value

(c) Approximate the class width as:

$$\text{Class width} = \frac{\text{Maximum value} - \text{Minimum value}}{\text{No. of classes}}$$

(d) Complete the following frequency distribution table:

Class limits (meters)	Class boundaries	Tally	Frequency (# of tiles)
01 - 10			
11 - 20			
21 - 30			
		Sum	250

(e) Construct a histogram.

(f) Construct a polygon on the above histogram. Is it symmetric like the front of the city entrance?

LUDI 3.18. (COUNTING FLIGHTS) Suppose that the number of flights from seven airports is distributed as follows:

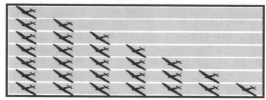

Fig. 3.62. Distribution of flights.

(a) The total number of flights is - - - - -
(b) The distribution of flights is - - - - -

LUDI 3.19. (COUNTING BUS DEPARTURES) Suppose that the number of buses departing from four bus stands is distributed as:

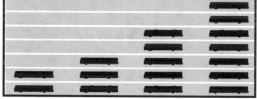

Fig. 3.63. Departure schedule of buses.

(a) The total number of buses is - - - - -
(b) The distribution of buses is - - - - -

LUDI 3.20. (GOAL DEPENDS ON THE JUDGEMENT OF THE HIT) A player hits a ball as shown below:

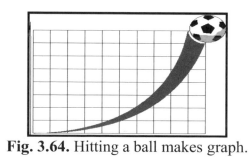

Fig. 3.64. Hitting a ball makes graph.

The path followed by the ball with time forms a - - - - -
(a) stem and leaf plot (b) dot plot (c) line plot (d) none of these.

LUDI 3.21. (ENJOY FIREWORKS NIGHT IN YOUR CITY)
Consider the following picture showing the path made by a cracker on fireworks night.

Fig. 3.65. Remembering a fire works night.

This is an illustration of - - - - -

(a) bar chart (b) line plot (c) histogram (d) none of these.

LUDI 3.22. (DIFFERENT GRAPHS SHOW THE SAME RESULTS) The following charts show a one-to-one correspondence between a - - - - -

(a) pie chart and stem and lcaf plot
(b) pie chart and histogram
(c) pie chart and bar chart
(d) pie chart and dot plot
(e) none of these

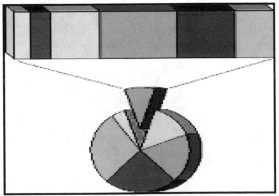

Fig. 3.66. Comparing results from different graphs.

LUDI 3.23. (GOOD GRADES HELP YOUR CAREER) Recorded here are the grades of 45 students in Stat 193 from the previous semester at St. Cloud State University

C, C, A, B, A, C, A, A, A, C, B, C, B, C, C, A, C, C, A, A, A, A, D, A, B, A, A, C, C, A, C, C, A, A, A, C, A, C, C, D, D, A, B, C, A.

(a) Summarize the data in a frequency distribution table:

Grades	Tally	Frequency	Relative Frequency	Angle
A				
B				
C				
D				
Sum				

(b) Complete the following circular graph to display the grades.

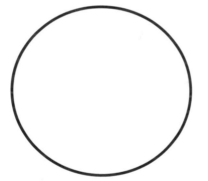

Fig. 3.67. Exam.

LUDI 3.24. (REMEMBER YOUR CHILDHOOD) The following slide is a good example of a --------- distribution.

(a) skewed to the left
(b) skewed to the right
(c) symmetric
(d) none of these

Fig. 3.68. Slide.

LUDI 3.25. (LEARNING ABOUT THE USA) Recorded here are the names of the 50 states in the USA.

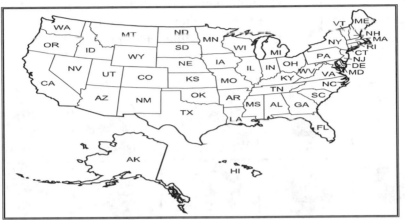

Fig. 3.69. Map of the USA.

(a) Complete the following frequency distribution table:

States names that start with	Tally	Frequency	Relative Frequency
A			
C			
D			
F			
G			
H			
I			
K			
L			
M			
N			
O			
P			
R			
S			
T			
U			
V			
W			
Sum			

(b) Construct a Pareto and an anti-Pareto chart.
(c) Which letter(s) have been used most to assign the names of different states? (**Optional**: Why?)

LUDI 3.26. (A VISIT TO A GARDEN) A class of 30 students visited a local garden. After a period of one month, the school teacher asked each student to write the name of one fruit on a piece of paper and submit it to the teacher. The teacher read the names of the fruit on the pieces of paper submitted by the students in the following sequence:

Apple	Apple	Watermelon	Raspberry	Grapes
Strawberry	Apple	Watermelon	Grapes	Grapes
Grapes	Watermelon	Apple	Grapes	Strawberry
Grapes	Grapes	Banana	Apple	Grapes
Grapes	Cherries	Watermelon	Banana	Apple
Raspberry	Cherries	Pomegranate	Raspberry	Raspberry

Fig. 3.70. Memorizing fruits.

(a) Construct a frequency distribution table using tally marking.
(b) Construct a Pareto chart and define it.
(c) Construct an anti-Pareto chart and define it.
(d) Construct a pie-chart.

LUDI 3.27. (A VISIT TO A FARM) A class of 30 students visited a local farm. After a period of one month, the school teacher asked each student to write the name of one vegetable on a piece of paper and submit it to the teacher. The teacher read the names of the vegetables on the pieces of paper submitted by the students in the following sequence:

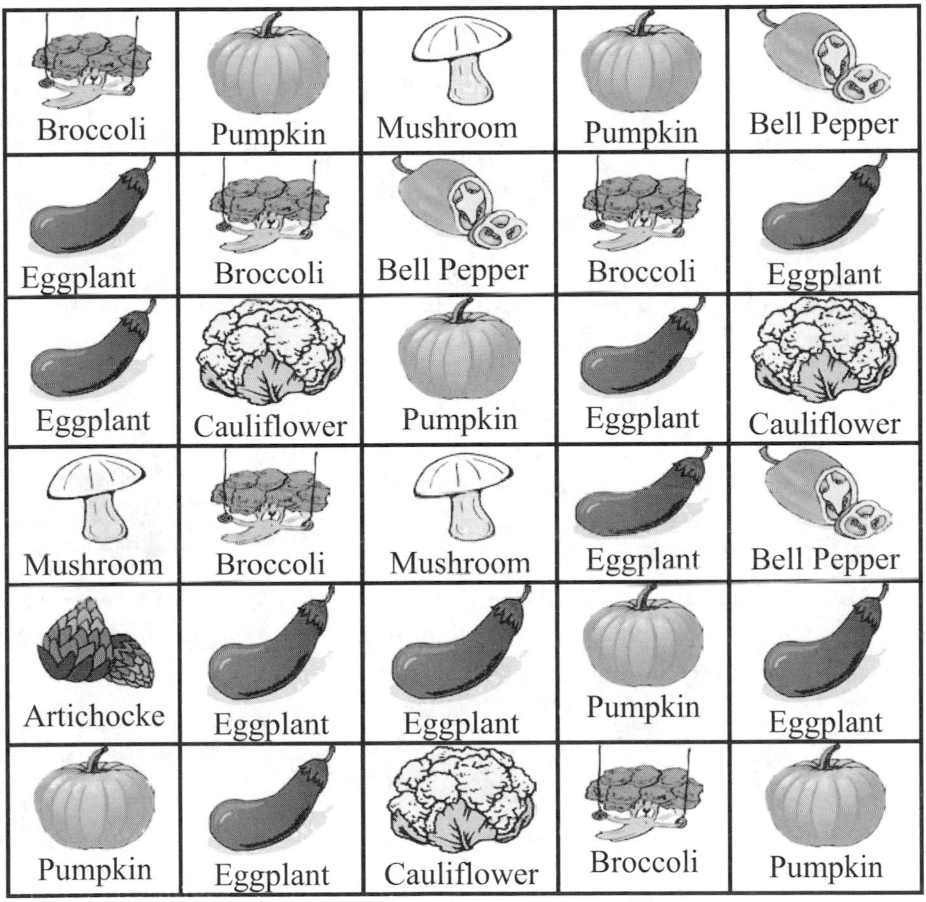

Broccoli	Pumpkin	Mushroom	Pumpkin	Bell Pepper
Eggplant	Broccoli	Bell Pepper	Broccoli	Eggplant
Eggplant	Cauliflower	Pumpkin	Eggplant	Cauliflower
Mushroom	Broccoli	Mushroom	Eggplant	Bell Pepper
Artichocke	Eggplant	Eggplant	Pumpkin	Eggplant
Pumpkin	Eggplant	Cauliflower	Broccoli	Pumpkin

Fig. 3.71. Memorizing vegetables.

(a) Construct a frequency distribution table using tally marking.
(b) Construct a Pareto chart and define it.
(c) Construct an anti-Pareto chart and define it.
(d) Construct a pie-chart.
(e) Which vegetable occurred most frequently?
(f) What is the proportion of students who remember eggplants?
(g) What is the percentage of students who remember broccoli?

LUDI 3.28. (ALWAYS LOOK FOR GOOD MODELS) The following table lists models of certain motor vehicles.

Vehicle	 Acura Legend	 Buick Century	 Toyota Camry	 Ford Escort
Type	Standard	Standard	Midsize	Large
Where Made	Foreign	Domestic	Foreign	Domestic
City MPG	20	23	18	21
Highway MPG	26	31	28	29

Fig. 3.72. Comparing cars.

Identify the individuals/units. Then list the variables recorded for each individual, and classify each variable as qualitative or quantitative. Further, classify the quantitative variables as discrete or continuous.

LUDI 3.29. (DO NOT DRINK AND DRIVE) Statistical abstracts of a country showed that during a year, there were 91,523 accidental deaths. Among these, 42,893 deaths were from motor vehicle accidents, 13,241 from falls, 3,807 from drowning, 3,900 from fires, and 7,382 from poisoning.

Fig. 3.73. Accident.

(a) Find the percentage of accidental deaths from each of these causes, rounded to the nearest percent.

Accident due to:	Frequency	Relative Frequency	Percent	Angle
Motor Vehicle				
Falls				
Drowning				
Fires				
Poisoning				
Other causes				
Total				

Hint: Complete the above table.

What percentage of accidental deaths was due to other causes?

(b) Make a well-labeled bar graph of the distribution of the causes of accidental deaths. Be sure to include an 'other causes' bar.

(c) Would it also be correct to use a pie chart to display this data set? Explain your answer.

LUDI 3.30. (ALWAYS REMEMBER TO FOLLOW GREAT PEOPLE) Professor Karl Pearson said, *"I rush from science to philosophy to our old friends the poets; and then, over wearied by too much idealism, I fancy I became practical in returning to science. Have you ever attempted to conceive all there is in the world worth knowing – that not one subject in the universe is unworthy of study? The giants of literature, the mysteries of many dimensional space, the attempt of Biltzmann and Crookes to penetrate Nature's*

Fig. 3.74.
Prof. Karl Pearson
(1857 - 1936).
Source: Refer to Bibliography

very laboratory, the Kantian theory of the universe, and the latest discoveries in embryology, with their wonderful tales of the development of life – what an immensity beyond our grasp! ... Mankind seems on the verge of a new and glorious discovery. What Newton did to simplify the planetary motions must now be done to unite in one whole the various isolated theories of mathematical physics."

Assume that the number of letters in a word is a measure of its length. For example, the word '*Mankind*' has a length of 7.

(a) Count the lengths of all the words in the above paragraph within quotes, and construct a frequency distribution table.

(b) Display the distribution of the length of words with a histogram.

(c) Describe the main features of the distribution. Is it symmetric, skewed to the right or to the left? What is the mode value? Are there any outliers? Is it uni-modal or bi-modal?

(d) Count the repeated words only once and repeat (a) to (c).

LUDI 3.31. (GOOD SCORES ARE ASSETS) Good score on a Mathematical Ability Test (MAT) is a must to enter a reputable college. A batch of 18 girls hoping to get into a college appeared for the MAT, and their scores were:

148 101 178 200 154 129 140 165 152

137 126 109 126 103 154 137 115 165

A batch of 20 boys wishing to enter the college also appeared for the MAT and their scores were:

104 146 187 169 92 115 126 70 115 151

180 113 91 88 114 75 140 132 109 108

Fig. 3.75. Girls and Boys competition.

(a) Make a back-to-back stem and leaf plot of the girls' and boys' scores. Are there any outliers?

(b) Compare the midpoints and the ranges of the two distributions. What is the most noticeable contrast between the girls and boys?

(c) Which distribution is more skewed: the girls' scores or the boys' scores?

(d) How many girls scored higher than the boys' maximum score?

(e) How many boys scored lower than the girls' minimum score?

(f) What is the distribution of the boys' scores?

(g) What is the distribution of the girls' scores?

(h) Who performed better: girls or boys?

LUDI 3.32. (GOOD CHARACTER ALWAYS PAYS)

The following paragraph describes the character of Sir R.A. Fisher:

"He was capable of tremendous charm, and warmth in friendship. But he also was the victim, as he himself recognised, of an uncontrollable temper; and his devotion to scientific truth as he saw it being literally passionate, he was an implacable enemy of those whom he judged guilty of propagating error."

Sir R.A. Fisher
1890-to-1962
Fig. 3.76.
Source: Refer to Bibliography

Assume that the number of letters in a word is a measure of its length. For example, the word '*devotion*' has a length of 8.

(a) Construct a dot plot to display the lengths of all the words in the above paragraph.
(b) Is the distribution symmetric?
(c) Are there any outliers?
(d) From the dot plot, can we say that Sir R.A. Fisher had good character?
(e) Count the repeated words only once and repeat (a) to (d).

LUDI 3.33. (TRY TO FIND AND REDUCE YOUR OWN WEAKNESSES) The following paragraph describes the "strengths and weaknesses" of Sir R.A. Fisher:

"As a penetrating thinker Fisher was outstanding; but his writings are difficult for many readers. Indeed, some of his teachings have been most effectively conveyed by the books of others who have been able to simplify their expression. As a lecturer also, Fisher was too difficult for the average student; his classes would rapidly fall away until only two or three students who could stand the pace remained as fascinated disciples. Nor was he particularly successful as an administrator; he perhaps failed to appreciate the limitations of the ordinary man. But with his wide interests and penetrating mind he was a most stimulating and sympathetic conversationalist."

Sir R.A. Fisher
(1890-1962)
Fig. 3.77.
Source: Refer to Bibliography

Assume that the number of letters in a word is a measure of its length. For example, the word 'Sympathetic' has a length of 11.

(a) Count the lengths of all the words in the above paragraph within quotes, and construct a frequency distribution table.
(b) Display the distribution of the length of words with a histogram.
(c) Describe the main features of the distribution. Is it symmetric, skewed to the right or to the left? What are the modal values? Are there any outliers? Is it uni-modal or bi-modal?
(d) Can we compare the "strengths and weaknesses" of Sir R.A. Fisher using the histogram?
(e) Count the repeated words only once and repeat (a) to (d).

LUDI 3.34. (USE SHARP TOOLS CAREFULLY) A construction company XYZ orders the following tools: an anvil, drill press, staple gun, vise, circular saw, and tool box.

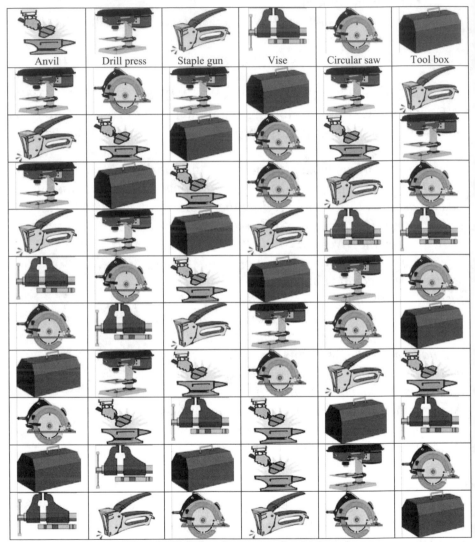

Fig. 3.78. Plumbers need good tools.

(a) Construct a Pareto diagram for this raw data.
(b) What proportion of the tools is the maximum?
(c) Use a pie chart to represent the same distribution of the tools.

4. NUMERICAL REPRESENTATION

4.1 INTRODUCTION

In this chapter, we introduce a way to write the sum of numbers with the help of mathematical symbols. Then we discuss the three measures of central value: mean, median and mode, and five measures of variation: range, mean absolute deviation, variance, standard deviation, and coefficient of variation. Steps to construct a box-plot, the empirical rule and the Tchebysheff's rule are discussed.

4.2 MEANING OF Σ

The symbol sigma (Σ) is an uppercase Greek letter corresponding to the uppercase letter **S** in the English language. As **S** stands for **Sum**, similarly the symbol sigma (Σ) stands for **Sum**.

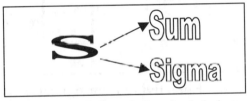

Fig. 4.1. English and Greek alphabets.

Consider adding three numbers $x_1 = 5$, $x_2 = 4$ and $x_3 = 7$. Thus, we have:

$$x_1 + x_2 + x_3 = 5 + 4 + 7 = 16$$

and this **sum** can be written as:

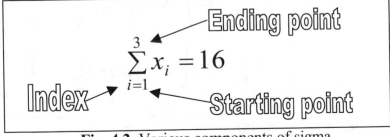

Fig. 4.2. Various components of sigma.

where the suffix i denotes the index, $i=1$ denotes the starting point of the index i which changes to 1, 2, and 3. Note that there are three terms involved in the **sum**. Number 3 on top of the sigma denotes the ending point. The letter x denotes the variable whose different values need to be added to get the required sum.

Thus we expand:

$$\sum_{i=1}^{3} x_i = x_1 + x_2 + x_3$$

Lets have some fun by looking at the following two figures to help you better understand this new symbol **SIGMA**.

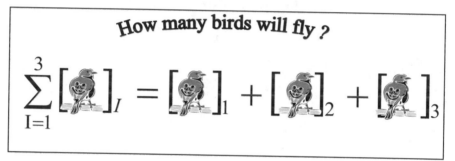

Fig. 4.3. Birds flying from sigma ($\sum$).

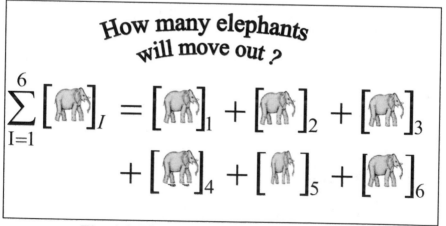

Fig. 4.4. Elephants moving from sigma ($\sum$).

4.3 CENTRAL VALUE OF A DATA SET

Note that a given data set may be a population or a sample according to how we define it. For example, all the students in a class can be considered a population or a sample according to our interest.

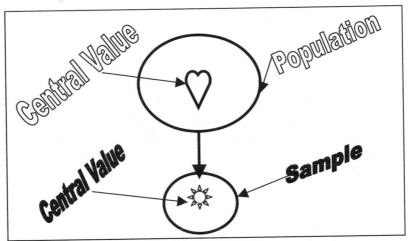

Fig. 4.5. Population and sample central values.

There are three measures of the middle value called: **mean, median,** and **mode**. Just add to your memory MMM like WWW.

Note that if a given data set represents a population, then we can find the **population mean, population median, or population mode**. If a given data set represents a sample, then we can find the **sample mean, sample median, or sample mode**.

4.4 POPULATION MEAN

Consider a population consisting of N units, say $X_1, X_2, ..., X_N$. Then the population mean is defined as the sum of the data values in the population divided by the total number of data values. It is denoted by the Greek letter: Mu (μ). Thus we have:

$$\text{Population Mean} (\mu) = \frac{X_1 + X_2 + ... + X_N}{N} = \frac{\sum\limits_{I=1}^{N} X_I}{N}$$

Example 4.1. (WEIGHING ELEPHANTS) Consider a circus having only five elephants with their weights in kg as:

Amy	Bob	Chris	Don	Eric
2,000	2,800	2,200	3,000	4,000

Fig. 4.6. All elephants in a zoo.

Find the average weight of all the elephants in the circus (or population mean).

Solution. We are given a population size of $N = 5$ and the weights of the elephants as: $X_1 = 2,000$ kg, $X_2 = 2,800$ kg, $X_3 = 2,200$ kg, $X_4 = 3,000$ kg and $X_5 = 4,000$ kg. Thus, the average weight of all the elephants is given by:

$$\text{Population Mean} \ (\mu) = \frac{\sum\limits_{i=I}^{N} X_I}{N} = \frac{\sum\limits_{I=1}^{5} X_I}{5}$$

$$= \frac{X_1 + X_2 + X_3 + X_4 + X_5}{5}$$

$$= \frac{2,000 + 2,800 + 2,200 + 3,000 + 4,000}{5}$$

$$= \frac{14,000}{5} = 2,800 \text{ kg} .$$

Note that if the original data is measured in **kg**, then the units of the population mean are also **kg**. In other words, the data values and the population mean have the same unit of measurement. Further note that the average weight of all the elephants in this population mean $\mu = 2,800$ kg is a **parameter,** which is an unknown and fixed quantity.

4.5 SAMPLE MEAN

Consider a sample consisting of n units, say $x_1, x_2, .., x_n$. Then the sample mean is defined as the sum of the data values in the sample divided by the total number of data values. It is denoted by x-bar ($\overline{x}$).

Thus we have:

$$\text{Sample Mean}\ (\bar{x}) = \frac{x_1 + x_2 + ... + x_n}{n} = \frac{\sum\limits_{i=1}^{n} x_i}{n}$$

Example 4.2. (SOMETIMES DOCTOR GIVES FREE SAMPLE)

Say we selected a sample of two elephants:

SRS Sample		
Elephants	Amy	Chris
Wt. (kg)	2,000	2,200

Fig. 4.7. Selected elephants.

(a) Find the sample mean.
(b) Interpret the sample mean.
Solution. (a) Here we are given the sample size $n = 2$ and sample values: $x_1 = 2,000$ kg and $x_2 = 2,200$ kg.

The average weight of the elephants in the sample is given by:

$$\text{Sample Mean}\ (\bar{x}) = \frac{\sum\limits_{i=1}^{n} x_i}{n} = \frac{\sum\limits_{i=1}^{2} x_i}{2} = \frac{x_1 + x_2}{2}$$

$$= \frac{2,000 + 2,200}{2} = \frac{4,200}{2} = 2100\ \text{kg}.$$

(b) Based on a sample of $n = 2$ elephants, we estimate that the true average weight, 2,800 kg, of all the $N = 5$ elephants is 2,100 kg.

Note that the sample mean $\bar{x} = 2,100$ kg is a **statistic**, which is known from a given sample, and varies from sample to sample. Again, note that if the original data values are in **kg,** then the sample mean is also in **kg.** The sample mean $\bar{x}$ is called an estimator of the population mean μ.

Example 4.3. (DANCING IS FUN) The following table gives the dancing time in minutes/day of eight elephants in a circus.

Elephants								
Time (x_i)	20	25	30	45	35	65	75	40

Fig. 4.8. Dancing elephants.

Compute the following:

$$(\text{i}) \sum_{i=1}^{8} x_i, \quad (\text{ii}) \left(\sum_{i=1}^{8} x_i \right)^2, \quad \text{and} \quad (\text{iii}) \sum_{i=1}^{8} x_i^2 \ .$$

Solution. We are given $n = 8$, $x_1 = 20$, $x_2 = 25$, $x_3 = 30$, $x_4 = 45$, $x_5 = 35$, $x_6 = 65$, $x_7 = 75$ and $x_8 = 40$.

$$(\text{i}) \ \sum_{i=1}^{8} x_i = x_1 + x_2 + x_3 + x_4 + x_5 + x_6 + x_7 + x_8$$
$$= 20 + 25 + 30 + 45 + 35 + 65 + 75 + 40 = 335.$$

$$(\text{ii}) \left(\sum_{i=1}^{8} x_i \right)^2 = (335)^2 = 112{,}225.$$

and

$$(\text{iii}) \ \sum_{i=1}^{8} x_i^2 = x_1^2 + x_2^2 + x_3^2 + x_4^2 + x_5^2 + x_6^2 + x_7^2 + x_8^2$$
$$= 20^2 + 25^2 + 30^2 + 45^2 + 35^2 + 65^2 + 75^2 + 40^2$$
$$= 400 + 625 + 900 + 2{,}025 + 1{,}225 + 4{,}225 + 5{,}625 + 1{,}600$$
$$= 16{,}625.$$

Remark: Note that the expressions $\left(\sum_{i=1}^{8} x_i \right)^2$ and $\sum_{i=1}^{8} x_i^2$ are different.

In the expression $\left(\sum_{i=1}^{8} x_i \right)^2$, we have to first find the sum of all terms and then square it, where as in the expression $\sum_{i=1}^{8} x_i^2$, we have to first square each term and then find the sum of all the squared terms.

4.6 OUTLIERS

If a data set has a few very large (or very small) values relative to the majority of the data values, such large (or small) values are called **outliers**.

For example, consider the following data set showing the GPA of four students as:

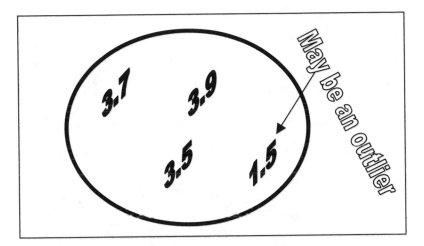

Fig. 4.9. Understanding outliers.

4.7 EFFECT OF OUTLIERS ON THE SAMPLE MEAN

Consider the three data values, $n = 3$, as: $x_1 = 2$, $x_2 = 3$, and $x_3 = 4$. Thus the sample mean is:

$$\bar{x} = \frac{\sum_{i=1}^{n} x_i}{n} = \frac{x_1 + x_2 + x_3}{3} = \frac{2+3+4}{3} = \frac{9}{3} = 3$$

which is clearly a central value of the data set: 2, 3, and 4.

Now say a few smaller values are included in the data set $n = 5$: $x_1 = 2$, $x_2 = 3$, $x_3 = 4$, $x_4 = 0$, and $x_5 = 0$, then we have:

$$\bar{x} = \frac{\sum\limits_{i=1}^{n} x_i}{n} = \frac{x_1 + x_2 + x_3 + x_4 + x_5}{5} = \frac{2+3+4+0+0}{5} = \frac{9}{5} = 1.8$$

Clearly, the sample mean $\bar{x} = 3.0$ is shifted towards the small value **0** and becomes $\bar{x} = 1.8$.

Similarly, consider what would happen if a few large values are included in the data set $n = 5$: $x_1 = 2$, $x_2 = 3$, $x_3 = 4$, $x_4 = 100$, and $x_5 = 100$, then we have:

$$\bar{x} = \frac{\sum\limits_{i=1}^{n} x_i}{n} = \frac{x_1 + x_2 + x_3 + x_4 + x_5}{5} = \frac{2+3+4+100+100}{5} = \frac{209}{5} = 41.8$$

Clearly the sample mean $\bar{x} = 3.0$ is shifted towards the large value **100** and becomes $\bar{x} = 41.8$.

We conclude that the sample (or population) mean is sensitive to the extreme values (or outliers) in the data set, thus it may or may not give the exact central value of the data set.

As an example, consider the situation of four rats and one elephant living together in a forest near a tree.

Fig. 4.10. Elephant is an outlier.

Clearly the elephant is an outlier among the rats if we mix them together while calculating the average weight or average diet of the rats.

Now, we are looking for a measure that can exactly divide the data into two equal parts. One such measure is called the **median**.

4.8 MEDIAN

The median of a set of n (or N) data values is the value that falls in the middle position when the data set is arranged in ascending order. The sample median is denoted by the lowercase letter m and the population median by the uppercase letter M.

4.8.1 RULES TO FIND SAMPLE MEDIAN

The following rules generally apply to find the sample median:
(a) Arrange the data set of n values in ascending order.
(b) There are two cases:

(i) If n is odd, the value at $\left(\dfrac{n+1}{2}\right)th$ position is the median.

(ii) If n is even, the average of the values at the $\dfrac{n}{2}th$ and

$\left(\dfrac{n}{2}+1\right)th$ position is the median.

Example 4.4. (WEIGHING CHICKENS) The following table shows the weight (lbs) of five chickens.

2	7	12	5	6

Fig. 4.11. Weight of chickens.

Find the median weight.
Solution. **Step I.** Arrange the data in ascending order.

2	5	6	7	12

Fig. 4.12. Chickens arranged by weight.

Step II. Here $n = 5 \, (\text{odd})$, so the value at

$$\frac{n+1}{2} \text{th} = \frac{5+1}{2} \text{th} = \frac{6}{2} \text{th} = \text{3rd position} = 6 \text{ lbs}$$

will be the median.

Thus $m = 6$ lbs is the median value. In other words, the chicken carrying the basket of eggs  is of median weight, that is, 50% of the chickens weigh less than 6 lbs and 50% of the chickens weigh more than 6 lbs. Note that the median has the same unit of measurement as the original data set.

Example 4.5. (CHICKEN LOVERS) The following table shows the weight (lbs) of six chickens.

| 2 | 9 | 10 | 5 | 6 | 12 |

Fig. 4.13. Weights of chickens.

Find the median weight.
Solution. **Step I.** Arrange the data in ascending order.

| 2 | 5 | 6 | 9 | 10 | 12 |

Fig. 4.14. Arrange chickens by weight.

Step II. Here $n = 6 \, (\text{even})$.

Now the value at

$$\frac{n}{2} \text{th} = \frac{6}{2} \text{th} = \text{3rd position} = 6 \text{ lbs}$$

and the value at

$$\left(\frac{n}{2}+1\right)\text{th} = \left(\frac{6}{2}+1\right)\text{th} = (3+1)\text{th} = 4\text{th position} = 9\,\text{lbs}.$$

The median weight is the average of these two values given by:

$$m = \frac{6+9}{2} = \frac{15}{2} = 7.5\,\text{lbs}.$$

Thus, $m = 7.5$ lbs is the median value. In other words, a new chicken with the weight between the weight of ![chicken] and ![chicken] will be the median chicken. Thus 50% of the chickens weigh less than 7.5 lbs and 50% of the chickens weigh more than 7.5 lbs.

4.8.2 EFFECT OF OUTLIERS ON MEDIAN

Note that the median is less sensitive to outliers than the mean. For example, the median of the data set 2, 3, 4, 5, 6 will be same as that of 2, 3, 4, 5, 100, but the mean values of these data sets will be different.

A layman can take the following example of a symmetric cylinder and an ice cream cone to differentiate between mean and median values.

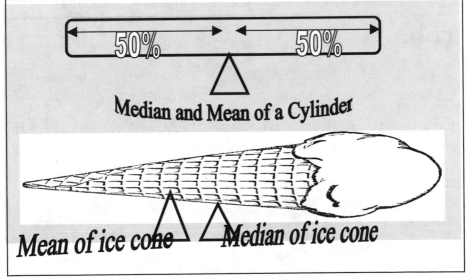

Fig. 4.15. Understanding mean and median.

Clearly the cylinder is equally heavy throughout and hence has the mean and the median at the same spot. The ice cream cone on the other hand is lighter from one side than the other thus it shifts the mean weight closer to the lighter side whereas the median remains at the same spot which divides it into two equal parts in length irrespective of weight. Thus if the ice cream cone has a long tail, then there will be less ice cream in it. This may be a reason why an ice cream cone seller makes long tails in ice cream cones.

Consider 1,000 shoes of different sizes bought by a shopkeeper as shown in **Figure 4.16**.

Shoe				
No.	5	6	7	8

Fig. 4.16. Shoes of different size.

At the end of the season, he found that size 8 had the maximum sales. It is obvious for the next season that the shopkeeper will buy the most shoes in size 8. What is size 8 for the shoe seller? The obvious answer is the mode or trend. Note that the mean value (or median value) of the shoe sizes, which is 6.5, is a useless measure for the shopkeeper. Thus, we have a new measure of central value called the **mode**.

4.9 MODE

The mode is the category or value that occurs the most frequently in the data set. It has same unit of measurement as the original data set.

Example 4.6. (GOOD CHICKENS LAY MORE EGGS)   The following table shows the number of eggs laid by seven hens during a 7 day week.

6	8	12	6	6	7	8

Fig. 4.17. Mode of eggs by different chickens.

What is the mode of the number of eggs laid/week?

Solution. In the given week, three chickens laid 6 eggs, thus the mode of the number of eggs laid/week is 6 eggs. Let us make a distribution table of the number of eggs laid by the chickens as follows:

Eggs	No. of chickens (f)
6	3
7	1
8	2
12	1
Sum	**7**

4.10 POSITIONS OF MEAN, MEDIAN, AND MODE

If the distribution is skewed to the left (the tail of the distribution goes to the left hand side like the thin side of the ice cream cone), then the mean, median and mode take the position as shown in **Figure 4.18**.

Thus for a distribution skewed to the left, we have:

$$\text{Mean} \leq \text{Median} \leq \text{Mode}$$

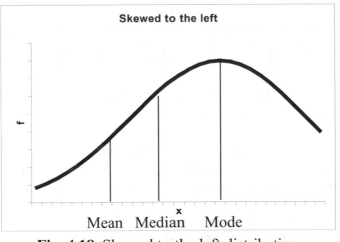

Fig. 4.18. Skewed to the left distribution.

In other words, if the distribution is skewed to the left then the mean is expected to be less than or equal to the median, and the median is expected to be less than or equal to the mode. The largest value will be the mode and the smallest value will be the mean.

If the distribution is symmetric and bell shaped (if a mirror can divide the distribution into two equal parts like a symmetric cylinder), then the mean, median, and mode take the position as shown in **Figure 4.19**.

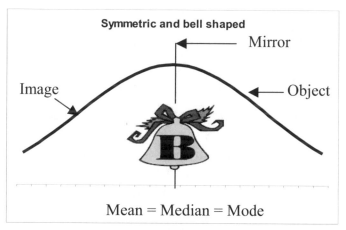

Fig. 4.19. Symmetric and bell shaped distribution.

Thus for a symmetric and bell shaped distribution, we have:

Mean = Median = Mode

In other words, if the distribution is symmetric and bell shaped then the mean, median, and mode values are equal.

If the distribution is skewed to the right (the tail of the distribution goes to the right side which is like the thin side of the ice cream cone), then the mean, median, and mode take the position as in **Figure 4.20**.

Now if a distribution is skewed to the right, we have:

Mode ≤ Median ≤ Mean

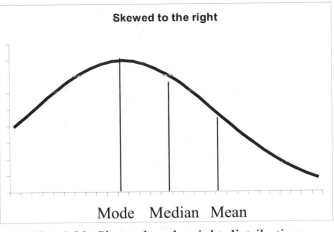

Fig. 4.20. Skewed to the right distribution.

In other words, if the distribution is skewed to the right, then the mode is expected to be less than or equal to the median, and the median is expected to be less than or equal to the mean. The largest value will be the mean and the smallest value will be the mode.

Note that whether the distribution is symmetric, skewed to the left, or skewed to the right, the median value always remains between the mean value and mode value.

4.10.1 MULTI-MODAL DATA SETS

Consider the data set: 5, 2, 4, 5, 9, 3, 5, 9, 6, and 9. In this data set, there are two modes: 5 and 9. It is called a bi-modal data set. A **dot plot** for such a data set is given by:

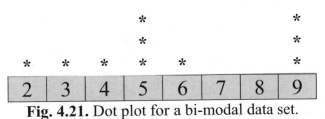

Fig. 4.21. Dot plot for a bi-modal data set.

Note that a mirror cannot divide the above dot plot into two equal parts: image and object. Thus, it is an example of an **asymmetric (or non-symmetric) and bi-modal** distribution.

The following figure shows a histogram made from a symmetric and bi-modal data set:

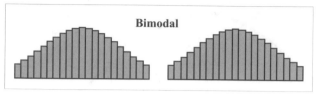

Fig. 4.22. Bi-modal histogram.

Note that mean and median are still same for the symmetric and bi-modal data, but **both modes** are different. Further note that mean, median, and mode are equal only for a **symmetric and bell shaped** distribution.

The following figure shows a symmetric and tri-modal data set:

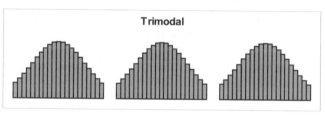

Fig. 4.23. Tri-modal histogram.

In the same way we may have *tetra-modal*, *penta-modal*, *hexa-modal*, etc. data sets. Such data sets are called multi-modal data sets.

Example 4.7. (FARMS NEED ATTENTION) Farmer Bob was very upset from the waste and disturbance created by rats in his field.

Fig. 4.24. A field surrounded by rats.

Then Bob hired a Statistician-cum-Rat detector to learn the average number of rats living per tunnel in the field. The Statistician-cum-Rat detector dug $n = 10$ tunnels in the field and noted the following number of rats living there.

Statistician-cum-Rat detector	Tunnel Name	Number of Rats
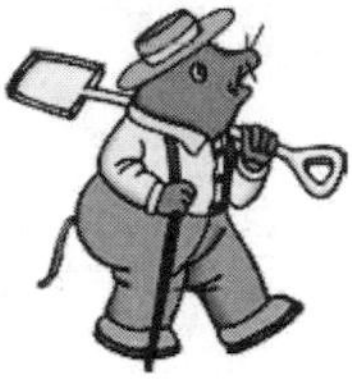	A	3
	B	2
	C	5
	D	6
	E	4
	F	4
	G	3
	H	5
	I	4
	J	4
Caution! data is not in any order.		

Fig. 4.25. Field experimental data.

At the end of the digging survey, the Statistician-cum-Rat detector reported to Bob that the distribution of rats in different tunnels in the field was symmetric and bell shaped. Justify the Statistician-cum-Rat detector's result using the following:

(a) Make a dot plot and comment.
(b) Find the sample mean.
(c) Find the sample median.
(d) Find the sample mode.
(e) Mean = Median = Mode implies symmetric and bell shaped.
(f) Discuss the Statistician-cum-Rat detector's opinion.

Solution. (a) Dot plot:

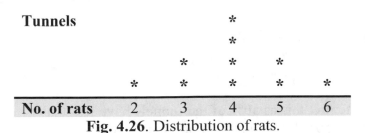

Fig. 4.26. Distribution of rats.

The dot plot in **Figure 4.26** shows that the distribution of rats in the tunnels in the field is symmetric and bell shaped.

(b) Sample mean:

Here we are given:

$$n = 10, \quad x_1 = 3, \quad x_2 = 2, \quad x_3 = 5, \quad x_4 = 6, \quad x_5 = 4, \quad x_6 = 4, \quad x_7 = 3,$$
$$x_8 = 5, \quad x_9 = 4, \text{ and } x_{10} = 4.$$

Thus, the sample mean is given by:

$$\bar{x} = \frac{\sum_{i=1}^{n} x_i}{n} = \frac{x_1 + x_2 + x_3 + x_4 + x_5 + x_6 + x_7 + x_8 + x_9 + x_{10}}{n}$$

$$= \frac{3 + 2 + 5 + 6 + 4 + 4 + 3 + 5 + 4 + 4}{10} = \frac{40}{10} = 4 \text{ rats}.$$

Thus, on average there are four rats per tunnel in the field.

(c) Sample median:

(I) Arrange the data in ascending order as:

2	3	3	4	4	4	4	5	5	6
B	A	G	E	F	I	J	C	H	D
Number of rats arranged in ascending order									

(II) Here $n = 10$ (even).

The value at $\frac{n}{2} th = \frac{10}{2} th = 5th$ position $= 4$ rats.

The value at $\left(\frac{n}{2} + 1\right) th = \left(\frac{10}{2} + 1\right) th = 6th$ position $= 4$ rats.

Thus, the median number of rats per tunnel is given by:

$$m = \frac{4 + 4}{2} = \frac{8}{2} = 4 \; .$$

We can interpret this as 50% of the tunnels have less than or equal to four rats and 50% have more than or equal to four rats.

(d) Sample mode:

The frequency distribution of tunnels having different numbers of rats is given by:

Rats	How many tunnels? (f)
2	1
3	2
4	4
5	2
6	1

Thus, one tunnel has either two or six rats, two tunnels have either three or five rats, and four tunnels have four rats. Thus, the maximum frequency of rats is four in four tunnels. In other words, the mode = 4. Note that here mode = 4 correspond to the number of rats and not the number of tunnels.

(e) Again note that:

Mean = Median = Mode = 4 rats.

Thus, the information by the Statistician-cum-Rat detector is correct.

(f) Opinion: In the Statistician-cum-Rat detector's opinion, farmer Bob is happy to know that the population of rats in the field follows symmetric distribution because he believes that nature loves symmetry. For example, the earth, sun, moon, eggs, trees, elephants, etc. are all almost symmetric.

4.11 IDEA OF SPREAD OF DATA

Data may have the same center but look different because of the spread of the numbers or variability in the numbers. For example, the data set: 5, 5, 5, 5, and 5 has no variation and its mean is 5; and the data set: 3, 4, 5, 6, and 7 has variation from 3 to 7 and its mean is also equal to 5. Thus, the two different data sets have the same mean value. Let us now enjoy an old story of a Camel who had refused to work in a zoo. Long ago, a camel was recruited by a zoo owner for the delivery of material to different destinations in the zoo. Monday through Friday, the camel had to walk to deliver a new hat to a monkey living nine miles away, a piece of bread to a dog living ten miles away and a new bottle of milk to a cat living eleven miles away. Thus, on Monday the average distance traveled by the camel was 10 miles. Although the total number of deliveries increases from Monday to Friday, the average distance traveled by the camel remains the same and is equal to 10 miles as shown in the last column of the following table. Assume the zoo owner knows only three measures: mean, median, and mode. The owner computed that the average distance traveled by the camel per day is the same, so the owner decided to pay the same amount of money to the camel every day. The camel argued with the owner and insisted that on Friday he was more tired than on Monday, so the owner should pay him more money on Friday.

 Camel & Owner	.	.	.	Lion	Mon-key	Dog	Cat	Rat	.	.	.	Distances in miles traveled by the Camel to make deliveries for different places / Ave-rage
Day												$\bar{x}$
Monday					9	10	11					10
Tuesday			8		9	10	11	12				10
Wednesday		7	8		9	10	11	12	13			10
Thursday	6	7	8		9	10	11	12	13	14		10
Friday	5	6	7	8	9	10	11	12	13	14	15	10

Fig. 4.27. Camel does delivery job.

Now the owner only knows how to find three measures: mean, median and mode, from a data set, and the camel refused to work. Then the owner went to see the zoo statistician to find a solution to his problem. The zoo statistician said to the owner, "Use of incomplete knowledge is dangerous." The zoo statistician suggested to look at the variation in the distances traveled by the camel.

The owner got excited and asked, "How?"

The zoo statistician told the owner that there are five measures of variation in a data set as listed below:

(a) Range
(b) Mean absolute deviation (MAD)
(c) Variance
(d) Standard deviation
(e) Coefficient of variation.

Fig. 4.28. Zoo Statistician.

The owner shouted, "Why so many measures?

Are you trying to confuse me?"

Fig. 4.29. Owner.

The zoo statistician nicely replied, "No. I will explain these to you one by one, along with their benefits and limitations."

4.12 RANGE

Range is the difference between the maximum value and the minimum value, that is:

$$\text{Range} = \text{Maximum value} - \text{Minimum value}$$
$$= X_{max} - X_{min}.$$

For example, if the prices of monkey hats are: $10, $16, $12, and $21. Then we have:

$$\text{Range} = \$21 - \$10 = \$11.$$

Note that the range has the same unit of measurement as that of the original data set. Its benefit is that it is easy to compute.

However, range has the following limitation:

The range depends only on the maximum and minimum values of the data set. For example, the data sets: 5, 7, 1, 2, and 4; and 5, 6, 2, 2, 4, and 8 have same range value 6. Thus, the range gives no importance to the other observations in the data set and hence may not be a good measure of variation.

4.13 MEAN ABSOLUTE DEVIATION

The mean absolute deviation (MAD) of a set of n data values, $x_1, x_2, ..., x_n$, is defined as:

$$MAD = \frac{\sum_{i=1}^{n}|x_i - \bar{x}|}{n}$$

where $\bar{x} = \dfrac{\sum_{i=1}^{n} x_i}{n}$ is the sample mean and $|\ |$ denotes the absolute value.

Note that the meaning of absolute value $|\ |$ is that $|5| = 5$ and $|-5| = 5$. In other words, $|\ |$ gives the magnitude of a number ignoring the sign. Further note that the unit of measurement of MAD is the same as that of the original data set.

Example 4.8. (THINK: IF WE HAD WINGS!) Consider five birds that can fly at different heights (km) as shown below.

B I R D	Chough	Eagle	Swan	Owl	Pigeon
Km	5	7	1	2	4

Fig. 4.30. Height of flying birds.

Find the mean absolute deviation (MAD) of the flying heights of these birds.

Solution. Here we have $n = 5$, $x_1 = 5$, $x_2 = 7$, $x_3 = 1$, $x_4 = 2$ and $x_5 = 4$.

Thus, the sample mean is given by:

$$\bar{x} = \frac{\sum_{i=1}^{n} x_i}{n} = \frac{\sum_{i=1}^{5} x_i}{5} = \frac{x_1 + x_2 + x_3 + x_4 + x_5}{5}$$

$$= \frac{5 + 7 + 1 + 2 + 4}{5} = \frac{19}{5} = 3.8 \text{ km}.$$

To find MAD, make the following table:

Birds	Height (km) x_i	Deviations from the mean $(x_i - \bar{x})$ (km)	Absolute values $\left| x_i - \bar{x} \right|$ (km)
Chough	5	$5 - 3.8 = 1.2$	$\left| 1.2 \right| = 1.2$
Eagle	7	$7 - 3.8 = 3.2$	$\left| 3.2 \right| = 3.2$
Swan	1	$1 - 3.8 = -2.8$	$\left| -2.8 \right| = 2.8$
Owl	2	$2 - 3.8 = -1.8$	$\left| -1.8 \right| = 1.8$
Pigeon	4	$4 - 3.8 = 0.2$	$\left| 0.2 \right| = 0.2$
Sum	**19**	**0**	**9.2**

From the above table:

$$n = 5, \ \sum_{i=1}^{n}(x_i - \bar{x}) = 0 \text{ km, and } \sum_{i=1}^{n} |x_i - \bar{x}| = 9.2 \text{ km.}$$

Note that the *sum of the deviations from the mean* is always *zero*.

Thus, the value of mean absolute deviation (MAD) is given by:

$$\text{MAD} = \frac{\sum_{i=1}^{n} |x_i - \bar{x}|}{n} = \frac{\sum_{i=1}^{5} |x_i - \bar{x}|}{5}$$

$$= \frac{|x_1 - \bar{x}| + |x_2 - \bar{x}| + |x_3 - \bar{x}| + |x_4 - \bar{x}| + |x_5 - \bar{x}|}{5}$$

$$= \frac{1.2 + 3.2 + 2.8 + 1.8 + 0.2}{5} = \frac{9.2}{5} = 1.84 \text{ km.}$$

Note that while calculating MAD, we have to take the absolute value of each deviation from the sample mean, which is not difficult using calculators or computers. However, its theory becomes too much complicated and is beyond the scope of this book.

Further note that the measures mean, median, mode, range, and MAD have the same unit of measurement as that of the original data set. For example, if the original data is in km, then the mean, median, mode, range and MAD will also be measured in km.

Let us now look at the unit of measurement of the next measure of variation, variance.

4.14 VARIANCE

In this section, we discuss two types of variance: (a) population variance and (b) sample variance.

4.14.1 POPULATION VARIANCE

The population variance of N data values is defined as the average of the square of the deviations of the data values about the population mean μ. It is denoted by the symbol sigma square (σ^2). Note that the symbol σ is a lowercase Greek letter *sigma*. If $X_1, X_2,..., X_N$ are the N values in the population, then the population variance is given by:

Read as
"Sigma Square"

$$\sigma^2 = \frac{\sum_{I=1}^{N}(X_I - \mu)^2}{N}, \quad \text{where} \quad \mu = \frac{\sum_{I=1}^{N}X_I}{N}$$

Note that if the original data is in *kg*, then the value of σ^2 will in kg^2 while that of the population mean μ will be in *kg*. Therefore, to keep consistency, we define a new term: **population standard deviation**.

4.14.1.1 POPULATION STANDARD DEVIATION

Population standard deviation is the positive square root of the population variance, σ^2. It is denoted by the lowercase Greek letter sigma (σ) as:

Read as: sigma

$$\sigma = \sqrt{\sigma^2}$$

Note that the unit of measurement of the population standard deviation will be the same as that of the original data set. For example, if the data is in *kg*, then σ will also be in *kg*.

Example 4.9. (WEIGHING ELEPHANTS) Consider a circus having only five elephants with weights (in kg) as shown below:

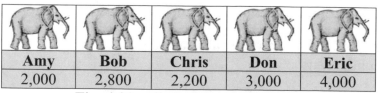

Amy	Bob	Chris	Don	Eric
2,000	2,800	2,200	3,000	4,000

Fig. 4.31. All elephants in a circus.

Find the population variance and population standard deviation of the weight of all the elephants in the circus (or in the population).

Solution. We are given a population of size $N = 5$ and the weights of the elephants as: $X_1 = 2,000$, $X_2 = 2,800$, $X_3 = 2,200$, $X_4 = 3,000$ and $X_5 = 4,000$.

Thus, we have the following table:

Elephant	Weight (Kg) X_I	Deviations from the mean (Kg) $(X_I - \mu)$	Squares of the deviations (Kg²) $(X_I - \mu)^2$
Amy	2,000	2,000-2,800 = -800	640,000
Bob	2,800	2,800-2,800 = 0	0
Chris	2,200	2,200-2,800 = -600	360,000
Don	3,000	3,000-2,800 = 200	40,000
Eric	4,000	4,000-2,800 = 1,200	1,440,000
Sum	**14,000**	**0**	**2,480,000**

From the above table we have:

$$\sum_{I=1}^{N} X_I - 14,000 \, \text{kg}, \quad \sum_{I=1}^{N} (X_I - \mu) = 0 \, \text{kg},$$

and $\quad \sum_{I=1}^{N} (X_I - \mu)^2 = 2,480,000 \, \text{kg}^2.$

The population mean is given by:

$$\mu = \frac{\sum\limits_{I=1}^{N} X_I}{N} = \frac{\sum\limits_{I=1}^{5} X_I}{5} = \frac{14,000}{5} = 2,800 \, \text{kg}.$$

Thus, the population variance is given by:

$$\sigma^2 = \frac{\sum\limits_{I=1}^{N} (X_I - \mu)^2}{N} = \frac{2,480,000}{5} = 496,000 \, \text{kg}^2$$

and the population standard deviation is given by:

$$\sigma = \sqrt{\sigma^2} = \sqrt{496,000 \, \text{kg}^2} = 704.27 \, \text{kg}.$$

4.14.2 SAMPLE VARIANCE

The sample variance of n data values, say $x_1, x_2, ..., x_n$, is defined as the sum of the squared deviations about the sample mean $\bar{x}$ divided by $(n-1)$ and is denoted by the symbol s^2 (read as lowercase English letter s-square). Note that there are two formulae to find the value of s^2. The first one is given by:

Read as:
s-square
$$s^2 = \frac{\sum\limits_{i=1}^{n}(x_i - \bar{x})^2}{n-1}$$

and the second one is given by:

Read as:
s-square
$$s^2 = \frac{n\left(\sum\limits_{i=1}^{n} x_i^2\right) - \left(\sum\limits_{i=1}^{n} x_i\right)^2}{n(n-1)}$$

The first formula is called a formula by definition and the second one is called a computing formula. Note that both formulae give the same result. The easy thing in the computing formula is that we do not need to calculate the sample mean while computing the sample variance. The first formula can also be considered as a two steps formula where in the first step we have to find the sample mean and in the second step we find the sample variance. Again note that if the original data is in kg, then the unit of measurement of the sample variance s^2 will be in kg^2. Thus, to make it consistent with the units of measurement of the original data, we define a new term called **sample standard deviation**.

4.14.2.1 SAMPLE STANDARD DEVIATION

Sample standard deviation is defined as the positive square root of the sample variance. In other words:

Read as: s
$$s = \sqrt{s^2}$$

The unit of measurement of the sample standard deviation will be the same as that of the original data. For example, if the data is in kg, then s will also be in kg.

4.15 COEFFICIENT OF VARIATION

The sample coefficient of variation (CV) is defined as:

$$CV = \frac{\text{sample standard deviation}}{\text{sample mean}} \times 100\%$$

$$= \frac{s}{\bar{x}} \times 100\%$$

Note that the coefficient of variation has the following properties:
(a) It is independent of the unit of measurement of the original data, but can be expressed in terms of a percentage (%).
(b) It is useful for comparing the variation of two sets of data having different mean values or measured in different units.
(c) The smaller the value of the CV, the more stable or consistent the data.

Example 4.10. (FLYING BIRDS) Consider five birds that can fly at different heights (km) as shown below.

B I R D	 Chough	 Eagle	 Swan	 Owl	 Pigeon
Km	5	7	1	2	4

Fig. 4.32. The height of the flying birds.

(a) Find the sample variance using the first formula based on the definition.
(b) Find the sample variance using the computing formula.
(c) Are the sample variances based on (a) and (b) the same?
(d) Derive the value of the sample standard deviation.
(e) Derive the value of the coefficient of variation.
Solution. Note that we are given: $n = 5$, $x_1 = 5$, $x_2 = 7$, $x_3 = 1$, $x_4 = 2$ and $x_5 = 4$.
(a) To use the first formula of sample variance, let us make the following table:

Bird	x_i km	$(x_i - \bar{x})$ km	$(x_i - \bar{x})^2$ km²
Chough	5	5-3.8 = 1.2	$(1.2)^2 = 1.44$
Eagle	7	7-3.8 = 3.2	$(3.2)^2 = 10.24$
Swan	1	1-3.8 = -2.8	$(-2.8)^2 = 7.84$
Owl	2	2-3.8 = -1.8	$(-1.8)^2 = 3.24$
Pigeon	4	4-3.8 = 0.2	$(0.2)^2 = 0.04$
Sum	19	0	22.80

From the above table, we have:

$$\sum_{i=1}^{n} x_i = 19\,\text{km},\ \sum_{i=1}^{n}(x_i - \bar{x}) = 0\,\text{km, and}\ \sum_{i=1}^{n}(x_i - \bar{x})^2 = 22.80\,\text{km}^2.$$

Note that the sample mean is given by:

$$\bar{x} = \frac{\sum_{i=1}^{n} x_i}{n} = \frac{19}{5} = 3.8\text{km}.$$

The sample variance based on the first method is given by:

$$s^2 = \frac{\sum_{i=1}^{n}(x_i - \bar{x})^2}{n-1} = \frac{22.80}{5-1} = \frac{22.80}{4} = 5.7\,\text{km}^2.$$

(b) For the computing formula, complete the following table:

Bird	x_i km	x_i^2 km²
Chough	5	$5^2 = 25$
Eagle	7	$7^2 = 49$
Swan	1	$1^2 = 1$
Owl	2	$2^2 = 4$
Pigeon	4	$4^2 = 16$
Sum	19	95

From the above table, we have:

$n = 5$, $\sum\limits_{i=1}^{n} x_i = 19 \, \text{km}$, and $\sum\limits_{i=1}^{n} x_i^2 = 95 \, \text{km}^2$. Thus, the sample variance by the computing formula is given by:

$$s^2 = \frac{n\left(\sum\limits_{i=1}^{n} x_i^2\right) - \left(\sum\limits_{i=1}^{n} x_i\right)^2}{n(n-1)} = \frac{5 \times 95 - (19)^2}{5(5-1)} = \frac{475 - 361}{20} = \frac{114}{20} = 5.7 \text{km}^2.$$

(c) Yes, from (a) and (b), we can see that the value of the sample variance s^2 is the same by both formulae. Thus, both formulae of the sample variance are equivalent.

(d) The sample standard deviation is given by:
$$s = \sqrt{s^2} = \sqrt{5.7} = 2.39 \, \text{km}.$$

(e) The coefficient of variation is given by:
$$CV = \frac{s}{\bar{x}} \times 100\% = \frac{2.39}{3.80} \times 100\% = 62.89\%$$

Example 4.11. (TRY TO BE CONSISTENT IN LIFE) A zoo statistician wishes to compare **monkey jumps** measured in *meters,* with the **diet of an elephant** measured in *kg*. The following data set is collected from an elephant on six randomly selected days and a monkey's jumps are measured on five randomly selected days.

Monkey's Jumps (meters)		4	5	2	6	8	
Elephant's diet (kg)		204	205	202	206	208	202

Fig. 4.33. Smaller CV is better.

(a) Which measure can be used to compare such data and why?
(b) Find the coefficient of variation for the monkey's jumps.
(c) Find the coefficient of variation for the elephant's diet.
(d) Which is more consistent: **Monkey's jumps or Elephant's diet**? Comment if possible.

Solution. (a) Coefficient of variation can be used to compare two data sets having different unit of measurements.

(b) To find the coefficient of variation for the monkey's jumps data set, complete the following table:

Monkey's Jumps	x_i	x_i^2
	4	$4^2 = 16$
	5	$5^2 = 25$
	2	$2^2 = 4$
	6	$6^2 = 36$
	8	$8^2 = 64$
Sum	25	145

Fig. 4.34. Sample variance by computing formula.

From the above table we have:

$$n = 5, \ \sum_{i=1}^{n} x_i = 25, \text{ and } \sum_{i=1}^{n} x_i^2 = 145.$$

Thus, the sample mean height of the monkey's jumps is given by:

$$\bar{x} = \frac{\sum\limits_{i=1}^{n} x_i}{n} = \frac{25}{5} = 5 \text{ meters.}$$

By the computing formula the sample variance is given by:

$$s^2 = \frac{n\left(\sum\limits_{i=1}^{n} x_i^2\right) - \left(\sum\limits_{i=1}^{n} x_i\right)^2}{n(n-1)} = \frac{5 \times 145 - (25)^2}{5(5-1)}$$

$$= \frac{725 - 625}{20} = \frac{100}{20} = 5 \text{ meters}^2.$$

Thus, the sample standard deviation of the monkey's jumps is given by:

$$s = \sqrt{s^2} = \sqrt{5} \approx 2.236 \text{ meters.}$$

Hence, the coefficient of variation of the monkey's jumps is given by:

$$CV(\text{Monkey's Jumps}) = \frac{s}{\bar{x}} \times 100\% = \frac{2.236}{5} \times 100\% = 44.72\%$$

(c) To find the coefficient of variation for the elephant's diet data set, complete the following table:

Elephant's diet	x_i	x_i^2
	204	$204^2 = 41{,}616$
	205	$205^2 = 42{,}025$
	202	$202^2 = 40{,}804$
	206	$206^2 = 42{,}436$
	208	$208^2 = 43{,}264$
	202	$202^2 = 40{,}804$
Sum	**1,227**	**250,949**

Fig. 4.35. Sample variance by the computing formula.

From the above table, we have:

$$n = 6, \ \sum_{i-1}^{n} x_i = 1{,}227, \text{ and } \sum_{i=1}^{n} x_i^2 = 250{,}949.$$

Thus, the sample mean of the elephant's diet is given by:

$$\bar{x} = \frac{\sum_{i=1}^{n} x_i}{n} = \frac{1{,}227}{6} = 204.5kg \ .$$

By the computing formula, the sample variance is given by:

$$s^2 = \frac{n\left(\sum_{i=1}^{n} x_i^2\right) - \left(\sum_{i=1}^{n} x_i\right)^2}{n(n-1)} = \frac{6 \times 250{,}949 - (1{,}227)^2}{6(6-1)}$$

$$= \frac{1{,}505{,}694 - 1{,}505{,}529}{30} = \frac{165}{30} = 5.5 \ kg^2 \ .$$

Thus, the sample standard deviation of the elephant's diet is given by:

$$s = \sqrt{s^2} = \sqrt{5.5} \approx 2.345 \, kg \ .$$

Hence, the coefficient of variation of the elephant's diet is given by:

$$\text{CV}\left(\text{Elephant's diet}\right) = \frac{s}{\overline{x}} \times 100\% = \frac{2.345}{204.5} \times 100\% = 1.15\%$$

(d) Clearly, the value of the coefficient of variation for the elephant's diet is much less than that of the monkey's jumps. Thus, the elephant's diet is more consistent than the monkey's jumps. This means the elephant cares about his health and likes to eat consistently. In contrast, the monkey makes jumps depending upon his mood and the jumps are inconsistent.

Example 4.12. (AN OLD STORY) A long time ago, an old lion was living under a tree and six rats were living in a nearby tunnel.

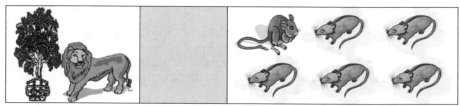

Fig. 4.36. Lion and six rats.

Out of the six rats, one rat had small whiskers. One day the lion was sleeping under a tree and all six rats made a plan to steal the lion's whiskers. They planned to replace the rat's whiskers with the lion's whiskers.

Fig. 4.37. Sleeping lion.

All six rats went up to the sleeping lion and removed his whiskers. In the meantime, the lion woke up and all the rats ran away in different directions of the field.

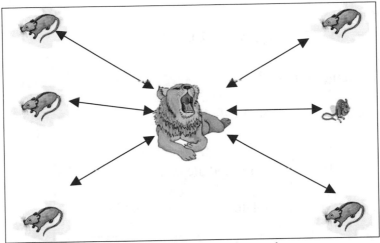

Fig. 4.38. The lion shouted.

As soon as the lion saw his face in a mirror without whiskers, he became very angry because his whiskers were a sign of pride. The lion called a statistician-cum-police inspector to learn the locations of the rats in the field. The statistician-cum-police inspector immediately suggested a **remote sensing survey** to locate the positions of the rats in the field and noted the distance (in meters) covered by the different rats as follows:

Rats						
Distance	125	204	178	298	320	120

Fig. 4.39. Data collected by the statistician.

Note that such data is sometimes also called **spatial data**. The lion wants to make sure that the rats are no longer living together so that his whiskers will remain safe in the future.

Find the spread of rats using:

(a) Range
(b) MAD
(c) Variance
(d) Standard deviation
(e) Coefficient of variation
(f) Comment on your answers.

Solution. We are given: $n = 6$, $x_1 = 125$, $x_2 = 204$, $x_3 = 178$, $x_4 = 298$, $x_5 = 320$, and $x_6 = 120$.

(a) The range is given by:

$$\text{Range} = \text{Maximum value} - \text{Minimum value}$$

$$= X_{max} - X_{min}$$

$$= 320 - 120 = 200 \text{ meters.}$$

(b) To find MAD complete the following table:

x_i	$(x_i - \bar{x})$	$|x_i - \bar{x}|$
125	-82.5	82.5
204	-3.5	3.5
178	-29.5	29.5
298	90.5	90.5
320	112.5	112.5
120	-87.5	87.5
Sum **1,245**	**0**	**406.0**

From the above table we have:

$$\sum_{i=1}^{n} x_i = 1,245 \quad \text{and} \quad \sum_{i=1}^{n} |x_i - \bar{x}| = 406.0$$

where the sample mean $\bar{x}$ is given by:

$$\bar{x} = \frac{\sum_{i=1}^{n} x_i}{n} = \frac{1,245}{6} = 207.50 \text{ meters.}$$

Thus, the mean absolute deviation (MAD) is given by:

$$\text{MAD} = \frac{\sum_{i=1}^{n} |x_i - \bar{x}|}{n} = \frac{406.0}{6} = 66.67 \text{ meters.}$$

(c) To find the sample variance, complete the following table:

x_i	$(x_i - \bar{x})$	$(x_i - \bar{x})^2$
125	-82.5	6,806.25
204	-3.5	12.25
178	-29.5	870.25
298	90.5	8,190.25
320	112.5	12,656.25
120	-87.5	7,656.25
Sum **1,245**	**0.0**	**36,191.50**

From the above table:

$$\sum_{i=1}^{n}(x_i - \bar{x})^2 = 36,191.50.$$

Thus the sample variance s^2 is given by:

$$s^2 = \frac{\sum_{i=1}^{n}(x_i - \bar{x})^2}{n-1} = \frac{36,191.50}{6-1} = \frac{36,191.50}{5} = 7,238.30 \text{ meters}^2.$$

(d) The sample standard deviation is given by:

$$s = \sqrt{s^2} = \sqrt{7,238.30} = 85.08 \text{ meters}.$$

(e) The coefficient of variation is given by:

$$CV = \frac{s}{\bar{x}} \times 100\% = \frac{85.08}{207.50} \times 100\% = 41.00\%$$

(f) Thus, the statistician-cum-police inspector told the old lion that all the measures of variation show that the rats are no longer living together. Thereafter, the old lion felt that his whiskers would be safe.

Remark: Someday you may learn that none of these measures work for **spatial data,** but this explanation is beyond the scope of this book.

4.16 TCHEBYSHEFF's RULE

Given a real number $k > 1$ and a set of N data values in a population, then at least $\left(1 - \dfrac{1}{k^2}\right)$ of the N data values lie within k times standard deviation, σ, from the population mean, μ, that is, in the interval given by:

$$\mu \pm k\sigma, \text{ or } (\mu - k\sigma, \ \mu + k\sigma).$$

For example, if $k = 2$, then at least

$$\left(1 - \frac{1}{k^2}\right) = \left(1 - \frac{1}{2^2}\right) = \left(1 - \frac{1}{4}\right) = \frac{3}{4} \text{ of the } N \text{ data values}$$

lie within the interval $\mu \pm 2\sigma$, or $(\mu - 2\sigma, \ \mu + 2\sigma)$.
In other words, **at least** 75% of the data values lie in the interval: $(\mu - 2\sigma, \ \mu + 2\sigma)$.

If $k = 3$, then at least

$$\left(1 - \frac{1}{k^2}\right) = \left(1 - \frac{1}{3^2}\right) = \left(1 - \frac{1}{9}\right) = \frac{8}{9} \text{ of the } N \text{ data values lie}$$

within the interval $\mu \pm 3\sigma$, or $(\mu - 3\sigma, \ \mu + 3\sigma)$.
In other words, **at least** 89% of the data values lie in the interval: $(\mu - 3\sigma, \ \mu + 3\sigma)$.

If $k = 4$, then at least

$$\left(1 - \frac{1}{k^2}\right) = \left(1 - \frac{1}{4^2}\right) = \left(1 - \frac{1}{16}\right) = \frac{15}{16} \text{ of the } N \text{ data values lie}$$

within the interval $\mu \pm 4\sigma$, or $(\mu - 4\sigma, \ \mu + 4\sigma)$.
In other words, **at least** 93% of the data values lie in the interval: $(\mu - 4\sigma, \ \mu + 4\sigma)$.

Tchebysheff's rule holds for any kind of distribution that is symmetric, skewed to the left, or skewed to the right. A pictorial representation of Tchebysheff's rule is shown in **Figure 4.40** with the help of a skewed to the right distribution.

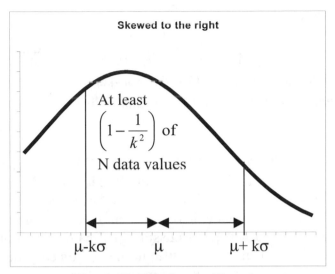

Fig. 4.40. Tchebysheff's rule.

Note that the same result is also true for a sample data set using sample mean and sample standard deviation. Given a real number k greater than one and a set of n data values in a sample, then at least $\left(1 - \dfrac{1}{k^2}\right)$ of the n data values lie within k times the sample standard deviation, s, from the sample mean, $\bar{x}$, that is in the interval:

$$\bar{x} \pm ks, \text{ or } \left(\bar{x} - ks, \ \bar{x} + ks\right).$$

For example, if $k = 2$, then at least
$$\left(1 - \frac{1}{k^2}\right) = \left(1 - \frac{1}{2^2}\right) = \left(1 - \frac{1}{4}\right) = \frac{3}{4} \text{ of the } n \text{ data values lies}$$
within the interval $\bar{x} \pm 2s$ or $\left(\bar{x} - 2s, \ \bar{x} + 2s\right)$.

In other words, **at least** 75% of the data values lie in the interval: $\left(\bar{x} - 2s, \ \bar{x} + 2s\right)$.

If $k = 3$, then at least
$$\left(1 - \frac{1}{k^2}\right) = \left(1 - \frac{1}{3^2}\right) = \left(1 - \frac{1}{9}\right) = \frac{8}{9} \text{ of the } n \text{ data values lie}$$
within the interval $\bar{x} \pm 3s$ or $\left(\bar{x} - 3s, \ \bar{x} + 3s\right)$.

In other words, **at least** 89% of the data values lie in the interval: $\left(\bar{x} - 3s, \ \bar{x} + 3s\right)$.

If $k = 4$, then at least

$$\left(1 - \frac{1}{k^2}\right) = \left(1 - \frac{1}{4^2}\right) = \left(1 - \frac{1}{16}\right) = \frac{15}{16}$$ of the n data values lie

within the interval $\bar{x} \pm 4s$ or $\left(\bar{x} - 4s,\ \bar{x} + 4s\right)$.

In other words, **at least** 93% of the data values lie in the interval: $\left(\bar{x} - 4s,\ \bar{x} + 4s\right)$.

Example 4.13. (THINK BEFORE YOU ACT) A long time ago twenty-five rats were living together and a cat was also living near them in a field of sugarcane. The rats were always afraid of the cat. One day all twenty-five rats called an urgent meeting to put a bell on the cat's neck. The rats were discussing this matter when an old rat raised a question to them, "Who will put the bell on cat's neck?" The meeting went a bit longer and in the meantime the cat came. As soon as the rats saw the cat they all ran away with an average speed of 75 meters/minute and a variance of 100 (meters/minute)^2, respectively.

Fig. 4.41. Cat and rats.

(a) What was the minimum number of rats running within the speed limit of 55 meters/minute to 95 meters/minute?

(b) What was minimum number of rats running within the speed limit of 45 meters/minute to 105 meters/minute?

Solution. Given a sample mean of $\bar{x} = 75$ and a sample variance of $s^2 = 100$, the sample standard deviation $s = \sqrt{s^2} = \sqrt{100} = 10$.

(a) If $k - 2$, then by Tchebysheff's rule, at least:

$$\left(1 - \frac{1}{k^2}\right) = \left(1 - \frac{1}{2^2}\right) = \left(1 - \frac{1}{4}\right) = \frac{3}{4} \text{ of the } n = 25 \text{ rats}$$

$$= \frac{3}{4} \times 25 = \frac{75}{4} \approx 19 \text{ rats}$$

had speed between the interval

$$\bar{x} \pm ks, \quad \text{or} \quad \bar{x} \pm 2s, \text{ or } 75 \pm 2 \times 10, \text{ or } 75 \pm 20, \text{ or}$$

$$\left(75 - 20, \ 75 + 20\right), \text{ or } \left(55, \ 95\right).$$

(b) If $k = 3$, then by Tchebysheff's rule, at least

$$\left(1 - \frac{1}{k^2}\right) = \left(1 - \frac{1}{3^2}\right) = \left(1 - \frac{1}{9}\right) = \frac{8}{9} \text{ of the } n = 25 \text{ rats}$$

$$= \frac{8}{9} \times 25 = \frac{200}{9} \approx 22 \text{ rats}$$

had speed between the interval

$$\bar{x} \pm ks, \quad \text{or} \quad \bar{x} \pm 3s, \text{ or } 75 \pm 3 \times 10, \text{ or } 75 \pm 30, \text{ or}$$

$$\left(75 - 30, \ 75 + 30\right), \text{ or } \left(45, \ 105\right).$$

4.17 EMPIRICAL RULE

Remember that distribution is said to be symmetric if a mirror can divide it into two equal parts, that is, if one side can be considered an object, then the other side will be its image as shown in the **Figure 4.43**. Given a distribution of data values that is approximately bell shaped and symmetric or normal as shown in **Figure 4.42**, then:

Fig. 4.42. Bell.

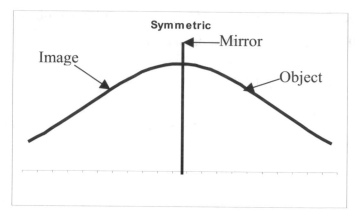

Fig. 4.43. Mirror divides symmetric distribution into two equal halves.

(a) the one sigma interval $\mu \pm \sigma$ or $(\mu - \sigma, \mu + \sigma)$ contains approximately 68% of the population data values, N.

(b) the two sigma interval $\mu \pm 2\sigma$ or $(\mu - 2\sigma, \mu + 2\sigma)$ contains approximately 95% of the population data values, N.

(c) the three sigma interval $\mu \pm 3\sigma$ or $(\mu - 3\sigma, \mu + 3\sigma)$ contains approximately 99% or almost all of the population data values, N.

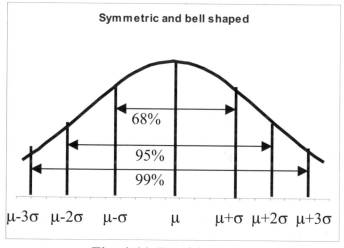

Fig. 4.44. Empirical rule.

Figure 4.44 provides a pictorial representation of the empirical rule. Further note that the sample analogous intervals $\bar{x} \pm s$, $\bar{x} \pm 2s$ and $\bar{x} \pm 3s$ also contain approximately 68%, 95% and 99% of the sample data values n, respectively.

Example 4.14. (SPEED THRILLS, BUT KILLS). In a race, the following 25 cars participated with an average speed of 75 miles/hr and a variance of 25 $(\text{miles/hr})^2$.

Fig. 4.45. Car race.

Assuming that the distribution of the speed of the cars is exactly bell shaped and symmetric, find:

(a) How many cars were going between 70 miles/hr and 80 miles/hr?
(b) How many cars were going between 65 miles/hr and 85 miles/hr?
(c) How many cars were going between 60 miles/hr and 90 miles/hr?
(d) How many cars were going more than 70 miles/hr?

Solution. Given $n = 25$, $\bar{x} = 75$ and $s = 5$, thus by the empirical rule, the interval:

(a) $\bar{x} \pm s$, or 75 ± 5, or $(75 - 5, 75 + 5)$, or $(70, 80)$ contains approximately 68% of the data values. Thus, 68% of 25 = 17 cars were going between 70 miles/hr and 80 miles/hr.
(b) $\bar{x} \pm 2s$, or $75 \pm 2 \times 5$, or $(75 - 10, 75 + 10)$, or $(65, 85)$ contains approximately 95% of the data values. Thus, 95% of 25 = 23.75 ≈ 24 cars were going between 65 miles/hr and 85 miles/hr.
(c) $\bar{x} \pm 3s$, or $75 \pm 3 \times 5$, or $(75 - 15, 75 + 15)$, or $(60, 90)$ contains approximately 99% of the data values. Thus, 99% of 25 = 24.75 ≈ 25 cars were going between 60 miles/hr and 90 miles/hr.
(d) Note that the distribution is exactly bell-shaped, thus from (a) the number of cars going more than 70 miles/hr = 17+4 = 21.

Example 4.15. (KEEP THE WASHROOM NEAT AND CLEAN)
A monkey sells different brands of toilet paper at the same price but with different lengths depending upon the quality of the paper as given below:

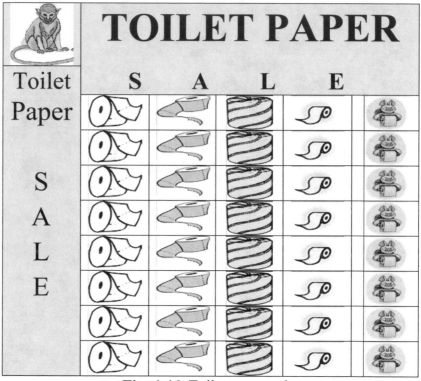

Fig. 4.46. Toilet paper sale.

The sales monkey claims that these $n = 40$ toilet paper rolls have an average length and a standard deviation of 12.8 meters and 1.7 meters, respectively.

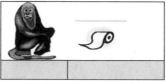

Fig. 4.47. Monkey in a washroom.

Another monkey found that his washroom is running out of toilet paper and he wants to know if he buys all 40 rolls of toilet paper, then:

(a) How many toilet paper rolls will have a length between 11.1 meters and 14.5 meters?
(b) How many toilet paper rolls will have a length between 9.4 meters and 16.2 meters?
(c) How many toilet paper rolls will have a length between 7.7 meters and 17.9 meters?

Solution. Given $n = 40$, $\bar{x} = 12.8$ meters and $s = 1.7$ meters.

From the empirical rule:

(a) the interval $\bar{x} \pm s$, or 12.8 ± 1.7, or $(12.8 - 1.7, \ 12.8 + 1.7)$, or $(11.1, 14.5)$ contains 68% of the $n = 40$ rolls.

Thus, approximately 68% of 40 or $68\% \times 40 = \dfrac{68}{100} \times 40 = 27.2 \approx 27$

toilet paper rolls will have a length between 11.1 meters and 14.5 meters.

(b) the interval $\bar{x} \pm 2s$, or $12.8 \pm 2 \times 1.7$, or $(12.8 - 3.4, \ 12.8 + 3.4)$, or $(9.4, 16.2)$ contains 95% of the $n = 40$ rolls.

Thus, approximately 95% of 40 or $95\% \times 40 = \dfrac{95}{100} \times 40 = 38$ toilet

paper rolls will have a length between 9.4 meters and 16.2 meters.

(c) the interval $\bar{x} \pm 3s$, or $12.8 \pm 3 \times 1.7$, or $(12.8 - 5.1, \ 12.8 + 5.1)$, or $(7.7, 17.9)$ contains 99% of the $n = 40$ rolls.

Thus, approximately 99% of 40 or $99\% \times 40 = \dfrac{99}{100} \times 40 = 39.6 \approx 40$

toilet paper rolls will have a length between 7.7 meters and 17.9 meters.

Example 4.16. (HIGH SCORES ARE EXCITING) Three owl instructors are comparing the scores of elephants in a dancing competition. Each owl instructor has seven elephants in class.

In class **A**, one elephant has a score of 20, one has a score of 80, and the rest have scores of 50.

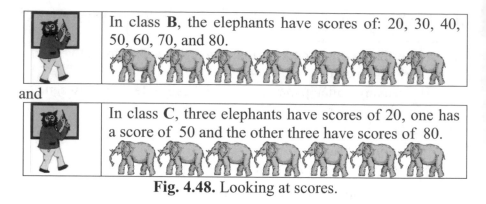

In class **B**, the elephants have scores of: 20, 30, 40, 50, 60, 70, and 80.

and

In class **C**, three elephants have scores of 20, one has a score of 50 and the other three have scores of 80.

Fig. 4.48. Looking at scores.

(a) Construct dot plots for classes **A**, **B** and **C**. Guess which class has the maximum variation.

(b) Find the variation by range. Which class has the maximum variation?

(c) Find the variation by sample standard deviation. Which class has the maximum variation?

(d) Comment on the guess.

Solution. (a) Three dot plots for the classes **A**, **B** and **C** are:

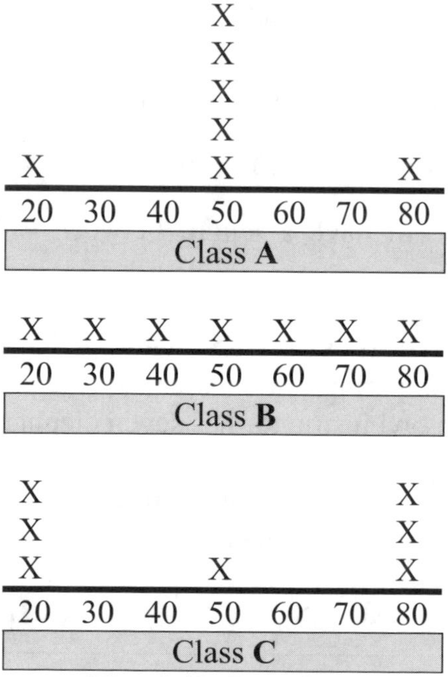

```
                        X
                        X
                        X
                        X
        X               X               X
      ─────────────────────────────────────
       20  30  40  50  60  70  80
      ┌─────────────────────────────────┐
      │             Class A              │
      └─────────────────────────────────┘


       X   X   X   X   X   X   X
      ─────────────────────────────────────
       20  30  40  50  60  70  80
      ┌─────────────────────────────────┐
      │             Class B              │
      └─────────────────────────────────┘


       X                               X
       X                               X
       X               X               X
      ─────────────────────────────────────
       20  30  40  50  60  70  80
      ┌─────────────────────────────────┐
      │             Class C              │
      └─────────────────────────────────┘
```

Fig. 4.49. Dot plots.

Our guess is that the maximum variation is in Class **C**.

(b) Range of the three classes:

Class A	20	50	50	50	50	50	80
Class B	20	30	40	50	60	70	80
Class C	20	20	20	50	80	80	80

Range for class **A** = Maximum value - Minimum value
$$= X_{max} - X_{min}$$
$$= 80 - 20 = 60.$$

Range for class **B** = Maximum value - Minimum value
$$= X_{max} - X_{min}$$
$$= 80 - 20 = 60.$$

Range for class **C** = Maximum value - Minimum value
$$= X_{max} - X_{min}$$
$$= 80 - 20 = 60.$$

Thus from the range point of view, all three classes have the same or equal variation.

(c) Standard deviations for the three classes:

Class A		Class B		Class C	
x_i	x_i^2	x_i	x_i^2	x_i	x_i^2
20	400	20	400	20	400
50	2,500	30	900	20	400
50	2,500	40	1,600	20	400
50	2,500	50	2,500	50	2,500
50	2,500	60	3,600	80	6,400
50	2,500	70	4,900	80	6,400
80	6,400	80	6,400	80	6,400
Sum **350**	**19,300**	**350**	**20,300**	**350**	**22,900**

$n = 7$	$n = 7$	$n = 7$
$\sum_{i=1}^{7} x_i = 350$	$\sum_{i=1}^{7} x_i = 350$	$\sum_{i=1}^{7} x_i = 350$
$\sum_{i=1}^{n} x_i^2 = 19,300$	$\sum_{i=1}^{n} x_i^2 = 20,300$	$\sum_{i=1}^{n} x_i^2 = 22,900$

The sample variance for class **A** is:

$$s^2 = \frac{n\left(\sum\limits_{i=1}^{n} x_i^2\right) - \left(\sum\limits_{i=1}^{n} x_i\right)^2}{n(n-1)} = \frac{7 \times 19{,}300 - (350)^2}{7(7-1)} = 300\,.$$

Thus, the sample standard deviation for class **A** is:

$$s = \sqrt{s^2} = \sqrt{300} = 17.32\,.$$

The sample variance for class **B** is:

$$s^2 = \frac{n\left(\sum\limits_{i=1}^{n} x_i^2\right) - \left(\sum\limits_{i=1}^{n} x_i\right)^2}{n(n-1)} = \frac{7 \times 20{,}300 - (350)^2}{7(7-1)} = 466.67\,.$$

Thus, the sample standard deviation for class **B** is:

$$s = \sqrt{s^2} = \sqrt{466.67} = 21.60\,.$$

The sample variance for class **C** is:

$$s^2 = \frac{n\left(\sum\limits_{i=1}^{n} x_i^2\right) - \left(\sum\limits_{i=1}^{n} x_i\right)^2}{n(n-1)} = \frac{7 \times 22{,}900 - (350)^2}{7(7-1)} = 900\,.$$

Thus, the sample standard deviation for class **C** is:

$$s = \sqrt{s^2} = \sqrt{900} = 30\,.$$

On the basis of sample standard deviation, we conclude that class **C** has the maximum variation.

(d) Therefore, our guess was right. Now we can smile because we learned that our formula produces the result we expected. It's good to know that although a mathematical formula cannot speak, it gives the message that we expect, and hence can be trusted in the future.

4.18 RELATIVE STANDING

Consider four students with GPAs of 2.5, 2.4, 2.6, and 2.5. We would like to see the relative standing of a 5th student with a GPA of 3.8, that is, we wish to know whether the 5th student belongs to the group of four students.

There are two methods: (a) Z-score (b) Box plot

Now we like to discuss these methods in detail as follows:

4.18.1 Z-SCORE

The sample Z-score is a measure of relative standing of the ith data value, say x_i in a sample of n data values defined as:

$$Z = \left(\frac{x_i - \bar{x}}{s} \right)$$

where $\bar{x}$ is the sample mean and s is the sample standard deviation. Note that the sample Z score has no unit of measurement.

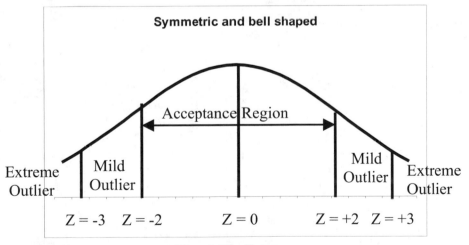

Fig. 4.50. Z-score.

If the value of Z lies between –2 and +2, then we say the x_i value belongs to the same group, that is, it is not an outlier. If the value of Z either lies between –3 and –2 or +2 and +3, then we say the x_i

value is a **mild outlier**. If the value of Z is either less than -3 or more than $+3$ then we say the x_i value is an **extreme outlier**.

Example 4.17. (WHO DID A GOOD JOB?) A farmer wishes to open a pig farm. The farmer selected a sample of 10 pigs of different breeds to see which breed gives maximum progenies.

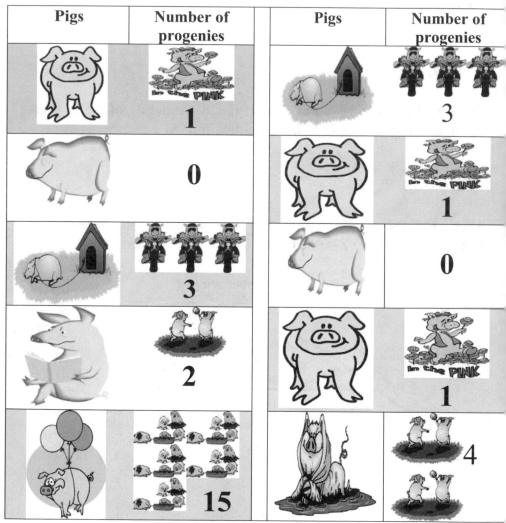

Pigs	Number of progenies	Pigs	Number of progenies
	1		3
	0		1
	3		0
	2		1
	15		4

Fig. 4.51. Pig farm.

The farmer found that the pig carrying balloons has the maximum number of progenies. The farmer thought that this pig might have bought so many balloons to make his/her kids happy and hence may be taking better care of his/her progenies. The farmer found that the pig carrying balloons was quite a different breed and may be a right choice to open a pig farm. Justify the farmer's claim that the pig with balloons is an outlier, and does not belong to the others.

Solution. To find the Z-score, complete the following table:

Pig #	Number of progenies x_i	x_i^2
1	1	1
2	1	1
3	0	0
4	15	225
5	2	4
6	3	9
7	4	16
8	0	0
9	1	1
10	3	9
Sum	30	266
Caution! data is not in any order		

From the above table, we have:

$$n = 10, \sum_{i=1}^{n} x_i = 30, \text{ and } \sum_{i=1}^{n} x_i^2 = 266.$$

Thus, the average number of progeny per pig is given by:

$$\bar{x} = \frac{\sum_{i=1}^{n} x_i}{n} = \frac{30}{10} = 3 \text{ progenies,}$$

and the sample variance of the number of progenies is given by:

$$s^2 = \frac{n\left(\sum_{i=1}^{n} x_i^2\right) - \left(\sum_{i=1}^{n} x_i\right)^2}{n(n-1)} = \frac{10(266) - (30)^2}{10(10-1)} = \frac{2,660 - 900}{90} = 19.556.$$

Thus, the sample standard deviation of the number of progenies is given by:

$$s = \sqrt{s^2} = \sqrt{19.556} \approx 4.42 \text{ progenies.}$$

Thus, the Z-score for the pig with balloons and 15 progenies is given by:

$$Z = \frac{x_i - \bar{x}}{s} = \frac{15 - 3}{4.42} = \frac{12}{4.42} = 2.71.$$

The Z score 2.71 is quite close to 3.0. Thus, the claim by the farmer that this particular pig with balloons is of a different breed than the others in the sample seems to be correct.

4.18.2 QUARTILES

Quartiles are three measures that divide a ranked data set into four equal parts.

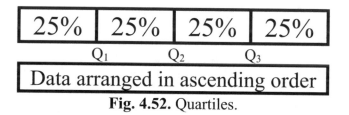

Fig. 4.52. Quartiles.

In **Figure 4.52** the values Q_1, Q_2 and Q_3 are called first, second and third quartiles of a data set when arranged in ascending order. Note that the second quartile Q_2 is the median of the whole data set, Q_1 is the median of the data set below Q_2, and Q_3 is the median of the data set above Q_2. Thus, if we know how to find the median of a data set then we can also find Q_1, Q_2 and Q_3. It is equivalent to finding a median three times. The first quartile and third quartile are sometimes called lower and upper quartiles. The unit of measurement of Q_1, Q_2 and Q_3 are the same as that of the original data. We can also present these three quartiles with a mound shaped graph as in **Figure 4.53**:

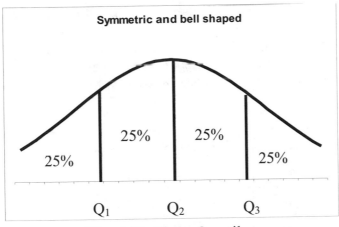

Fig. 4.53. Three Quartiles.

4.18.3 INTERQUARTILE RANGE

The interquartile range is the range between the third and the first quartile. It is denoted by IQR and is defined as:

$$IQR = Q_3 - Q_1$$

The unit of measurement of the IQR is also the same as that of the original data set.

Example 4.18. (DEER LOOK DEAR) The following table shows the distance traveled in miles by 12 deer in a race lasting an hour:

Fig. 4.54. Deer in a race.

(a) Find the values of the three quartiles and interpret them.
(b) If a dear runs 88 miles in the race, where does this particular deer stand in relation to these quartiles?
(c) Find the interquartile range (IQR) and interpret it.

Solution. (a) Arrange the deer in ascending order of their running distance as follows:

Deer	Distance	Quartiles
	53	
	58	
	68	
	$Q_1 = \dfrac{68 + 73}{2} = 70.5$	
	73	
	75	
	76	
	$Q_2 = \dfrac{76 + 79}{2} = 77.5$	
	79	
	80	
	85	
	$Q_3 = \dfrac{85 + 88}{2} = 86.5$	
	88	**88**
	91	
	99	

Fig. 4.55. Deer arranged in ascending order.

Note that every time we have an even number of deer while calculating three medians corresponding to three quartiles. Thus, 3 deer have a running distance below 70.5 miles, 3 deer have a running distance between 70.5 miles and 77.5 miles, 3 deer have a running distance between 77.5 miles and 86.5 miles, and 3 deer have a running distance above 86.5 miles. Therefore we divided the entire dataset into four equal parts.

(b) A deer whose running distance is 88 miles can be classified among the top 25% of deer participating in the race.

(c) The value of the interquartile range (IQR) is given by:

$$IQR = Q_3 - Q_1 = 86.5 - 70.5 = 16 \text{ miles.}$$

This means the difference between the running distance of a deer at the 75th position and the 25th position is 16 miles.

Example 4.19. (CHEETAHS RUN FAST) The following are the times in minutes taken by nine cheetahs in a race of 10 miles:

47	28	39	51	33	37	59	24	33

Fig. 4.56. Cheetahs in a race.

(a) Find the three quartiles and interpret them.
(b) If a cheetah completes the race in 28 mins, where does he fall?
(c) Find the interquartile range.
Solution. (a) Arrange the data in ascending order as follows:

Cheetah	Time	
	24	
	28	**28**
$Q_1 = \dfrac{28+33}{2} = 30.5 \text{ minutes}$		
	33	
	33	

 🐆	37	$Q_2 = 37$ minutes
 🐆	39	
 🐆	47	
	$Q_3 = \dfrac{47 + 51}{2} = 49$ minutes	
 🐆	51	
 🐆	59	

Fig. 4.57. Cheetahs arranged in ascending order.

This means two cheetahs have a running time less than 30.5 minutes, two cheetahs have a running time between 30.5 minutes and 37 minutes, two cheetahs have a running time between 37 minutes and 49 minutes, and two cheetahs have a running time of more than 49 minutes. Thus, we divided the entire dataset into four equal parts.

(b) If a cheetah takes 28 minutes to complete the race, then he lies among the fastest 25% of cheetahs.

(c) The interquartile range (IQR) is given by:

$$\text{IQR} = Q_3 - Q_1 = 49.0 - 30.5 = 18.5 \text{ minutes.}$$

4.19 BOX PLOT

A box plot is also called a box whisker plot, and it makes use of the seven measures listed below:

(a) First quartile, Q_1.

(b) Second quartile, Q_2 or median.

(c) Third quartile, Q_3.

(d) Smallest data value, x_s, between lower and upper inner fences.

(e) Largest data value, x_l, between lower and upper inner fences.

(f) Maximum data value, x_{max}.

(g) Minimum data value, x_{min}.

4.19.1 USE OF A BOX PLOT

A box plot can be used to see:

(a) the center of a data set
(b) the spread of a data set
(c) the skewness of a data set
(d) any outliers in a dataset.

4.19.2 STEPS TO MAKE A BOX PLOT

The main steps to construct a box plot are as follows:

(a) Arrange the data in ascending order (from lowest to highest).
(b) Find the three quartiles Q_1, Q_2, and Q_3; and the interquartile range:

$$IQR = Q_3 - Q_1$$

Sometimes Q_1 and Q_3 are also called **Hinges**.

(c) Find the STEP as:

$$STEP = 1.5 \times IQR$$

(d) Find the lower inner fence (**LIF**) and upper inner fence (**UIF**) as:

$$LIF = Q_1 - STEP$$

and

$$UIF = Q_3 + STEP$$

(e) Draw a **box** consisting of two boxes as shown below:

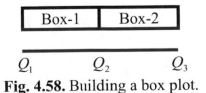

Fig. 4.58. Building a box plot.

If box-1 and box-2 are of the same length, the distribution is symmetric; if box-1 is longer than box-2, the distribution is skewed to the left; and if box-2 is longer than box-1, the distribution is skewed to the right. Note that Q_2 is not always midpoint of Q_1 and Q_3.

(f) Let x_{min} and x_{max} denote the minimum and maximum data values; and x_s and x_l be the smallest and largest data values between the lower inner fence (LIF) and the upper inner fence (UIF). Then, complete the box plot as follows:

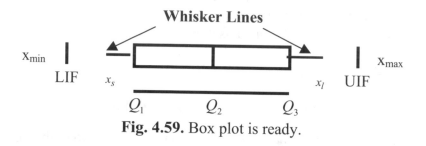

Fig. 4.59. Box plot is ready.

4.19.3 DETECTION OF OUTLIERS USING A BOX PLOT

A box plot can be used to find outliers from a given data set. It is most frequently used in engineering science to find defective items. For example, consider a factory manufacturing bulbs where the engineers are generally interested in the lifetime of bulbs. If most of the bulbs' lifetime remains within the LIF and UIF, then the engineers believe that their factory output is in a good position. In other words, if $x_{min} = x_s$ and $x_{max} = x_l$, then there are no outliers. As discussed before, outliers are of two types: (a) mild outliers, and (b) extreme outliers.

To differentiate between them, we define two more fences as follows:

Lower outer fence (LOF) $= Q_1 - 2(\text{STEP})$

and

Upper outer fence (UOF) $= Q_3 + 2(\text{STEP})$

Then we complete the following box plot:

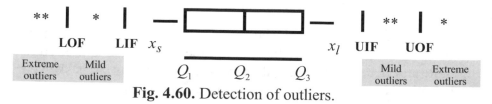

Fig. 4.60. Detection of outliers.

Thus, all the values lying between the LIF and the UIF can be considered real data values. Note that any data value between the UIF (or LIF) and the UOF (or LOF), are called **mild outliers**, and any value less than the LOF (or more than the UOF), are called **extreme outliers**.

4.19.4 SEVEN MEASURES FOR A BOX PLOT

Note that X_{min}, X_s, Q_1, Q_2, Q_3, X_l, and X_{max} are also called the seven measures used in the construction of a box plot. In general, we say there are five measures if there are no outliers, or if the outliers are dropped from the data set before analyzing it.

Example 4.20. (LEARNING THE BOX PLOT) The following data set gives the ages in months of 12 monkeys living in a zoo.

Fig. 4.61. Monkeys of different ages.

(a) Construct a box plot of their ages.
(b) Find a monkey who can be classified as an outlier.

Solution. (a) To construct a box plot, arrange the monkeys in ascending order of their age as follows:

Monkey	Age	Quartiles
	17	
	22	
	23	
	$Q_1 = \dfrac{23 + 27}{2} = 25$	
	27	
	29	
	32	
	$Q_2 = \dfrac{32 + 38}{2} = 35$	
	38	
	42	
	46	
	$Q_3 = \dfrac{46 + 52}{2} = 49$	
	52	
	60	
	92	**92**

Fig. 4.62. Monkeys arranged in ascending order of age.

(i) The values of the three quartiles and the interquartile range are:

$Q_1 = 25$, $Q_2 = 35$, $Q_3 = 49$, and IQR $= Q_3 - Q_1 = 49 - 25 = 24$.

(ii) STEP $= 1.5 \times$ IQR $= 1.5 \times (24) = 36$.

Thus, the values of the lower inner fence (LIF) and the upper inner fence (UIF) are:

$$\text{LIF} = Q_1 - \text{STEP} = 25 - 36 = -11$$

and

$$\text{UIF} = Q_3 + \text{STEP} = 49 + 36 = +85.$$

(iii) To find x_s and x_l values:

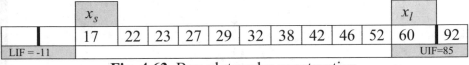

Fig. 4.63. Box plot under construction.

Thus, the smallest and largest values between LIF and UIF are:

$$x_s = 17 \text{ and } x_l = 60.$$

(iv) Using the above information, the box plot is given by:

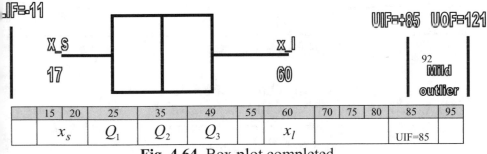

Fig. 4.64. Box plot completed.

(b) Note that the values of the lower outer fence (LOF) and the upper outer fence (UOF) are given by:

$$\text{LOF} = Q_1 - 2(\text{STEP}) = 25 - 2(36) = -47$$

and

$$\text{UOF} = Q_3 + 2(\text{STEP}) = 49 + 2(36) = +121.$$

Note that the age of one monkey is $x_l = 92$ months, which lies between the $\text{UIF} = +85$ months and the $\text{UOF} = +121$ months. In other words, there is one monkey who can be considered an outlier, or aged.

Example 4.21. (READING BOX PLOTS) The following two box plots give the ages of children in two schools. The number of children in school 1 and school 2 is 500 and 800 respectively.

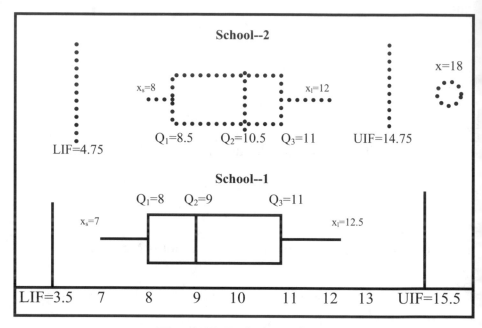

Fig. 4.65. Twin box plots.

(a) In school 1, how many children are older than 9?
(b) In school 1, how many children are between 8 and 11 years old?
(c) In school 1, how many children are more than 11 years old?
(d) In school 1, what is the minimum age?
(e) In school 1, what is the maximum age?
(f) What is the interquartile range for school 1?
(g) A child can enter school 1 at age 6 and leave at age 14. Are there any children who are registered below the age of 6?
(h) A child can enter school 2 at age 7 and leave at age 13. Are there any children who are not going to graduate on time? What may be the reason?
(i) What is the interquartile range for school 2?
(j) How many children in school 2 are older than 8.5 years?
(k) What is the minimum age in school 2?
(l) What is the maximum age in school 2?
(m) How many children in both schools are 11 or older?

(n) How many children in school 1 are less than the minimum age of a child in school 2?

(o) What is the range of age for school 1 and school 2?

Solution.

(a) Since $Q_2 = 9$ is the median, 50% of the children in school 1 are older than 9. The total number of children in school 1 is 500, therefore the number of children older than 9 is:

$$= 50\% \text{ of } 500 = \frac{50}{100} \times 500 = 250.$$

(b) Since $Q_1 = 8$ and $Q_3 = 11$, 50% of the children are between 8 and 11 years old. Again, because the total number of children in school 1 is 500, the number of children between 8 and 11 is:

$$= 50\% \text{ of } 500 = \frac{50}{100} \times 500 = 250.$$

(c) Since $Q_3 = 11$ is the upper quartile (or third quartile), 25% of the children will be more than 11 years old.

Number of such children $= 25\% \text{ of } 500 = = \frac{25}{100} \times 500 = 125.$

(d) The minimum age will be: $X_s = 7$ years.

(e) The maximum age will be: $X_l = 12.5$ years.

(f)The value of the IQR for school 1 is given by:

$$\text{IQR} = Q_3 - Q_1 = 11 - 8 = 3 \text{ years.}$$

(g) No, there are no outliers in school 1.

(h) Yes, there is one 18 year old who is still going to school 2. The reason may be that this student is not spending enough time on homework or may not be attending class.

(**i**) The value of the IQR for school 2 is given by:

$$\text{IQR} = Q_3 - Q_1 = 11 - 8.5 = 2.5 \text{ years.}$$

(**j**) Since $Q_1 = 8.5$ is a first quartile (or lower quartile) for school 2, 75% of the children will be more than 8.5 years old. Note that there are 800 children in school 2.

Thus, the number of children older than 8.5 years will be:

$$= 75\% \text{ of } 800 \text{ children} = \frac{75}{100} \times 800 = 600 \text{ children.}$$

(**k**) The minimum age in school 2 is $X_s = 8$ years.

(**l**) The maximum age in school 2 is $X_l = 18$ years, which is an outlier.

(**m**) The value $Q_3 = 11$ years is a third quartile for both schools, thus 25% of the children in school 1 and 25% of the children in school 2 are older than 11.

> Now, 25% of the children in school 1 $= \dfrac{25}{100} \times 500 = 125$
> and, 25% of the children in school 2 $= \dfrac{25}{100} \times 800 = 200$
> Total children in both schools $= 125 + 200 = 325$.

(**n**) The minimum age in school 2 is $X_s = 8$. Also, note that $Q_1 = 8$ years is the first quartile for school 1, thus 25% of the children in school 1 are younger than the minimum age of a child in school 2. The number of such children $= \dfrac{25}{100} \times 500 = 125$.

(**o**) For school 1, the range = maximum value – minimum value

$$= 12.5 - 7 = 5.5 \text{ years.}$$

For school 2, the range = maximum value – minimum value

$$= 18 - 8 = 10 \text{ years.}$$

Example 4.22. (PIZZA LOVERS) An inspector took a random sample of 900 pepperoni pizzas and 600 vegetable pizzas from Pizza Stores in a local area. The following two box plots give the number of calories in these two types of pizza.

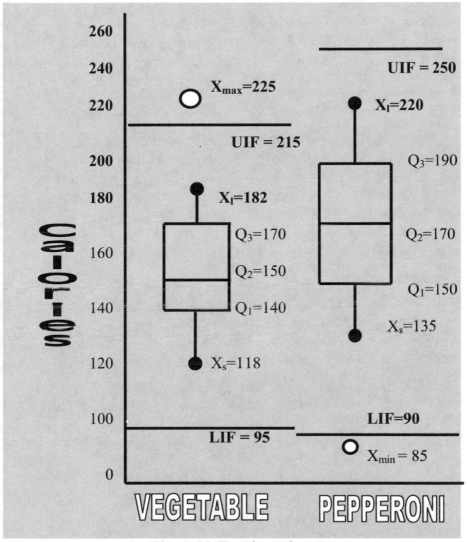

Fig. 4.66. Two box plots.

(a) How many vegetable pizzas have more than 150 calories?
Answer: Note that $Q_2 = 150$ is a median, thus 50% of the 600 pizzas have more than 150 calories. The total number of vegetable pizzas is 600.

Therefore, the number of vegetable pizzas having more than 150 calories is:

$$= 50\% \text{ of } 600 = \frac{50}{100} \times 600 = 300.$$

(b) How many pepperoni pizzas have between 170 and 190 calories?
Answer: Note that $Q_2 = 170$ and $Q_3 = 190$ are the median and upper quartiles. Thus, 25% of the pepperoni pizzas have between 170 and 190 calories. The total number of pepperoni pizzas is 900. Thus, the number of pepperoni pizzas having calories between 170 and 190 is:

$$= 25\% \text{ of } 900 = \frac{25}{100} \times 900 = 225.$$

(c) How many pizzas have 85 or more calories?
Answer: All 600 + 900=1500 pizzas have more than 85 calories.
(d) Are there any outliers? If so, should these be kept or dropped?
Answer: There are two outliers. One vegetable pizza shows 225 calories and one pepperoni pizza shows only 85 calories. If in the vegetable pizza, meat was added by mistake resulting in higher calories, then this pizza should be dropped from the analysis. Similarly, if someone forgot to add pepperoni to the pizza showing only 85 calories, then it must be dropped from the analysis. Otherwise, these outlier pizzas should be kept in the analysis.
(e) How many pizzas have more than 150 calories?
Answer: Note that 150 calories is the median for vegetable pizzas and the first quartile for pepperoni pizzas. Thus, we have:

$$= 50\% \text{ of } 600 = \frac{50}{100} \times 600 = 300 \text{ vegetable pizzas,}$$

and

$$= 75\% \text{ of } 900 = \frac{75}{100} \times 900 = 675 \text{ pepperoni pizzas.}$$

Thus, the total number of pizzas with more than 150 calories is:
$$= 300 + 675 = 975.$$

(f) What is the distribution of calories among vegetable pizzas?
Answer: Skewed to the right.
(g) What is the distribution of calories among pepperoni pizzas?
Answer: Almost symmetric.
(h) What is the IQR of calories for vegetable pizzas?
Answer: IQR = $Q_3 - Q_1$ = 170-140 = 30.
(i) What is the IQR of calories for pepperoni pizzas?
Answer: IQR = $Q_3 - Q_1$ = 190-150 = 40.

(j) What is the range of calories for vegetable pizzas?
Answer: Assuming there is no outlier, the range is:
$$= X_{max} - X_{min} = 225\text{-}118 = 107.$$
(k) What is the range of calories for pepperoni pizzas?
Answer: Assuming there is no outlier, the range is:
$$= X_{max} - X_{min} = 220\text{-}85 = 135.$$
(l) What is range of calories for both types of pizzas?
Answer: Assuming there is no outlier, the range is:
$$= X_{max} - X_{min} = 225\text{-}85 = 140.$$

Example 4.23. (ALWAYS WATCH THE OUTSIDE TEMP) The following data set shows the daily temperature at a zoo in New York over a period of two weeks:

°C	48	50	47	54	49	53	48	48	51	58	53	54	49	52

Find the following:
(a) Sample size
(b) Sample mean
(c) Median
(d) Mode
(e) On the basis of mean, median, and mode, is the distribution symmetric or skewed?
(f) Range
(g) Sample variance with two formulae
(h) Sample standard deviation
(i) First quartile
(j) Second quartile
(k) Third quartile
(l) Minimum value (X_{min})
(m) Maximum value (X_{max})
(n) Interquartile range
(o) STEP = 1.5*IQR
(p) Lower Inner Fence (LIF)
(q) Upper Inner Fence (UIF)
(r) X_s, smallest data value between LIF and UIF
(s) X_l, largest data value between LIF and UIF
(t) Make a box plot
(u) Are there any outliers in the data set?

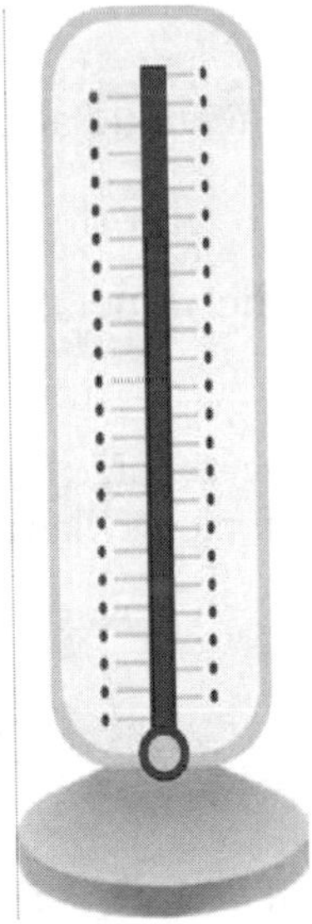

Fig. 4.67. Temp.

Solution. (a) Sample size:

Total number of data values, $n = 14$.

(b) Sample mean :

$$\bar{x} = \frac{\sum\limits_{i=1}^{n} x_i}{n} = \frac{\sum\limits_{i=1}^{14} x_i}{14}$$

$$= \frac{x_1 + x_2 + x_3 + x_4 + x_5 + x_6 + x_7 + x_8 + x_9 + x_{10} + x_{11} + x_{12} + x_{13} + x_{14}}{14}$$

$$= \frac{48 + 50 + 47 + 54 + 49 + 53 + 48 + 48 + 51 + 58 + 53 + 54 + 49 + 52}{14}$$

$$= \frac{714}{14} = 51°C.$$

(c) Median: Let us do both steps here:

(i)Arrange the data in ascending order:

°C	47	48	48	48	49	49	50	51	52	53	53	54	54	58
							7th	8th						
				Median=			$\frac{50+51}{2} = 50.5$							

(ii) Here $n = 14$ (even)

The value at the $\frac{n}{2}th = \frac{14}{2}th = 7th$ position $= 50°C$.

The value at the $\left(\frac{n}{2}+1\right)th = \left(\frac{14}{2}+1\right)th = 8th$ position $= 51 °C$.

The average value of the values at the 7th and 8th positions gives us the sample median as:

$$m = \frac{\text{Value at 7th position} + \text{Value at 8th position}}{2}$$

$$= \frac{50+51}{2} = 50.5°C$$

(d) Mode: The mode is the value which occurs most frequently, so:

Mode $= 48 °C$.

(e) Here the mean, median and mode are given by:

Mean $= 51°C$, Median $= 50.5 °C$, and Mode $= 48 °C$.

```
    *
    *    *                      *    *
*   *    *    *    *    *    *   *    *
47  48  49  50  51  52  53  54  58
```
Mode $<$ Median $<$ Mean

Fig. 4.68. Dot plot.

Thus, the distribution is skewed to the right.

(f) Range:

Range $=$ Maximum value - Minimum value $= 58 - 47 = 11°C$.

(g) Sample variance: We apply both methods of computation of the sample variance as follows:

Method I: (By definition)

x_i	$(x_i - \bar{x})$	$(x_i - \bar{x})^2$
48	-3	9
50	-1	1
47	-4	16
54	3	9
49	-2	4
53	2	4
48	-3	9
48	-3	9
51	0	0
58	7	49
53	2	4
54	3	9
49	-2	4
52	1	1
Sum 714	0	128

Using results found at the bottom of the left hand side table, we have:

Sample mean:

$$\bar{x} = \frac{\sum\limits_{i=1}^{n} x_i}{n} = \frac{714}{14} = 51 \ °C$$

Sample variance, by definition, is given by:

$$s^2 = \frac{\sum\limits_{i=1}^{n}(x_i - \bar{x})^2}{14 - 1}$$

$$= \frac{128}{14 - 1} = \frac{128}{13} = 9.85°C^2.$$

From the above table, we have:

Sum of x_i values $= \sum\limits_{i=1}^{n} x_i = 714$.

Sum of the deviations from the mean value $= \sum\limits_{i=1}^{n} (x_i - \bar{x}) = 0$.

Sum of the square of the deviations from the mean value $= \sum\limits_{i=1}^{n} (x_i - \bar{x})^2 = 128$.

Method II: (Computing formula)

x_i	x_i^2
48	2,304
50	2,500
47	2,209
54	2,916
49	2,401
53	2,809
48	2,304
48	2,304
51	2,601
58	3,364
53	2,809
54	2,916
49	2,401
52	2,704
Sum 714	**36,542**

The computing formula of sample variance given by:

$$s^2 = \frac{n\left(\sum\limits_{i=1}^{n} x_i^2\right) - \left(\sum\limits_{i=1}^{n} x_i\right)^2}{n(n-1)}$$

$$= \frac{14 \times 36{,}542 - (714)^2}{14(14-1)}$$

$$= \frac{511{,}588 - 509{,}796}{14 \times 13}$$

$$= \frac{1{,}792}{182} = 9.85 \ {}^{\circ}C^2.$$

From the above table, we have:

Sum of x_i values $= \sum\limits_{i=1}^{n} x_i = 714$ and Sum of x_i^2 values $= \sum\limits_{i=1}^{n} x_i^2 = 36{,}542$.

(**h**) Sample standard deviation:

$$s = \sqrt{s^2} = \sqrt{9.85} = 3.14 \ {}^{\circ}C.$$

(i), (j), and (k): Q_1, Q_2, and Q_3

To find Q_2:

Arrange the data in ascending order:

°C	47	48	48	48	49	49	50	51	52	53	53	54	54	58
							7th	8th						
			Median=			$\dfrac{50+51}{2}=50.5$							

$$Q_2 = 50.5 \,^\circ C.$$

To find Q_1: The number of data values below the median value 50.5 °C are $n = 7$ and are given in the following table:

°C	Data below Q_2 values (n =7)						
	47	48	48	48	49	49	50
				Q_1			

Here now $n = 7 \,(\text{odd})$,

so the value at $\dfrac{n+1}{2}th = \dfrac{7+1}{2}th = \dfrac{8}{2}th = 4th$ position $= 48 = Q_1$

To find Q_3: The number of data values above the median value 50.5 °C are $n = 7$ and are given in the following table:

°C	Data above Q_2 value (n = 7)						
	51	52	53	53	54	54	58
				Q_3			

Here again, $n = 7 \,(\text{odd})$,

so the value at the $\dfrac{n+1}{2}th = \dfrac{7+1}{2}th = \dfrac{8}{2}th = 4th$ position $= 53 = Q_3$

Thus, the values of Q_1, Q_2 and Q_3 are as follows:

$$Q_1 = 48 \,^\circ C, \quad Q_2 = 50.5 \,^\circ C, \text{ and } Q_3 = 53 \,^\circ C.$$

(l) Minimum value: $X_{min} = 47 \,^\circ C$.

(m) Maximum value: $X_{max} = 58 \,^\circ C$.

(n) Interquartile range: $IQR = Q_3 - Q_1 = 53 - 48 = 5\,^{\circ}C.$

(o) $STEP = 1.5 \times IQR = 1.5 \times 5 = 7.5\,^{\circ}C.$

(p) Lower inner fence : $LIF = Q_1 - STEP = 48 - 7.5 = 40.5\,^{\circ}C.$

(q) Upper inner fence : $UIF = Q_3 + STEP = 53 + 7.5 = 60.5\,^{\circ}C.$

(r) $X_s = 47^{\circ}C.$

(s) $X_1 = 58^{\circ}C.$

(t) Box plot:

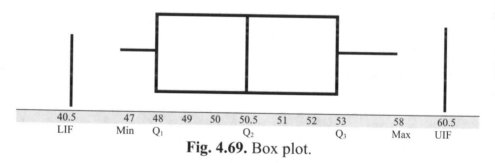

Fig. 4.69. Box plot.

(u) Are there any outliers in the data set?

All of the values in the data set lie between the lower inner fence (LIF) = 40.5 °C and the upper inner fence (UIF) = 60.5 °C, hence there are no outliers.

Example 4.24. (CAT LOVERS) Consider the following data giving the number of cats in 10 households.

Number of cats in ten households									
3	4	0	3	1	0	3	2	1	3

Fig. 4.70. Cat lovers.

Find the sample standard deviation using two methods.

Solution. We will show that both of the following methods used to find the sample variance are the same.

Method I	Method II
$s^2 = \dfrac{\sum_{i=1}^{n}(x_i - \bar{x})^2}{n-1}$	$s^2 = \dfrac{n\left(\sum_{i=1}^{n}x_i^2\right) - \left(\sum_{i=1}^{n}x_i\right)^2}{n(n-1)}$

Method I: Here, we show the following table:

Method I	x_i	$(x_i - \bar{x})$	$(x_i - \bar{x})^2$
	3	3 - 2 = 1	1
	4	4 - 2 = 2	4
	0	0 - 2 = -2	4
	3	3 - 2 = 1	1
	1	1 - 2 = -1	1
	0	0 - 2 = -2	4
	3	3 - 2 = 1	1
	2	2 - 2 = 0	0
	1	1 - 2 = -1	1
	3	3 - 2 = 1	1
Sum	**20**	**0**	**18**

Fig. 4.71. Steps to find variance by definition.

From the above table, we have:

$$n = 10, \ \sum_{i=1}^{10} x_i = 20, \ \sum_{i=1}^{10}(x_i - \bar{x}) = 0, \text{ and } \sum_{i=1}^{10}(x_i - \bar{x})^2 = 18.$$

Using the first method, the sample variance of the number of cats is given by:

$$s^2 = \frac{\sum_{i=1}^{n}(x_i - \bar{x})^2}{n-1} = \frac{18}{10-1} = \frac{18}{9} = 2, \text{ where } \bar{x} = \frac{\sum_{i=1}^{n} x_i}{n} = \frac{20}{10} = 2.$$

and the sample standard deviation is given by:

$$s = \sqrt{s^2} = \sqrt{2} \approx 1.414.$$

Method II: Here, we show the following table:

Method II	x_i	x_i^2
	3	$3^2 = 9$
	4	$4^2 = 16$
	0	$0^2 = 0$
	3	$3^2 = 9$
	1	$1^2 = 1$
	0	$0^2 = 0$
	3	$3^2 = 9$
	2	$2^2 = 4$
	1	$1^2 = 1$
	3	$3^2 = 9$
Sum	**20**	**58**

Fig. 4.72. Steps to use the computing formula.

From the above table, we have:

$$n = 10, \quad \sum_{i=1}^{10} x_i = 20, \quad \text{and} \quad \sum_{i=1}^{10} x_i^2 = 58.$$

Using the second method, the sample variance of the number of cats is:

$$s^2 = \frac{n\left(\sum_{i=1}^{n} x_i^2\right) - \left(\sum_{i=1}^{n} x_i\right)^2}{n(n-1)} = \frac{10 \times 58 - 20^2}{10(10-1)} = \frac{580 - 400}{90} = \frac{180}{90} = 2$$

and the sample standard deviation is given by:

$$s = \sqrt{s^2} = \sqrt{2} \approx 1.414.$$

Note that both methods give the same sample variance as well as the same sample standard deviation.

Example 4.25. (GOLF LOVERS) The following are the golf scores of 12 members of a men's golf team in a tournament: 89, 90, 87, 95, 86, 81, 102, 105, 83, 88, 91, and 79. Compute the sample mean, sample variance and sample standard deviation of these scores.

Fig. 4.73. Golf scores.

Solution. Here $n = 12$, $x_1 = 89$, $x_2 = 90$, $x_3 = 87$, $x_4 = 95$, $x_5 = 86$, $x_6 = 81$, $x_7 = 102$, $x_8 = 105$, $x_9 = 83$, $x_{10} = 88$, $x_{11} = 91$, and $x_{12} = 79$.

We have the following table:

	x_i	$(x_i-\bar{x})$	$(x_i-\bar{x})^2$
	89	-0.67	0.45
	90	0.33	0.11
	87	-2.67	7.13
	95	5.33	28.41
	86	-3.67	13.47
	81	-8.67	75.17
	102	12.33	152.03
	105	15.33	235.01
	83	-6.67	44.49
	88	-1.67	2.79
	91	1.33	1.77
	79	-10.67	113.85
Sum	1,076	0.0	674.67

From the above table:

$$\sum_{i=1}^{n} x_i = 1,076, \quad \sum_{i=1}^{n}(x_i - \bar{x}) = 0.0, \quad \text{and} \quad \sum_{i=1}^{n}(x_i - \bar{x})^2 = 674.67.$$

The sample mean is given by:

$$\bar{x} = \frac{\sum_{i=1}^{n} x_i}{n} = \frac{x_1 + x_2 + \dots\dots + x_{12}}{12} = \frac{1,076}{12} = 89.67 \text{ scores.}$$

The sample variance is given by:

$$s^2 = \frac{\sum_{i=1}^{n}(x_i - \bar{x})^2}{n-1} = \frac{674.67}{12-1} = 61.33 \text{ (scores)}^2,$$

and the sample standard deviation is given by:

$$s = \sqrt{s^2} = \sqrt{61.33} = 7.83 \text{ scores.}$$

Example 4.26. (BOX PLOT EXPERTS) Given that the LIF = 20 and the UIF = 50, find Q_1 and Q_3. Can we find Q_2?

Solution. Note that the $IQR = \dfrac{UIF - LIF}{4} = \dfrac{50 - 20}{4} = 7.5$.

Thus,

$$STEP = 1.5 \times IQR = 1.5 \times 7.5 = 11.25$$
$$Q_1 = LIF + STEP = 20 + 11.25 = 31.25$$
$$Q_3 = UIF - STEP = 50 - 11.25 = 38.75.$$

No, it is not possible to find Q_2, because Q_2 is not necessarily the midpoint of Q_1 and Q_3.

Example 4.27. (HORSE RACING IS FUN) A horse rider made 24 trips in different times (in minutes) as shown below:

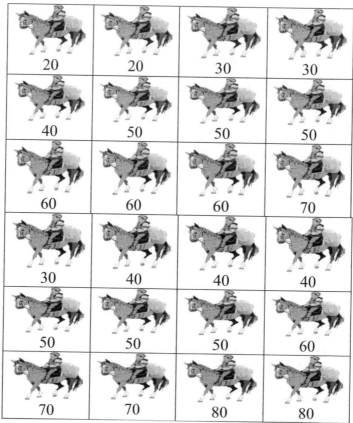

Fig. 4.74. Horse rider.

Did he follow the empirical rule?

Solution. From the data values we have:

	x_i	$(x_i - \bar{x})$	$(x_i - \bar{x})^2$
	20	-30	900
	20	-30	900
	30	-20	400
	30	-20	400
	30	-20	400
	40	-10	100
	40	-10	100
	40	-10	100
	40	-10	100
	50	0	0
	50	0	0
	50	0	0
	50	0	0
	50	0	0
	50	0	0
	60	10	100
	60	10	100
	60	10	100
	60	10	100
	70	20	400
	70	20	400
	70	20	400
	80	30	900
	80	30	900
Sum	**1,200**	**0**	**6,800**

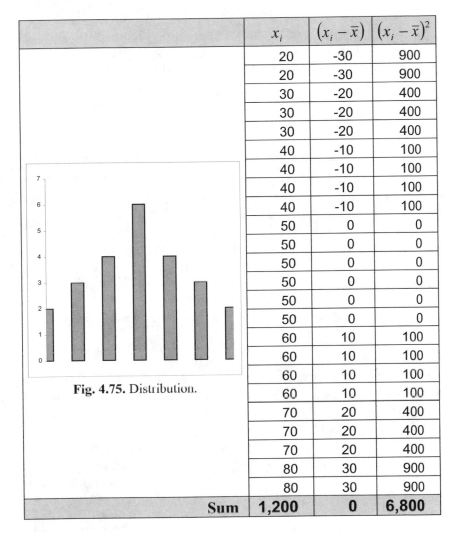

Fig. 4.75. Distribution.

From the above table:

$$n = 24, \ \sum_{i=1}^{n} x_i = 1{,}200, \ \sum_{i=1}^{n}(x_i - \bar{x}) = 0, \ \sum_{i=1}^{n}(x_i - \bar{x})^2 = 6{,}800.$$

Thus,

$$\bar{x} = \frac{\sum_{i=1}^{n} x_i}{n} = \frac{1{,}200}{24} = 50, \ s^2 = \frac{\sum_{i=1}^{n}(x_i - \bar{x})^2}{n-1} = \frac{6{,}800}{23} = 295.65$$

and

$$s = \sqrt{s^2} = \sqrt{295.65} = 17.19.$$

(a) The interval $\bar{x} \pm s$, or 50 ± 17.19, or $(50 - 17.19, 50 + 17.19)$, or $(32.81, 67.19)$ contains approximately 68% of $24 = 16.32 \approx 16$ data values. If we count, there are 14 values within this interval.

(b) The interval $\bar{x} \pm 2s$ or $50 \pm 2 \times 17.19$ or 50 ± 34.38 or $(50 - 34.38, 50 + 34.38)$ or $(15.62, 84.38)$ contains approximately 95% of $24 = 22.8 \approx 23$ data values. If we count, there are 24 values within this interval.

(c) The interval $\bar{x} \pm s$ or $50 \pm 3 \times 17.19$ or 50 ± 51.57 or $(50 - 51.57, 50 + 51.57)$ or $(-1.57, 101.57)$ contains approximately 99% of $24 = 23.76 \approx 24$ data values. If we count, there are indeed 24 values within this interval as expected.

Although the distribution is mound shaped, but due to small sample size, the empirical rule is approximately followed in (a) and (b) but it is exactly followed in case (c).

Example 4.28. (SPEED THRILLS, BUT KILLS) Assuming the empirical rule holds, find the mean and standard deviation of the speeds of the cars in each of the following two cases:
(a) 68% of the cars are going between 40 miles/hr and 60 miles/hr.
(b) 95% of the cars are going between 40 miles/hr and 60miles/hr.
Solution. (a) Note that 68% of the data values lie between $\bar{x} \pm s$. Thus we have:

$$\bar{x} - s = 40 \quad \text{and} \quad \bar{x} + s = 60$$

Solving for $\bar{x}$ and s, we have:

$$\bar{x} = 50 \quad \text{and} \quad s = 10.$$

Fig. 4.76. Car.

(b) Note that 95% of the data values lie between $\bar{x} \pm 2s$. Thus we have:

$$\bar{x} - 2s = 40 \quad \text{and} \quad \bar{x} + 2s = 60.$$

Solving for $\bar{x}$ and s, we have:

$$\bar{x} = 50 \quad \text{and} \quad s = 5.$$

Fig. 4.77. Car.

LUDI 4.1. (STEPS FOR STANDARD DEVIATION)

(a) Calculate the sample standard deviation $s = \sqrt{s^2}$ for the data set:

	x_i	x_i^2
	2	
	5	
	9	
	7	
	8	
	5	
Sum		

Math is fun

$99 \times 99 + 99 =$

Fig. 4.78. Let me try.

From the above table, find:

$$n = \text{------}, \quad \sum_{i=1}^{n} x_i = \text{--------}, \quad \sum_{i=1}^{n} x_i^2 = \text{--------}$$

Use the above results to find s^2:

$$s^2 = \frac{n\left(\sum_{i=1}^{n} x_i^2\right) - \left(\sum_{i=1}^{n} x_i\right)^2}{n(n-1)} = \text{-----------}$$

Then find the sample standard deviation:

$$s = \sqrt{s^2} = \text{--------}$$

(b) Multiply all the data points in (a) with a number, say 4, to complete the following table:

	$x_i = 4x_i(a)$	x_i^2
	?	
	?	
	?	
	?	
	?	
	?	
Sum		

Math needs time

$99 \times 99 + 99 =$

Fig. 4.79. It was good.

From the above table:

$$n = \text{--------}, \quad \sum_{i=1}^{n} x_i = \text{--------}, \quad \sum_{i=1}^{n} x_i^2 = \text{--------}$$

Use the above results to find s^2 :

$$s^2 = \frac{n\left(\sum\limits_{i=1}^{n} x_i^2\right) - \left(\sum\limits_{i=1}^{n} x_i\right)^2}{n(n-1)} = \text{----------}$$

Then find the sample standard deviation:

$$s = \sqrt{s^2} = \text{--------}$$

Divide the sample standard deviation in (b) with the sample standard deviation in (a), and the result $= \text{--------}$

(c) Add a number, say 4, to all the data points in (a) to complete the following table:

	$x_i = x_i(a) + 4$	x_i^2
	?	
	?	
	?	
	?	
	?	
	?	
Sum		

Secret of Math

99×99+99=

Fig. 4.80. Do more enjoy more.

From the above table:

$$n = \text{--------}, \quad \sum_{i=1}^{n} x_i = \text{--------}, \quad \sum_{i=1}^{n} x_i^2 = \text{--------}$$

Use the above results to find s^2 :

$$s^2 = \frac{n\left(\sum\limits_{i=1}^{n} x_i^2\right) - \left(\sum\limits_{i=1}^{n} x_i\right)^2}{n(n-1)} = \text{----------}$$

Then find the sample standard deviation:

$$s = \sqrt{s^2} = --------$$

Divide the sample standard deviation derived in (c) with the sample standard deviation in (a), and the result = --------

Can you develop some relationship between the results in (a), (b) and (c)?   If yes, please write it here:

LUDI 4.2. (PRACTICING SAMPLE STANDARD DEVIATION)
The following data shows the scores of 8 students in a statistics class:

| 42 | 52 | 67 | 42 | 52 | 52 | 52 | 36 |

Fig. 4.81. Students with their scores.

Complete the following table:

	x_i	x_i^2
Sum		

Fig. 4.82. Students.

Find the following:

(a) $n = $ ------- , (b) $\sum\limits_{i=1}^{n} x_i = $ ------- , (c) $\sum\limits_{i=1}^{n} x_i^2 = $ -------

(d) $\left(\sum\limits_{i=1}^{n} x_i \right)^2 = $ -------

(e) Sample mean :

$$\bar{x} = \frac{\sum\limits_{i=1}^{n} x_i}{n} = \text{-------}$$

(f) Sample variance by using the computing formula:

$$s^2 = \frac{n\left(\sum\limits_{i=1}^{n} x_i^2 \right) - \left(\sum\limits_{i=1}^{n} x_i \right)^2}{n(n-1)} = \text{-----------}$$

(g) Complete the following table:

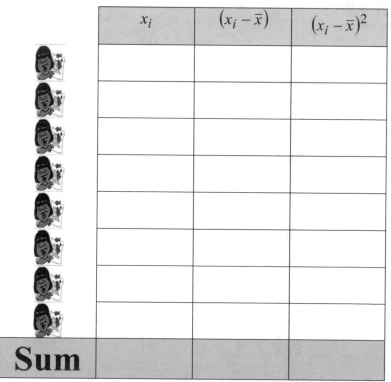

	x_i	$(x_i - \bar{x})$	$(x_i - \bar{x})^2$
Sum			

Fig. 4.83. Students.

(h) Sample variance by using the definition formula:

$$s^2 = \frac{\sum_{i=1}^{n}(x_i - \bar{x})^2}{n-1} = \text{_____}$$

(i) Find the sample standard deviation:

$$s = \sqrt{s^2} = \text{_____}$$

LUDI 4.3. (READING BOX PLOTS). The following two box plots give the scores of students in two colleges. The number of students in college 1 and college 2 is 800 and 1200, respectively.

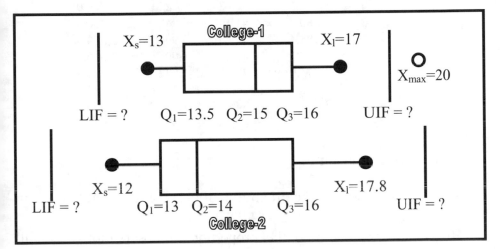

Fig. 4.84. Box plots.

(**a**) Complete the box plots by finding the lower inner fences and the upper inner fences.
(**b**) In college 2, how many students have scores higher than 14?
(**c**) In college 2, how many students have scores between 14 and 16?
(**d**) In college 1, what is the minimum score?
(**e**) In college 2, what is the maximum score?
(**f**) How many students in both colleges have scores higher than 16?
(**g**) What is the range of scores in college 1?
(**h**) The distribution of scores in college 1 is skewed to…..?
(**i**) The difference between the median marks of the two colleges is…..?
(**j**) Are there any outliers?

LUDI 4.4. (REMEMBER YOUR FRIENDS) How can sigma be used to send new years greetings?

Hint: For example Happy New Year (2007) can be written as:

$$\text{HNY}(2007) = \sum_{I=1}^{365} (\text{Happy day})_I$$

LUDI 4.5. (EMPIRICAL RULE) The following table gives a sample of 50 data values:

0.2	51.5	388.8	571.4	1,519.9
0.4	56.4	405.7	635.7	1,692.8
3.4	57.6	426.2	722.0	1,716.0
4.3	80.7	431.4	848.3	2,466.8
16.7	188.4	440.5	906.2	2,580.3
19.3	197.2	464.5	1,006.0	2,610.5
27.5	274.0	494.7	1,022.7	3,520.3
29.2	298.3	540.6	1,228.6	3,585.4
38.0	348.3	549.5	1,241.3	3,909.7
43.2	386.4	557.6	1,372.4	3,928.7

Compute the following:

(a) $\bar{x} \pm s$, and find the proportion of observations that fall within this interval.

(b) $\bar{x} \pm 2s$, and find the proportion of observations that fall within this interval.

(c) $\bar{x} \pm 3s$, and find the proportion of observations that fall within this interval.

(d) Comment on the results related to the **empirical rule** in each case.

LUDI 4.6. (WEIGHING ELEPHANTS) The weight of four elephants in kg is given below:

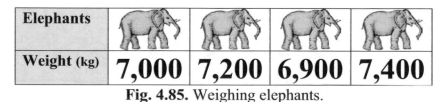

Elephants				
Weight (kg)	7,000	7,200	6,900	7,400

Fig. 4.85. Weighing elephants.

(**a**) Find the mean and standard deviation of their weights.

(**b**) A special herbal medicine is given to these four elephants to reduce their weight by 500 kg. Construct a new data set. Find the average weight and the standard deviation of the new data set.

(**c**) Another herbal medicine reduces their weight to 8/10 of their original weight. Construct a new data set. Find the average weight and standard deviation from the new data.

LUDI 4.7. (PROPERTIES OF STANDARD DEVIATION)

(**a**) Calculate the value of the sample standard deviation for the data set:

	x_i	$(x_i - \bar{x})$	$(x_i - \bar{x})^2$
	12		
	15		
	19		
	17		
	18		
	15		
Sum			

Math is fun

$99 \times 99 + 99 =$

Fig. 4.86. Let me try.

From the above table:

$$n = \text{---------}, \quad \sum_{i=1}^{n}(x_i - \bar{x}) = \text{---------}, \quad \sum_{i=1}^{n}(x_i - \bar{x})^2 = \text{---------}$$

Use the above results to find s^2 :

$$s^2 = \frac{\sum_{i=1}^{n}(x_i - \bar{x})^2}{n-1} = \text{-----------}$$

Then, find the sample standard deviation:

$$s = \sqrt{s^2} = \text{-------}$$

(**b**) Multiply all the data points in (**a**) with a number, say 3.2, to complete the following table:

	$x_i = 3.2x_{i(a)}$	$(x_i - \bar{x})$	$(x_i - \bar{x})^2$
Sum			

Math needs patience

$99 \times 99 + 99 =$

Fig. 4.87. It was great fun.

From the above table:

$$n = --------, \quad \sum_{i=1}^{n}(x_i - \bar{x}) = --------, \quad \sum_{i=1}^{n}(x_i - \bar{x})^2 = --------$$

Use the above results to find s^2:

$$s^2 = \frac{\sum_{i=1}^{n}(x_i - \bar{x})^2}{n-1} = -----------$$

Then find the sample standard deviation:

$$s = \sqrt{s^2} = -------$$

Divide the sample standard deviation derived in (b) with the sample standard deviation in (a) and the result $= --------$

Read this after finishing your calculations:
Note that we multiplied the original data by 3.2, and we observed that the new standard deviation became 3.2 times the old standard deviation.

(c) Add a number, say 5, to all the data points in (a) to complete the following table:

	$x_i = x_i(a) + 5$	$(x_i - \bar{x})$	$(x_i - \bar{x})^2$
Sum			

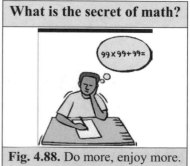

What is the secret of math?

$99 \times 99 + 99 =$

Fig. 4.88. Do more, enjoy more.

From the above table, find:

$$n = \text{--------}, \quad \sum_{i=1}^{n} (x_i - \bar{x}) = \text{--------}, \quad \sum_{i=1}^{n} (x_i - \bar{x})^2 = \text{--------}$$

Use the above results to find s^2:

$$s^2 = \frac{\sum_{i=1}^{n} (x_i - \bar{x})^2}{n-1} = \text{----------}$$

Then find the sample standard deviation:

$$s = \sqrt{s^2} = \text{--------}$$

Divide the sample standard deviation derived in (c) with the sample standard deviation in (a), and the result = --------

Read this after finishing your calculations:
Note that we added the constant value of 5 to the original data, but we observed that the new standard deviation remains the same as the old standard deviation. In other words, the standard deviation in (c) = the standard deviation in (a).

Lesson: Note that if we add a constant value to a data set, the standard deviation will not change. But, if we multiply data values by a constant, then the new standard deviation will be equal to the same constant multiplied by the old standard deviation. Addition of a constant or multiplication of a constant to all the data values is also called a **linear transformation**.

LUDI 4.8. (DEALING WITH LONG DATA SETS) For the following data set: 8, 7, 13, 3, 6, 4, 8, 6, 3, 4, 0, 1, 11, 7, 1, 8, 6, 12, 13, 10, 5, 5, 2, 8, 17, 8, 5, 2, 15, 10, 2, 3, 16, 1, 7, 9, 2, 3, 11, 9.
(a) Find the sample mean $\bar{x}$.
(b) Find the sample variance s^2 by the computing formula.
(c) Find the sample standard deviation s.
(d) Find the percentage of data values in the interval $\bar{x} \pm 2s$.
(e) Compare your above percentage with the empirical rule.

LUDI 4.9. (A SINGLE WORD DECISION MAY BE TOUGH)
True / False statements:

	Statement	Circle one	
a	If the standard deviation of a set of values is 0, then the mean must be zero.	True	False
b	The median is the midpoint between Q_1 and Q_3	True	False
c	If two data sets have the same mean and variance, their polygons must be the same.	True	False
d	The variance and standard deviation of the data set: 5,5,5,5,5 is zero.	True	False
e	If a distribution is skewed to the right, then: Mode ≤ Median ≤ Mean .	True	False
f	If a distribution is skewed to the left, then: Mode ≤ Median ≤ Mean .	True	False
g	If a distribution is symmetric and bell shaped, then: Mode = Median = Mean	True	False
h	If n is even, then the average of the two middle values from the arranged data is the median.	True	False
i	If n is odd, then the single middle value from the ordered data will be the median.	True	False
j	Range depends only upon extreme values.	True	False
k	The mean value shifts toward the outlier.	True	False
l	The median is effected less by outliers.	True	False
m	The interquartile range may be negative.	True	False
n	Sample variance may be negative.	True	False
o	Sample mean may be negative.	True	False
p	Sample standard deviation may be negative.	True	False
q	The data set 2, 3, 4, 4, 5, 6, 6, 7 has two modes and is called a bimodal data set.	True	False
r	The sum of the deviations from the mean is always zero.	True	False
s	A box plot can be used to detect outliers.	True	False
t	The range of 5,5,5,5,5 is zero.	True	False
u	A bimodal histogram can never be symmetric.	True	False
v	A tri-modal histogram can never be symmetric.	True	False
w	A tetra-modal histogram has four modes.	True	False
x	If a distribution is symmetric then: Mode = Median = Mean	True	False

LUDI 4.10. (THINK BEFORE YOU CIRCLE) Choose one correct answer.

a	The mean value of 10, 10, 10 and 10 is---	10	100	40
b	If the mean of 5 values is 20.2, then the total was ---	20.2	101	Cannot be found
c	The variance of 7,7,7,7, and 7 is ---	7	0	49
d	If $\sum_{i=2}^{5} x_i = 20$ then the sample mean $\bar{x} =$ ---	20	5	4
	In the following figure, we are given a ---			
e	400 450 500 550 600 650 700 750 800 850 900	Dot plot	Box plot	Bar Chart
	From the graph in (e), the value of the IQR is ---	300	600	800
	From the graph in (e), the values of $Q_1=$ ---, $Q_2 =$ --- and $Q_3=$ ---	800	500	600
	From the graph in (e): STEP = ---, the maximum value = --- and the minimum value = ---	400	800	450
f	If $\sum_{i=1}^{6}(x_i - \bar{x})^2 = 125$, then the sample standard deviation is ---	25	5	125
g	If $\bar{x} = 20$ and $\sum_{i=1}^{n} x_i = 140$, then the value of $n =$ ---	Cannot be found	7	20
h	From a sample of $n = 7$ data values, we found the sample mean $\bar{x} = 45$ and $\sum_{i=1}^{7} x_i^2 = 16{,}109$. Then the value of the sample standard deviation is, $s =$ ---	17.95	16109	315

LUDI 4.11. (SEVEN MAY BE A LUCKY NUMBER) A data set for a sample of 14 data values is given by

21, 92, 26, 24, 17, 24, 18, 22, 23, 02, 15, 10, 17 and 19

Find the **seven numbers summary** for the above data.
Hint: Find Q_1, Q_2, Q_3, X_{min}, X_{max}, X_s and X_l.

LUDI 4.12. (WHY DOES THE DOCTOR WEIGH INFANTS?)
The following are the weights (in pounds) of five randomly selected infants:

Infants					
Weight (lbs)	12	15	11	13	9

Fig. 4.89. Weighing infants is important.

(a) Determine the following values for this sample:
(i) The sample median
(ii) The sample mean
(iii) The sample standard deviation
(b) List the following:
(i) Which quantities in (a) are measures of variability?
(ii) Which quantities in (a) are measures of central value?

LUDI 4.13. (READING A DOT PLOT) Consider the following two graphs:

```
                        *
        *               *
        *               *             *
   ┌─────────────────────────────────────┐
      10    20    30    40    50
              Graph I

        *                             *
        *                             *
        *                             *
   ┌─────────────────────────────────────┐
      10    20    30    40    50
              Graph II
```

Fig. 4.90. Two dot plots.

(a)Which data has more variation based on range?

Range of data I = - - - - - - -	Range of data II = - - - - - - -
State your result here:	

(b) Which data has more variation based on sample variance? (**Hint:** compute s^2 in each case using the computing formula, and show all work)

Data I			Data II	
x_i	x_i^2		x_i	x_i^2
Sum			**Sum**	
From above:			From above:	
$n = ...,\ \sum x_i = ...,\ \sum x_i^2 = ...$			$n = ...,\ \sum x_i = ...,\ \sum x_i^2 = ...$	

Then complete the following table:

Use this space to compute sample variance for data I	Use this space to compute sample variance for data II
State your result here:	

LUDI 4.14. (READING A BIMODAL HISTOGRAM) A local manufacturing company is producing two types of bolts. The manufacturer is interested in the average weight of the bolts produced. He takes a simple random sample of 40 and records the weight for each bolt. A histogram of the recorded weights is shown below:

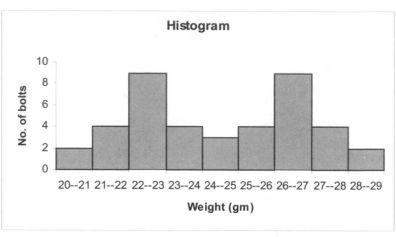

Fig. 4.91. Histogram of the weight of bolts.

(i) The shape of the distribution of weight is best described by - - -
    (a) Skewed to the left
    (b) Symmetric and uni-modal
    (c) Bi-modal and symmetric
    (d) None of these.

(ii) The modal classes are - - -   and - - -

(iii)The difference between the midpoints of the modal classes is - - -

(iv) The median class is - - -

(v) The mean, median and mode(s) are ....
    (a) equal because it is a symmetric distribution.
    (b) unequal because it is a bi-modal and symmetric distribution.
    (c) none of theses.

LUDI 4.15. (READING HISTOGRAMS) The following graph is an
example of - - - data
(a) uni-modal
(b) bi-modal
(c) tri-modal
(d) tetra-modal
(e) hexa-modal
(f) none of these

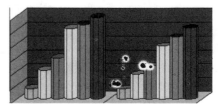

Fig. 4.92. Reading graphs.

LUDI 4.16. (BOAT RACE). In the American gold cup race, the following boats participated. A statistician made an observation that the speeds of the boats followed a mound shaped distribution with an average speed of 14 knots and a variance of 4 knots^2.

Fig. 4.93. Boat race.

Apply the **empirical rule** to find out how many boats were:

(a) racing between 12 knots and 16 knots.
(b) racing between 10 knots and 18 knots.
(c) racing between 8 knots and 20 knots.

LUDI 4.17. (ANTIQUE LOVERS) Assume that at an annual tournament, the following antique vehicles participated in a race in a particular city in the United States of America.

Fig. 4.94. Antiques are valuable.

A statistician made an observation that the average speed of all the vehicles was 16.5 miles/hour with a variance of 6.25 (miles/hour)^2. Apply the empirical rule to find:

(a) How many antique vehicles were racing between 14 miles/hour and 19 miles/hour?

(b) How many antique vehicles were racing between 11.5 miles/hour and 21.5 miles/hour?

(c) How many antique vehicles were racing between 9 miles/hour and 24 miles/hour?

LUDI 4.18. (SAVE ELECTRICITY FOR WINTER) Once a **squirrel** was living under a tree. She was always worried about the coming winter and cold weather.

Fig. 4.95. Squirrel with nuts.

Every year she collected special kinds of nuts to survive during the coming winter season. One summer, although she was sick, she collected $N = 1,000$ nuts to get through winter inside her tunnel. She called a statistician to measure the total weight of all the nuts collected and to determine how much she should eat daily so that she would have enough to get through the winter.

Fig. 4.96. Voluntary statistician.

The statistician took an SRSWOR sample of $n = 5$ nuts, and recorded their weights as given below:

| Weight of nuts (gm) | 6.2 | 5.3 | 7.4 | 9.1 | 3.5 |

(a) Complete the following steps to estimate the total weight of all 1,000 nuts. From the sample information, we have:

$$x_1 = \text{---}, \quad x_2 = \text{---}, \quad x_3 = \text{---}, \quad x_4 = \text{---}, \quad \text{and} \quad x_5 = \text{---}.$$

An estimate of the average weight of all the nuts based on the sample information will be:

$$\bar{x} = \frac{\sum\limits_{i=1}^{n} x_i}{n} = \text{--------} \, \text{gm}.$$

Note that an estimate of the **total** weight of all the nuts based on the sample information will be:

$$\hat{x} = N \bar{x} = \text{--------} \, \text{gm}.$$

(b) How much should she eat daily so that she will not have to leave her tunnel during the winter?

The statistician assumed that the period of winter is generally 90 days, then:

$$\text{daily nuts diet} = \frac{\hat{x}}{90} = \text{--------} \, \text{gm}.$$

Thus, the statistician told the squirrel that if she eats - - - - - - - - gm of nuts daily, then she can get through winter without making any trips from her tunnel during the coming winter. Then the squirrel smiled and thanked the statistician for the help.

LUDI 4.19. (PRACTICE GRAPHS) Draw three graphs showing bell shaped, skewed to the right, and skewed to the left distribution. Show the expected positions of the mean, median, and mode on the three graphs.

LUDI 4.20. (MAY YOU LIVE A LONG LIFE) A boy named Bob is playing with soap bubbles. The average lifetime of a bubble is 52 milliseconds with a variance of 49 milliseconds^2.

Fig. 4.97. Bob blows bubbles.

(I) Empirical Rule:

(a) What approximate proportion of bubbles has the age of 45 to 59 milliseconds?

(b) What approximate proportion of bubbles has the age of 38 to 66 milliseconds?

(c) What approximate proportion of bubbles has the age of 31 to 73 milliseconds?

(II) Tchebysheff's Rule:

(a) What is the minimum proportion of bubbles that have the age of 38 to 66 milliseconds?

(b) What is the minimum proportion of bubbles that have the age of 31 to 73 milliseconds?

(c) What is the least proportion of bubbles that have the age of 38.28 to 65.72 milliseconds?

(d) What is the least proportion of bubbles that have the age of 33.94 to 70.06 milliseconds?

(e) What is the least proportion of bubbles that have the age of 41.5 to 62.5 milliseconds?

LUDI 4.21. (THINKING ON BOX PLOTS) Given that the lower inner fence is 40 and the upper inner fence is 90, find the first and the third quartile. Can we find the median?

LUDI 4.22. (SCOOTER RACING IS FUN) In a race of 40 scooters, the participants show the following performance in km/hr:

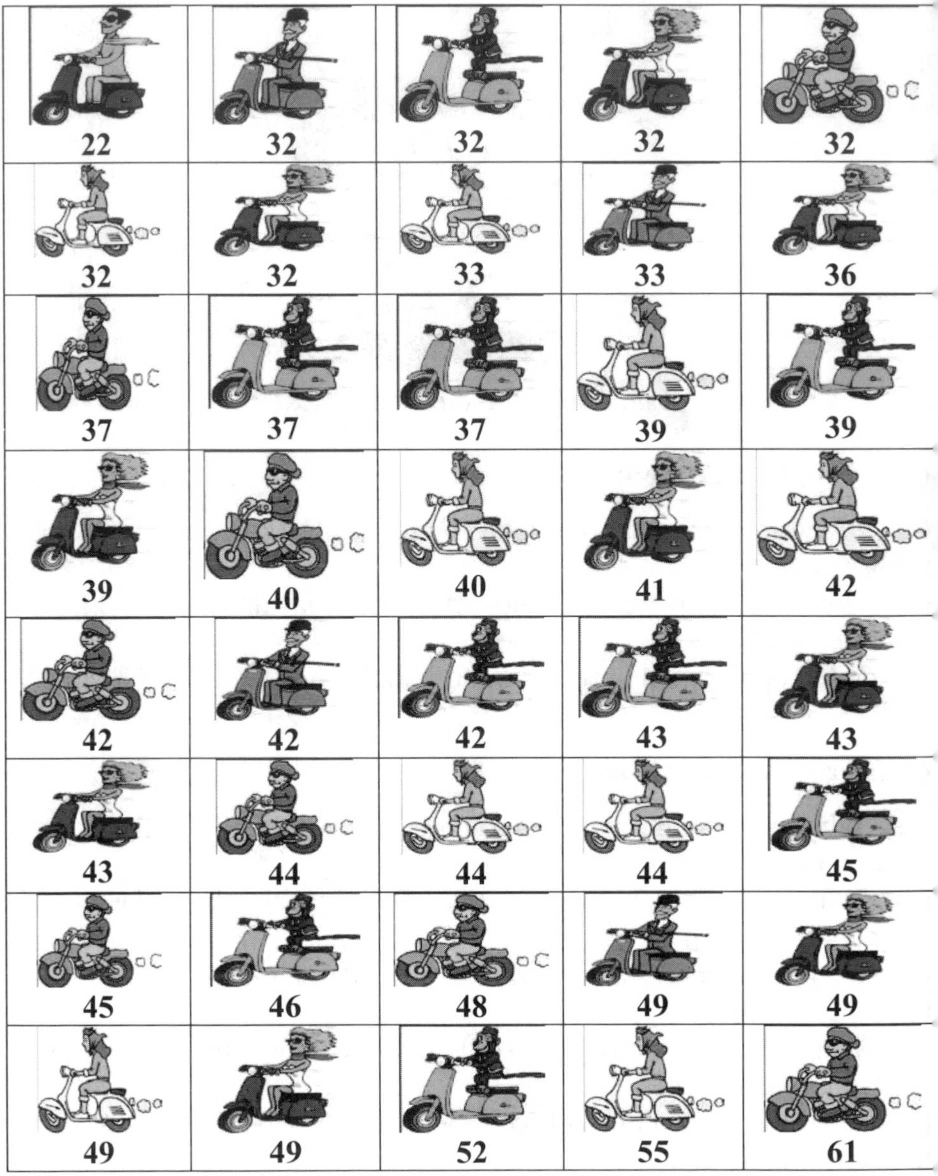

22	32	32	32	32
32	32	33	33	36
37	37	37	39	39
39	40	40	41	42
42	42	42	43	43
43	44	44	44	45
45	46	48	49	49
49	49	52	55	61

Fig. 4.98. Always use a helmet while driving a scooter.

Did the participants follow the empirical rule? Justify your claim by counting the number of participants in each one of the three standard intervals based on the empirical rule.

LUDI 4.23. (TYPES OF UNIVERSITIES) Two universities are providing the following information on their faculty salaries:

University	X_s	Q_1	Median	Q_3	X_l
Undergraduate	66,000	74,000	78,000	80,000	86,000
Research	70,000	76,500	82,000	85,000	100,000

(a) Construct a box plot for salaries in the undergraduate university.
(b) Construct a box plot for salaries in the research university.
(c) Are there larger differences at the lower or higher salary levels? Explain.

LUDI 4.24. (DEALING WITH NEGATIVE NUMBERS) For each one of the following sample data sets:

(i) -2, -5, -4, +2, -3, -5
(ii) -1, +2, 0, -2, 1, -1, -2, -2
(iii) -12, -11, -11, -12, -15, -20

Find the:

(a) mean, (b) median, (c) mode,
(d) range, (e) variance, and
(f) standard deviation.

Fig. 4.99. Math is fun.

LUDI 4.25. (GOOD SCORES NEED PRACTICE) The following are the golf scores of 11 members of a men's golf team:

90	91	88	99	87	82	103	106	84	89	92

Find and do the following:

(a) Arrange the data in ascending order:

Fig. 4.100. Golf.

(b) $X_{min} =$

(c) $X_{max} =$

(d) $Q_2 =$

(e) $Q_1 =$

(f) $Q_3 =$

(g) $IQR =$

(h) $STEP =$

(i) $LIF =$

Fig. 4.101. Get used to your calculator before using it in an examination.

(j) $UIF =$

(k) $X_s =$

(l) $X_l =$

(m) Make a box plot:

_____

(n) What is the distribution of the scores?

(o) Find the lower outer fence (LOF).

(p) Find the upper outer fence (UOF).

(q) Are there any outliers?

LUDI 4.26. (CIRCLE YOUR CHOICE) (I) The following histogram shows the distribution of income of employees in a factory:

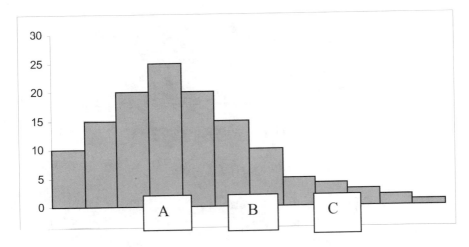

Fig. 4.102. Learning a histogram.

(a) What is A expected to be? (i) Mean (ii) Median (iii) Mode
(b) What is B expected to be? (i) Mean (ii) Median (iii) Mode
(c) What is C expected to be? (i) Mean (ii) Median (iii) Mode
(d) Which of the following is expected to be true?
     (i) A > B > C (ii) A < B < C (iii) B < C > A (iv) C < B < A

(II) The following histogram is an example of - - - - - - -

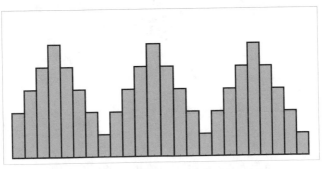

Fig. 4.103. Multi-modal situation.

(i) Uni-modal (ii) Bi-modal (iii) Tri-modal

(III) The following histogram is an example of ……

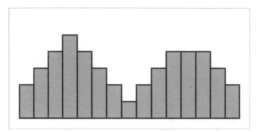

Fig. 4.104. Counting modes.

(i) Uni-modal (ii) Bi-modal (iii) Tri-modal

(IV) Let X represent the income (in $) of the employees at a university. Assume the median income of this population is $65,000. The following histogram is a sketch of the density curve for the distribution of income.

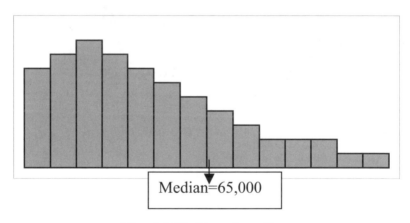

Median=65,000

Fig. 4.105. Use of median.

Circle one choice:
(i) What is the distribution of X?
(a) Symmetric (b) Skewed to the right (c) Skewed to the left

(ii) What can you say about the mean salary?
(a) Less than $65,000 (b) Equal to $65,000 (c) More that $65,000

(iii) What can you say about the mode of the distribution of X ?
(a) Less than $65,000 (b) Equal to $65,000 (c) More than $65,000

(V) Let X represent the income (in $) of the employees in a factory. Assume the mean income of this population is $65,000 with a standard deviation of $10,000. The following histogram is a sketch of the density curve for the distribution of income.

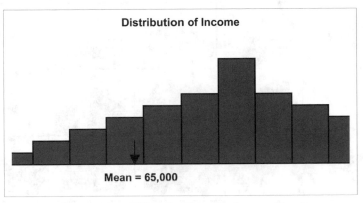

Fig. 4.106. Why do we need a median?

Circle one choice:

(i) What is the distribution of X?
(a) Symmetric (b) Skewed to the right (c) Skewed to the left

(ii) What can you say about the median salary?
(a) Less than $65,000 (b) Equal to $65,000 (c) More than $65,000

LUDI 4.27. (HANDLING NEGATIVE NUMBERS)
For each one of the following sample data sets:

(i) -22, -52, -42, +22, -32, -52
(ii) -14, +24, +14, -24, +14, -14, -24, -24
(iii) -122, -112, -112, -122, -152, -202

find:

(a) mean, (b) median, (c) mode, (d) range, (e) variance, and (f) standard deviation.

Fig. 4.107. Doing math.

LUDI 4.28. (SPPED THRILLS, BUT KILLS) The distribution of the speeds of 42 motorcyclists has been found to be exactly bell-shaped with an average speed of 80 miles/hr and a variance of 100 $(miles/hr)^2$.

Fig. 4.108. Motorcycle lovers.

Apply the empirical rule to find:
(a) How many motorcyclists were going between 70 miles/hr and 90 miles/hr?
(b) How many motorcyclists were going between 60 miles/hr and 100 miles/hr?
(c) How many motorcyclists were going between 50 miles/hr and 110 miles/hr?
(d) How many motorcyclists were going more than 60 miles/hr?

LUDI 4.29. (CYCLE RACES ARE FUN) In a race of 40 cyclists, the participants showed the following performance in km/hr:

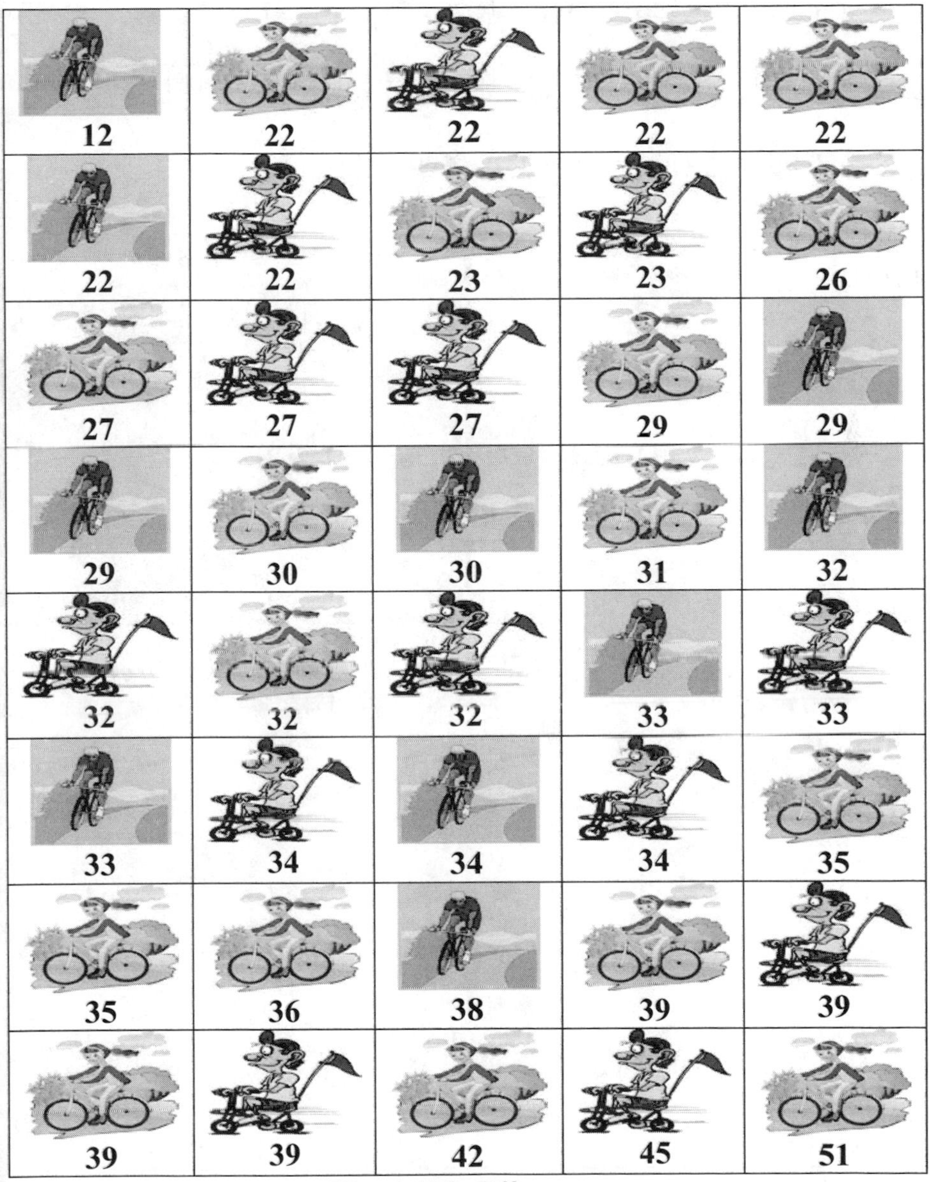

12	22	22	22	22
22	22	23	23	26
27	27	27	29	29
29	30	30	31	32
32	32	32	33	33
33	34	34	34	35
35	36	38	39	39
39	39	42	45	51

Fig. 4.109. Bike race.

Did the participants follow the empirical rule? Justify your claim by counting the number of participants in each one of the three standard intervals based on the empirical rule.

LUDI 4.30. (GOOD GRADES NEED HARD WORK) The following are the scores of 11 students in Stat 229 after adding their extra credit during the last semester:

| 80 | 81 | 78 | 89 | 77 | 72 | 105 | 107 | 74 | 35 | 82 |

(a) Arrange the data in ascending order:

| | | | | | | | | | | |

(b) $X_{min} =$

(c) $X_{max} =$

(d) $Q_2 =$

(e) $Q_1 =$

(f) $Q_3 =$

(g) $IQR =$

(h) $STEP =$

(i) $LIF =$

(j) $UIF =$

(k) $X_s =$

(l) $X_l =$

Fig. 4.110. Learn how to use your calculator before sitting in an exam.

(m) Make a box plot here:

(n) Are there any outliers?
(o) What is the distribution of the scores?

5. TOUCHING PROBABILITY

5.1 INTRODUCTION

In this chapter we discuss basic terminology and rules of probability theory, marginal and conditional probabilities, Venn diagrams, additive and multiplicative laws of probability, and simple examples associated with some fun. In layman's language the term probability is defined as the chance of the occurrence of an **event** in an **experiment**.

What is an experiment and an event?

5.2 EXPERIMENT

An **experiment** is any process by which an observation (or data value or measurement) can be obtained. For example:

(a) tossing a fair coin is an experiment.

Fig. 5.1. A fair coin.

(b) trying to hit a flying bird is an experiment.

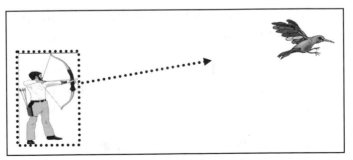

Fig. 5.2. Hunting is not fun.

(c) tossing a fair die is an experiment.

Fig. 5.3. A fair die.

Each experiment may result in one or more outcomes, which we call **events** and denote by capital letters E_1, E_2 etc.

5.3 EVENT

An **event** is an outcome of an experiment.

For example:
(a) while tossing a fair coin, the output will either be a head (H) or a tail (T), and we define these two possible events as:

    Event E_1 : Observe a head (H)
    Event E_2 : Observe a tail (T)

Here a fair coin means that it cannot land on the edge after flipping.

(b) while trying to hit a flying bird, the possible events can be:

    Event E_1 : Hit the flying bird (H)
    Event E_2 : Miss the flying bird (M)

(c) while tossing a die, the output will be any one of the six sides facing up, and we define these six events as:

    Event E_1 : Observe a 1,   Event E_2 : Observe a 2,
    Event E_3 : Observe a 3,   Event E_4 : Observe a 4,
    Event E_5 : Observe a 5,   Event E_6 : Observe a 6.

We can also define events while rolling a die as follows:

    Event A : Observe an odd number $\{1, 3, 5\} = \{E_1, E_3, E_5\}$
    Event B : Observe a number less than four $\{1, 2, 3\} = \{E_1, E_2, E_3\}$
    Event C: Observe an even number $\{2, 4, 6\} = \{E_2, E_4, E_6\}$

5.4 MUTUALLY EXCLUSIVE EVENTS

Two or more events are said to be **mutually exclusive** or disjoint if they cannot occur together.

For example:
(a) while tossing a fair coin, the events E_1 and E_2 are mutually exclusive (ME) because the appearance of a head (H) discards the appearance of a tail (T) and vice versa.
(b) while hitting a flying bird, the events E_1 and E_2 are also mutually exclusive because either the bird will get hit or it won't get hit.
(c) while rolling a die the events E_1, E_2, E_3, E_4, E_5 and E_6 are also mutually exclusive, that is, either we will observe a 1, 2, 3, 4, 5, or 6. All six sides of a die cannot face up together.

In contrast, note the events A and B defined as:

$$A = \text{Observe an odd number} = \{1, 3, 5\} = \{E_1, E_3, E_5\}$$

and

$$B = \text{Observe a number less than four} = \{1, 2, 3\} = \{E_1, E_2, E_3\}$$

are **not** mutually exclusive events, because the events E_1 and E_3 are common among A and B. However, the events $A = \{1, 3, 5\}$ and $C = \{2, 4, 6\}$ are mutually exclusive events.

5.5 COMPOUND EVENTS

If an event can be decomposed (or broken) into several events, it is called a **compound event**.

For example, the event $A = \text{observe an odd number} = \{1, 3, 5\} = \{E_1, E_3, E_5\}$ is a compound event because it can be split into several events such as: $\{E_1\}$, $\{E_2\}$, $\{E_3\}$, $\{E_1, E_2\}$, $\{E_1, E_3\}$, $\{E_2, E_3\}$ etc. Similarly, the events $B = \{1, 2, 3\}$ and $C = \{2, 4, 6\}$ are also compound events.

5.6 SIMPLE EVENTS

Any event that cannot be decomposed (or broken) is called a **simple event**. For example, while rolling a die, the events E_1, E_2, E_3, E_4, E_5, and E_6 are simple events.

5.7 SAMPLE SPACE

A list of all possible simple events of an experiment is called a **sample space**. A sample space is denoted by the uppercase letter, **S**.

For example, while tossing a fair coin the sample space, **S**, consists of two simple events: Head (H) and Tail (T), that is $S = \{H, T\}$.

5.8 TREE DIAGRAM

Any diagram showing all possible outcomes (simple events) of an experiment as branches of a tree is called a **tree diagram**.

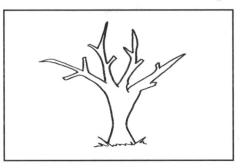

Fig. 5.4. A tree showing a trunk and branches.

The roots represent an experiment and the branches show the total number of simple events.

Example 5.1. (COIN) Toss a fair coin. Construct a tree diagram and derive the sample space.
Solution. We have:

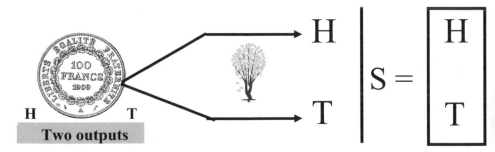

Fig. 5.5. A tree diagram and sample space with one coin.

Example 5.2. (TOSS TWO COINS) Construct a sample space while tossing two fair coins.

Solution. If we toss two coins, then the total number of outputs (events) is given by:

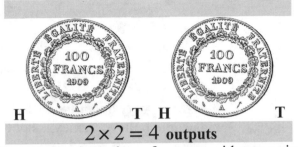

H   T H   T

$2 \times 2 = 4$ **outputs**

Fig. 5.6. Total number of outputs with two coins.

Thus, while tossing two coins there will be a total of four outcomes:

(Head, Head), (Head, Tail), (Tail, Head), and (Tail, Tail).

A tree diagram to present the sample space while tossing two coins:

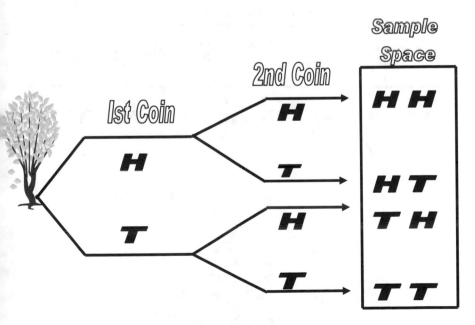

Fig. 5.7. A tree diagram and sample space with two coins.

Example 5.3. (MAN MADE MONEY: A BIG MISTAKE)
Construct a sample space for tossing three fair coins.
Solution. If we toss three coins, then the total number of outputs (events) is given by:

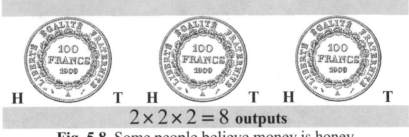

$2 \times 2 \times 2 = 8$ **outputs**

Fig. 5.8. Some people believe money is honey.

Thus, if tossing three coins, there will be **eight** outcomes:

(Head, Head, Head), (Head, Head, Tail), (Head, Tail, Head),
(Head, Tail, Tail), (Tail, Head, Head), (Tail, Head, Tail),
(Tail, Tail, Head), and (Tail, Tail, Tail).

A tree diagram and the sample space for three coins is in **Figure 5.9**.

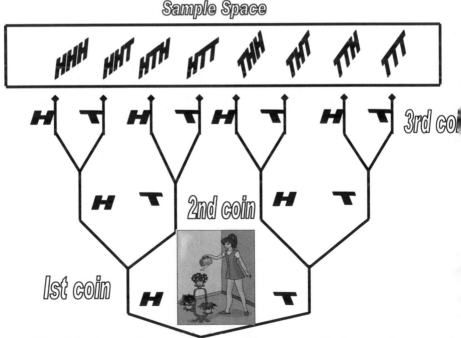

Fig. 5.9. A tree diagram and sample space with three coins.

Example 5.4. (KIDS' FAVORITE GAME: LUDO) Roll a fair die. Construct a tree diagram and derive the sample space.

Solution. We have six possible outputs when rolling a die as follows:

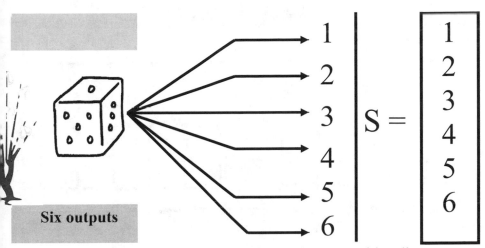

Six outputs

Fig. 5.10. A tree diagram and sample space with a die.

Example 5.5. (COIN AND DIE) Construct a sample space for tossing a fair coin and rolling a fair die.

Solution. If we toss a coin and roll a die, then the total number of outputs (events) is given by:

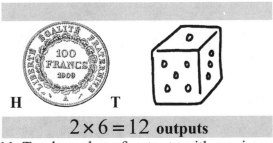

H   T

$2 \times 6 = 12$ **outputs**

Fig. 5.11. Total number of outputs with a coin and a die.

Thus, when tossing a fair coin and rolling a fair die, there will be a total of twelve outcomes:

(Head, 1), (Head, 2), (Head, 3), (Head, 4), (Head, 5), (Head, 6),
(Tail, 1), (Tail, 2), (Tail, 3), (Tail, 4), (Tail, 5), and (Tail, 6).

A tree diagram to present a sample space while tossing a coin and rolling a die is given in **Figure 5.12**.

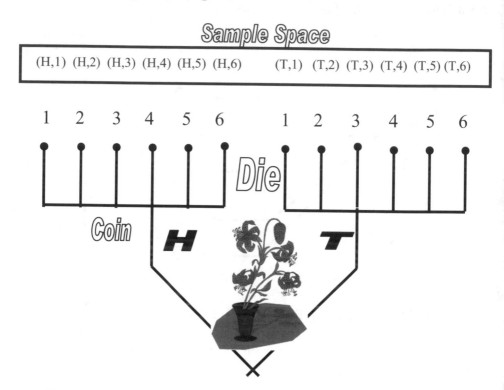

Fig. 5.12. A tree diagram and sample space with a coin and a die.

Example 5.6. (PAIR OF DICE) Construct a sample space for rolling two fair dice.
Solution. If we roll two dice, then the total number of outputs (events) is given by:

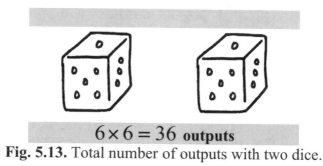

$6 \times 6 = 36$ **outputs**
Fig. 5.13. Total number of outputs with two dice.

Thus, while tossing two dice there will be thirty-six outcomes:

$$(1, 1), (1, 2), (1, 3), (1, 4), (1, 5), (1, 6)$$
$$(2, 1), (2, 2), (2, 3), (2, 4), (2, 5), (2, 6)$$
$$(3, 1), (3, 2), (3, 3), (3, 4), (3, 5), (3, 6)$$
$$(4, 1), (4, 2), (4, 3), (4, 4), (4, 5), (4, 6)$$
$$(5, 1), (5, 2), (5, 3), (5, 4), (5, 5), (5, 6)$$
$$(6, 1), (6, 2), (6, 3), (6, 4), (6, 5), (6, 6)$$

This sample space can also be represented with a tree diagram as shown in **Figure 5.14**.

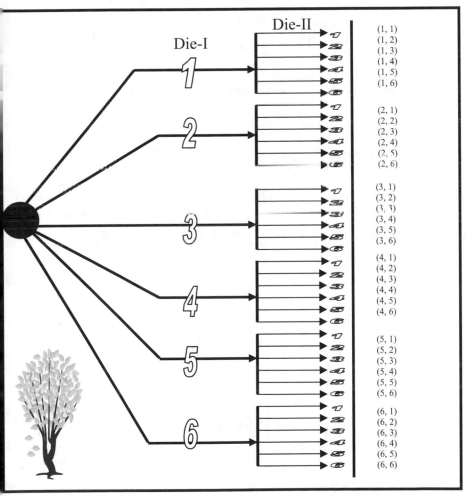

Fig. 5.14. A tree diagram and sample space for rolling two dice.

5.9 VENN DIAGRAM

A Venn diagram can be used to display events and sample space.

For example, rolling a die will result in six outcomes: E_1, E_2, E_3, E_4, E_5 and E_6. Consider the event A consisting of all odd numbers:

$$A = \{E_1, E_3, E_5\}.$$

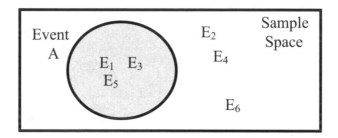

Fig. 5.15. Venn diagram.

5.10 NULL EVENT

An event having no element in it is called a **null** event. It is generally denoted by the Greek letter, phi (Φ). For example, observing a number 7 when rolling a die is a null event. It is also called an impossible event.

5.11 SURE EVENT

An event having all the elements in it is called a **sure** event or a sample space. It is denoted by uppercase letter **S**.

For example, observing a number between 1 and 6, both inclusive, when rolling a die.

5.12 INTERSECTION OF EVENTS

The intersection of two events, A and B, is an event where both the events A and B occur together. It is denoted by $A \cap B$.

For example, while rolling a die, consider the events:

A = Observe an odd number = $\{E_1, E_3, E_5\}$

B = Observe a number less than 4 = $\{E_1, E_2, E_3\}$

and

C = Observe an even number = $\{E_2, E_4, E_6\}$.

Then the intersection of A and B is given by the common events:

$$A \cap B = \{E_1, E_3\}.$$

A Venn diagram to present the intersection of A and B is given by:

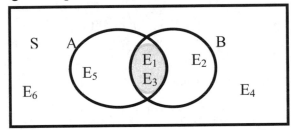

Fig. 5.16. The shaded area represents the intersection of two events.

Similarly, the intersection of A and C is given by the common events:

$$A \cap C = \Phi.$$

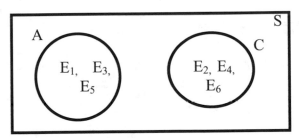

Fig. 5.17. For two disjoint events, the Venn diagram has no shaded area.

Note that if $A \cap C = \Phi$, then the events A and C are called disjoint events or mutually exclusive events.

The intersection of three events A, B and C is an event where all three events occur together. It is denoted by $A \cap B \cap C$. In this case A, B and C are mutually exclusive as shown below:

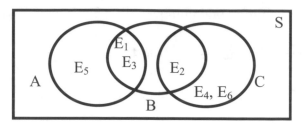

Fig. 5.18. A Venn diagram with three events.

Here $A \cap B \cap C = \Phi$.

5.13 UNION OF EVENTS

The union of two events A and B is the event that either A, B, or both occur. It is denoted by $A \cup B$.

For example, when rolling a die, if

$$A = \text{Observe an odd number} = \{E_1, E_3, E_5\}$$

and

$$B = \text{Observe a number less than } 4 = \{E_1, E_2, E_3\}.$$

Then the union of A and B is given by:

$$A \cup B = \{E_1, E_2, E_3, E_5\}.$$

A Venn diagram to present the above union of events is given by:

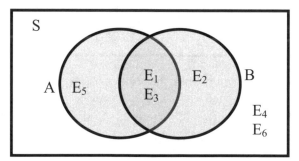

Fig. 5.19. A Venn diagram for the union of two events.

The union of three events A, B and C is an event where either A, B, C, $A \cap B$, $A \cap C$, $B \cap C$, or $A \cap B \cap C$ will occur. It is denoted by $A \cup B \cup C$.

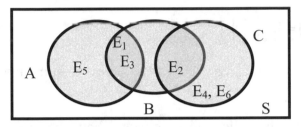

Fig. 5.20. A Venn diagram with three events.

5.14 COMPLEMENT OF AN EVENT

The complement of an event A consists of all the simple events in the sample space S that are not in A. It is denoted by A^c.

For example, while rolling a die if

$$A = \text{Observe an odd number} = \{E_1, E_3, E_5\}$$

then

$$A^c = \text{Observe an even number} = \{E_2, E_4, E_6\}.$$

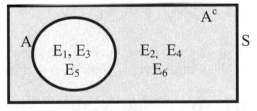

Fig. 5.21. Complement of an event.

5.15 DEFINITION OF PROBABILITY

Consider an experiment performed n times. Let A denote the event of interest. Out of n trials, the event A occurred x times, that is, the frequency of occurrence of event A is $f = x$. Thus, the relative frequency (RF) of the event A is given by:

$$RF = \frac{\text{Frequency}}{n} = \frac{x}{n}$$

If we assume that the number of trials approaches infinity ($n \rightarrow \infty$), then we are generating the whole population and the relative frequency becomes the probability of the event, A, that is:

$$P(A) = \lim_{n \to \infty}\left(\frac{x}{n}\right)$$

Such an experiment repeated many times is sometimes called a **simulation**.

Thus, we have the following statements about the probability of an event:

(a) $P(A)$ behaves like relative frequency.

(b) $P(A)$ must be a proportion between 0 and 1.

(c) $P(A) = 0$ if the event A never occurs, that is $A = \Phi$ (Null event). For example, observing a seven when rolling a fair die.

(d) $P(A) = 1$ if the event A is sure to occur. For example, the sun will rise tomorrow.

(e) The sum of probabilities of all the simple events in a sample space is equal to one, that is, $P(S) = 1$ where S stands for a sample space or a **sure** event.

5.16 PROBABILITY OF AN EVENT

The probability of an event of interest, say A, is equal to the sum of the probabilities of the simple events contained in the event A.

5.16.1 STEPS TO FIND THE PROBABILITY OF AN EVENT

The following steps need to be carried out while computing the probability of an event.

Step 1. List all the simple events in the sample space, S.

Step 2. Assign relative frequency (probability) to each simple event in the sample space, S.

Step 3. Find the event of interest, say A, in terms of simple events.

Step 4. Add the probabilities of all the simple events within the event of interest, A.

Example 5.7. (TWO COINS) Toss two fair coins and record the outcome. Find the probability of observing exactly one head.

Solution. Step 1. List all the simple events in the sample space. The experiment is made by tossing two coins as shown below:

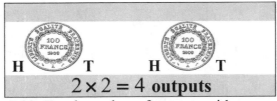

$$2 \times 2 = 4 \text{ outputs}$$

Fig. 5.22. Total number of outputs with two coins.

Thus, the sample space is given by:

$$S = [HH, HT, TH, TT]$$

Step 2. Assign the relative frequency (probability) to each simple event in the sample space.

1st coin	2nd coin	Event	Frequency (f)	$RF = f/n$ $= P(E_i)$
Head	Head	E_1: HH	1	$P(E_1) = 1/4$
Head	Tail	E_2: HT	1	$P(E_2) = 1/4$
Tail	Head	E_3: TH	1	$P(E_3) = 1/4$
Tail	Tail	E_4: TT	1	$P(E_4) = 1/4$
		Sum	$n = 4$	**1.00**

Step 3. Find the event of interest. Only two simple events E_2 and E_3 have exactly one head. Thus, the event of interest A is given by:

$$A = \{E_2, E_3\}.$$

Step 4. Add the probabilities of the simple events within the event of interest.

The required probability of getting exactly one head is given by:

$$P(A) = P(E_2) + P(E_3) = \frac{1}{4} + \frac{1}{4} = \frac{2}{4} = \frac{1}{2} = 0.5.$$

Example 5.8. (PAIR OF DICE) Two fair dice are tossed.
(a) How many simple events are there?
(b) Construct a sample space, **S**.
(c) What is the probability that the sum of the numbers shown above is equal to 7?
(d) What is the probability that the same number will show up on both dice?
(e) What is the probability that the sum of the numbers shown above will be either 7 or 11?
(f) What is the probability that the sum is neither 7 nor 11?
Solution. (a) If we roll two dice, then the total number of outputs (events) will be:

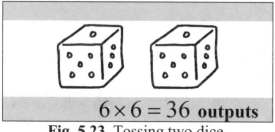

$6 \times 6 = 36$ **outputs**

Fig. 5.23. Tossing two dice.

(b) The sample space is given below:

(1,1)	(1,2)	(1,3)	(1,4)	(1,5)	(1,6)
(2,1)	(2,2)	(2,3)	(2,4)	(2,5)	(2,6)
(3,1)	(3,2)	(3,3)	(3,4)	(3,5)	(3,6)
(4,1)	(4,2)	(4,3)	(4,4)	(4,5)	(4,6)
(5,1)	(5,2)	(5,3)	(5,4)	(5,5)	(5,6)
(6,1)	(6,2)	(6,3)	(6,4)	(6,5)	(6,6)

(c) There are six simple events which show a sum = 7.

(1,1)	(1,2)	(1,3)	(1,4)	(1,5)	(1,6)
(2,1)	(2,2)	(2,3)	(2,4)	(2,5)	(2,6)
(3,1)	(3,2)	(3,3)	(3,4)	(3,5)	(3,6)
(4,1)	(4,2)	(4,3)	(4,4)	(4,5)	(4,6)
(5,1)	(5,2)	(5,3)	(5,4)	(5,5)	(5,6)
(6,1)	(6,2)	(6,3)	(6,4)	(6,5)	(6,6)

Thus, the probability that the sum $= 7$ is:

$$P(A) = \frac{6}{36}.$$

(d) Again there are six simple events where the same number will appear on both dice as shown below:

(1,1)	(1,2)	(1,3)	(1,4)	(1,5)	(1,6)
(2,1)	(2,2)	(2,3)	(2,4)	(2,5)	(2,6)
(3,1)	(3,2)	(3,3)	(3,4)	(3,5)	(3,6)
(4,1)	(4,2)	(4,3)	(4,4)	(4,5)	(4,6)
(5,1)	(5,2)	(5,3)	(5,4)	(5,5)	(5,6)
(6,1)	(6,2)	(6,3)	(6,4)	(6,5)	(6,6)

Thus, the probability that the same number will appear is:

$$P(A) = \frac{6}{36}.$$

(e) There are eight possibilities that the sum will either be 7 or 11.

(1,1)	(1,2)	(1,3)	(1,4)	(1,5)	(1,6)
(2,1)	(2,2)	(2,3)	(2,4)	(2,5)	(2,6)
(3,1)	(3,2)	(3,3)	(3,4)	(3,5)	(3,6)
(4,1)	(4,2)	(4,3)	(4,4)	(4,5)	(4,6)
(5,1)	(5,2)	(5,3)	(5,4)	(5,5)	(5,6)
(6,1)	(6,2)	(6,3)	(6,4)	(6,5)	(6,6)

Thus, the required probability that the sum will either be 7 or 11 is:

$$P(A) = \frac{8}{36}.$$

(f) Using the complement rule in (e), we have the probability that the sum is neither 7 or 11 as:

$$P(A^c) = 1 - P(A) = 1 - \frac{8}{36} = \frac{28}{36}.$$

5.17 MARGINAL AND CONDITIONAL PROBABILITIES

(RATING OF A PROFESSOR): Suppose a class consisting of 100 students (several males and several females) was asked about the teaching style of a professor.

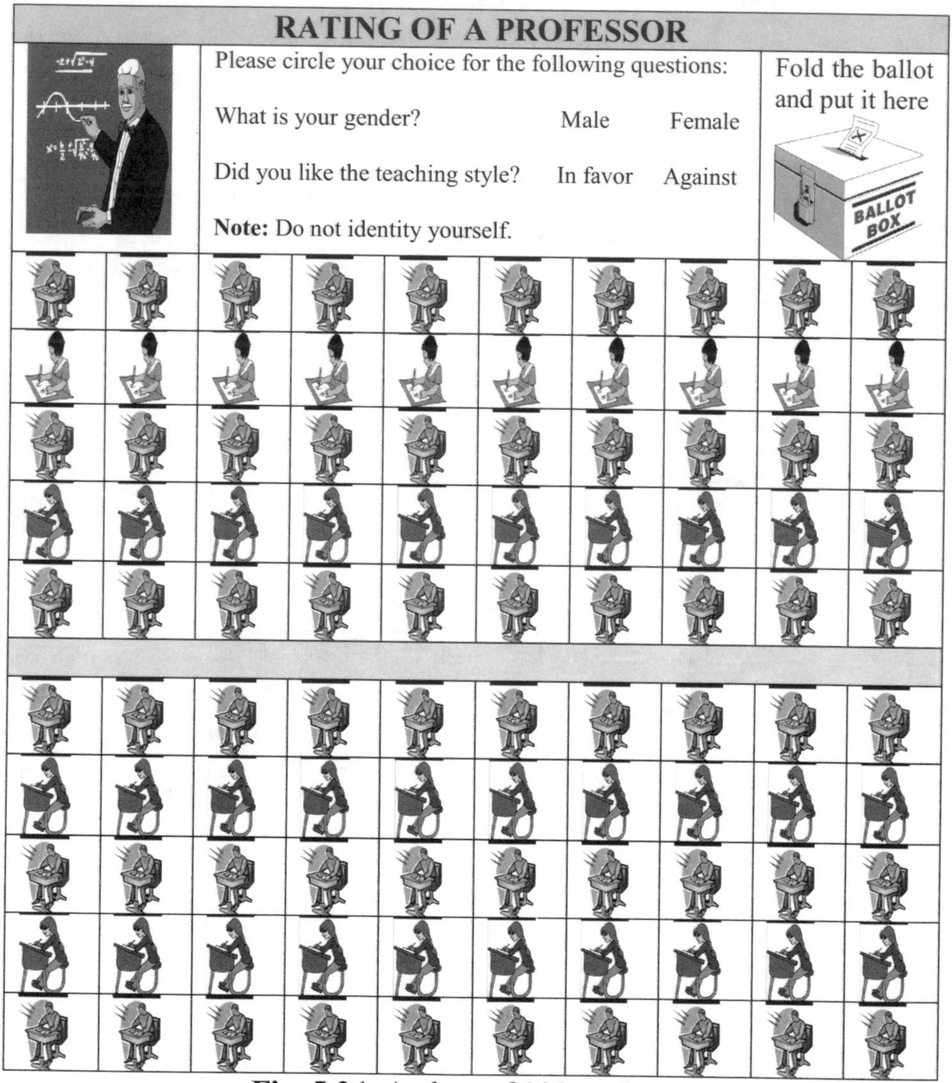

Fig. 5.24. A class of 100 students.

At the end of the survey, a two-way classification of the responses of 100 students is given below:

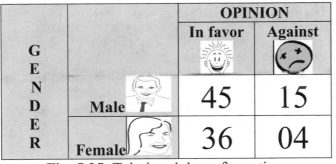

		OPINION		
		In favor	**Against**
G **E** **N** **D** **E** **R**	**Male**	45	15
	Female	36	04

Fig. 5.25. Tabulated data after rating.

Such a table is called 2×2 contingency table. Suppose one secret ballot is selected at random from these 100 responses from the students. Then the selected student may be classified either (a) on the basis of gender, or (b) on the basis of opinion.

If only one characteristic is considered at a time, the student selected can be a:

(i) Male-M, (ii) Female-F, (iii) In favor-R, or (iv) Against-T

The probability of each of these four events is called marginal probability or simple probability.

5.17.1 MARGINAL PROBABILITY

Marginal probability is the probability of a single event without consideration of any other event.

		Opinion		Total
		In favor	**Against**	
G **E** **N** **D** **E** **R**	**Male** 	45	15	60
	Female 	36	04	40
	Total	81	19	100

Fig. 5.26. Analyzing data after tabulating.

Thus, the four marginal probabilities are as follows:

$$P(\text{M}) = P(\text{Male}) = \frac{\#\ \text{of male students}}{\text{Total}\ \#\ \text{of students}} = \frac{60}{100} = 0.60$$

$$P(\text{F}) = P(\text{Female}) = \frac{\#\ \text{of female students}}{\text{Total}\ \#\ \text{of students}} = \frac{40}{100} = 0.40$$

$$P(\text{R}) = P(\text{In favor}) = \frac{\#\ \text{of in favor students}}{\text{Total}\ \#\ \text{of students}} = \frac{81}{100} = 0.81$$

and

$$P(\text{T}) = P(\text{Against}) = \frac{\#\ \text{of against students}}{\text{Total}\ \#\ \text{of students}} = \frac{19}{100} = 0.19.$$

Now assume that one student is selected at random from these 100 students. Further, assume that the selected student is a male. This implies that the event M has already occurred.

What is the probability that the selected male student is in favor of the teaching style of the professor?

5.17.2 CONDITIONAL PROBABILITY

Conditional probability is the probability that an event will occur given that another event has already occurred.

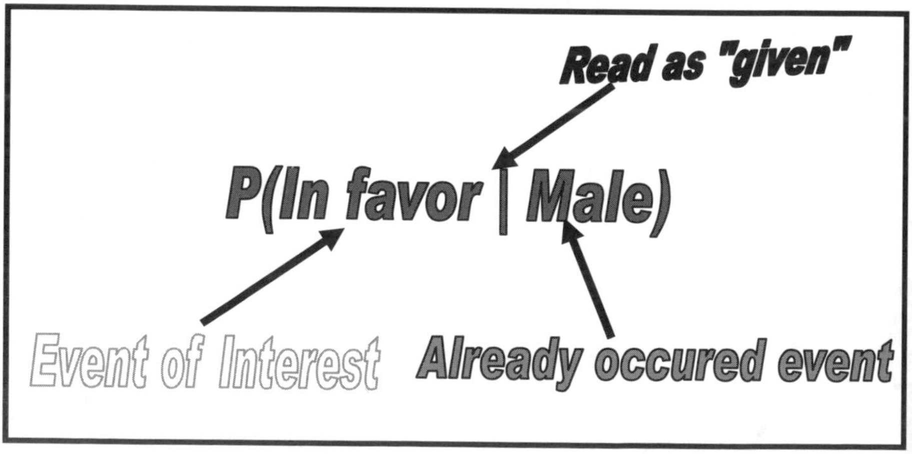

Fig. 5.27. Conditional probability.

If A and B are two events, then the conditional probability of **A given B** is written as:

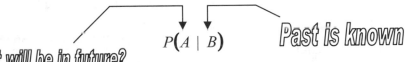

$$P(A \mid B)$$

For example, if we know that there is an accident in an intersection, what is the probability that the police are there to help?

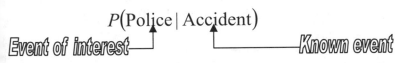

$$P(\text{Police} \mid \text{Accident})$$

Example 5.9. (HISTORY NEVER DIES) Find the probability that:
(a) the selected male student will be in favor of the teaching style.
(b) the selected male student will be against the teaching style.
(c) the selected female student will be in favor of the teaching style.
(d) the selected female student will be against the teaching style.

Solution. We are given:

	In favor 	Against 	Total
Male 	45	15	60
Female 	36	04	40
Total	81	19	100

Fig. 5.28. Finding conditional probabilities.

(**a**) The conditional probability that the selected male student will be in favor is given by:

$$P(\text{In favor} \mid \text{Male}) = P(\text{R} \mid \text{M}) = \frac{\text{\# of in favor male students}}{\text{Total \# of male students}} = \frac{45}{60} = 0.75.$$

Similarly, we read the other probabilities given by:

(b)

$$P(\text{Against} \mid \text{Male}) = P(T \mid M) = \frac{\#\,\text{of against male students}}{\text{Total}\,\#\,\text{of male students}} = \frac{15}{60} = 0.25$$

(c)

$$P(\text{In favor} \mid \text{Female}) = P(R \mid F) = \frac{\#\,\text{of in favor female students}}{\text{Total}\,\#\,\text{of female students}} = \frac{36}{40} = 0.90$$

(d)

$$P(\text{Against} \mid \text{Female}) = P(T \mid F) = \frac{\#\,\text{of against female students}}{\text{Total}\,\#\,\text{of female students}} = \frac{4}{40} = 0.10\,.$$

The conditional probabilities can be described with the help of a tree diagram as shown below:

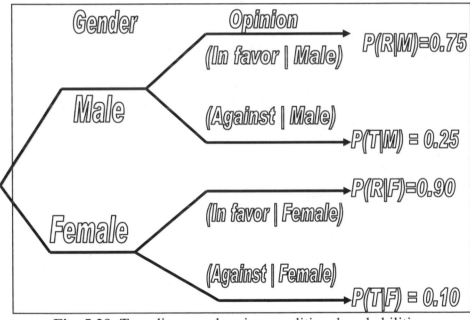

Fig. 5.29. Tree diagram showing conditional probabilities.

5.17.3 INDEPENDENT EVENTS

Two events, A and B, are said to be **independent** if

$$P(A \mid B) = P(A)$$

That is, the occurrence of event B does not effect the occurrence of event A, or

$$P(B \mid A) = P(B)$$

That is, the occurrence of event A does not effect the occurrence of event B.

On the other hand, if

$$P(A \mid B) \neq P(A), \quad \text{or} \quad P(B \mid A) \neq P(B)$$

then the events A and B are said to be **dependent**.

Example 5.10. (EVERYONE ENJOYS INDEPENDENCE) Are the events "Female" and "In favor" independent?
Solution. The events Female (F) and In favor (R) will be independent if

$$P(F \mid R) = P(F)$$

We have

$$P(F) = \frac{\#\, \text{of females}}{\text{Total}\, \#\, \text{of students}} = \frac{40}{100} = 0.40$$

and

$$P(F \mid R) = \frac{\#\, \text{of females in favor}}{\text{Total}\, \#\, \text{of in favor students}} = \frac{36}{81} = 0.44\,.$$

Thus, $P(F \mid R) \neq P(F)$ and the events **F** and **R** are not independent.

5.18 MATHEMATICAL LAWS OF PROBABILITY

If you can learn and remember these formulae, then higher courses in statistics will be much easier for you.

Fig. 5.30. Math: Do more, enjoy more.

If S is a sample space, and A and B are two events in it, then we have the following laws:

(a) Additive law of probability: The probability of occurrence of either **A** or **B** or **both** is given by:

$$P(A \cup B) = P(A) + P(B) - P(A \cap B)$$

(b) Probability of a sample space is one: The probability of all the simple events in a sample space is given by:

$$P(S) = 1$$

(c) Complement law: The probability that event **A** will not happen is given by:

$$P(A^c) = 1 - P(A)$$

(d) The probability of a null event is zero: The probability of an impossible event is given by:

$$P(\Phi) = 0$$

(e) Mutually exclusive events: If $P(A \cap B) = 0$, then the events A and B are mutually exclusive. Here, mutually exclusive means that there is nothing in common between these two events. In other words, $A \cap B = \Phi$. The additive law of probability becomes:

$$P(A \cup B) = P(A) + P(B)$$

Note that if there are several mutually exclusive events, say A, B, C, D,…etc., the probability of the occurrence of any of them is:

$$P(A \cup B \cup C \cup D....) = P(A) + P(B) + P(C) + P(D) + ...$$

(f) Multiplicative law of probability: In general, the multiplicative law is given by:

$$P(A \cap B) = P(A \mid B) P(B)$$

Note that if A and B are **independent**, then:

$$P(A \mid B) = P(A)$$

and the general multiplicative law becomes:

$$P(A \cap B) = P(A) P(B)$$

Note that if there are several independent events, say A, B, C, D,.. etc., the probability of the occurrence of all the events together is:

$$P(A \cap B \cap C \cap D....) = P(A) P(B) P(C) P(D)........$$

Example 5.11. (VENN DIAGRAM) Consider the following Venn diagram:

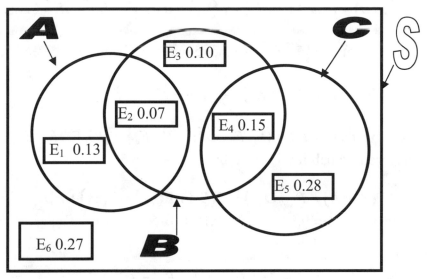

Fig. 5.31. Venn diagram showing the events A, B, C and S.

Find the following probabilities:
(i) $P(A)$ (ii) $P(B)$ (iii) $P(C)$ (iv) $P(A \cup B)$
(v) $P(A \cup C)$ (vi) $P(B \cup C)$ (vii) $P(A \cap B)$ (viii) $P(A \cap C)$
(ix) $P(B \cap C)$ (x) $P(A \cap B \cap C)$ (xi) $P(A \cup B \cup C)$
(xii) $P(B \cap C^c)$ (xiii) $P(A^c \cap C)$ (xiv) $P(A \cap B^c \cap C^c)$
(xv) $P(A^c \cap B \cap C^c)$ (xvi) $P(A^c \cap B^c \cap C)$ (xvii) $P(A^c \cup B^c)$.
Solution. Note that E_1, E_2, E_3, E_4, E_5, and E_6 are simple events, whereas A, B, C, and S are compound events.

We are given:

$P(E_1) = 0.13$, $P(E_2) = 0.07$, $P(E_3) = 0.10$ $P(E_4) = 0.15$,
$P(E_5) = 0.28$, and $P(E_6) = 0.27$.

(i) Note that $A = \{E_1, E_2\}$, so the required probability is given by:

$$P(A) = P(E_1) + P(E_2) = 0.13 + 0.07 = 0.20.$$

(ii) Note that $B = \{E_2, E_3, E_4\}$, so the required probability is given by:

$$P(B) = P(E_2) + P(E_3) + P(E_4) = 0.07 + 0.10 + 0.15 = 0.32.$$

(iii) Note that $C = \{E_4, E_5\}$, so the required probability is given by:

$$P(C) = P(E_4) + P(E_5) = 0.15 + 0.28 = 0.43.$$

(iv) Note that $A \cup B = \{E_1, E_2\} \cup \{E_2, E_3, E_4\} = \{E_1, E_2, E_3, E_4\}$, so the required probability is given by:

$$P(A \text{ or } B) = P(A \cup B) = P(E_1) + P(E_2) + P(E_3) + P(E_4)$$
$$= 0.13 + 0.07 + 0.10 + 0.15$$
$$= 0.45.$$

(v) Note that $A \cup C = \{E_1, E_2\} \cup \{E_4, E_5\} = \{E_1, E_2, E_4, E_5\}$, so the required probability is given by:

$$P(A \text{ or } C) = P(A \cup C) = P(E_1) + P(E_2) + P(E_4) + P(E_5)$$
$$= 0.13 + 0.07 + 0.15 + 0.28$$
$$= 0.63.$$

(vi) Note that $B \cup C = \{E_2, E_3, E_4\} \cup \{E_4, E_5\} = \{E_2, E_3, E_4, E_5\}$, so the required probability is given by:

$$P(B \text{ or } C) = P(B \cup C) = P(E_2) + P(E_3) + P(E_4) + P(E_5)$$
$$= 0.07 + 0.10 + 0.15 + 0.28$$
$$= 0.60.$$

(vii) Note that $A \cap B = \{E_1, E_2\} \cap \{E_2, E_3, E_4\} = \{E_2\}$, so the required probability is given by:

$$P(A \text{ and } B) = P(A \cap B) = P(E_2) = 0.07.$$

(viii) Note that $A \cap C = \{E_1, E_2\} \cap \{E_4, E_5\} = \{\Phi\}$, so the required probability is given by:

$$P(A \text{ and } C) = P(A \cap C) = P(\Phi) = 0.$$

(ix) Note that $B \cap C = \{E_2, E_3, E_4\} \cap \{E_4, E_5\} = \{E_4\}$, so the required probability is given by:

$$P(B \text{ and } C) = P(B \cap C) = P(E_4) = 0.15.$$

(x) Note that $A \cap B \cap C = \{E_1, E_2\} \cap \{E_2, E_3, E_4\} \cap \{E_4, E_5\} = \{\Phi\}$, so the required probability is given by:

$$P(A \text{ and } B \text{ and } C) = P(A \cap B \cap C) = P(\Phi) = 0.$$

(xi) Note that

$$A \cup B \cup C = \{E_1, E_2\} \cup \{E_2, E_3, E_4\} \cup \{E_4, E_5\}$$
$$= \{E_1, E_2, E_3, E_4, E_5\},$$

so the required probability is given by:

$$P(A \text{ or } B \text{ or } C) = P(A \cup B \cup C) = P(E_1, E_2, E_3, E_4, E_5)$$
$$= P(E_1) + P(E_2) + P(E_3) + P(E_4) + P(E_5)$$
$$= 0.13 + 0.07 + 0.10 + 0.15 + 0.28$$
$$= 0.73.$$

(xii) Note that $B \cap C^c = \{E_2, E_3, E_4\} \cap \{E_1, E_2, E_3, E_6\} = \{E_2, E_3\}$, so the required probability is given by:

$$P(B \text{ and } C^c) = P(B \cap C^c) = P(E_2) + P(E_3)$$
$$= 0.07 + 0.10 = 0.17.$$

(xiii) Note that $A^c \cap C = \{E_3, E_4, E_5, E_6\} \cap \{E_4, E_5\} = \{E_4, E_5\}$, so the required probability is given by:

$$P(A^c \text{ and } C) = P(A^c \cap C) = P(E_4) + P(E_5)$$

$$= 0.15 + 0.28 = 0.43.$$

(xiv) Note that

$$A \cap B^c \cap C^c = \{E_1, E_2\} \cap \{E_1, E_5, E_6\} \cap \{E_1, E_2, E_3, E_6\} = \{E_1\},$$

so the required probability is given by:

$$P(A \cap B^c \cap C^c) = P(E_1) = 0.13.$$

(xv) Note that

$$A^c \cap B \cap C^c = \{E_3, E_4, E_5, E_6\} \cap \{E_2, E_3, E_4\} \cap \{E_1, E_2, E_3, E_6\} = \{E_3\},$$

so the required probability is given by:

$$P(A^c \cap B \cap C^c) = P(E_3) = 0.10.$$

(xvi) Note that

$$A^c \cap B^c \cap C = \{E_3, E_4, E_5, E_6\} \cap \{E_1, E_5, E_6\} \cap \{E_4, E_5\} = \{E_5\},$$

so the required probability is given by:

$$P(A^c \cap B^c \cap C) = P(E_5) = 0.28.$$

(xvii) Note that

$$A^c \cup B^c = \{E_3, E_4, E_5, E_6\} \cup \{E_1, E_5, E_6\} = \{E_1, E_3, E_4, E_5, E_6\},$$

so the required probability is given by:

$$P(A^c \cup B^c) = P(E_1) + P(E_3) + P(E_4) + P(E_5) + P(E_6)$$

$$= 0.13 + 0.10 + 0.15 + 0.28 + 0.27$$

$$= 0.93.$$

Example 5.12. (HITTING BIRDS IS NOT FUN) The probability that a hunter can hit a flying bird is 0.7. Two hunters shoot two independent arrows from their bows to hit a flying bird. Find the probability that the bird will get hit exactly once.

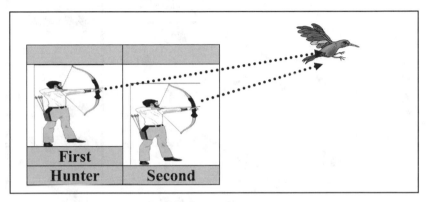

Fig. 5.32. Hunting a bird.

Solution. If two hunters try hitting the bird with two independent arrows, then the total number of outputs (events) is given by:

Fig. 5.33. Total number of outputs with two arrows.

Thus, while shooting two independent arrows at a flying bird, there will be a total of four outcomes given by:

(Hit, Hit), (Hit, Miss), (Miss, Hit), and (Miss, Miss).

A tree diagram to present the sample space for two hunters trying to hit a flying bird is given in **Figure 5.34**.

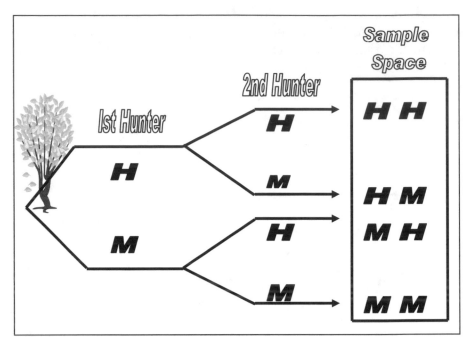

Fig. 5.34. Tree diagram and sample space.

We can assign the relative frequencies (probabilities) to each simple event in the sample space. Note that the probability that a given hunter will hit a bird is 0.7 and will miss a bird is 0.3. Thus, we have:

1st Arrow	2nd Arrow	Event	RF = P(E$_i$) (Independent trials)
H	H	E$_1$: HH	$P(E_1) = 0.7 \times 0.7 = 0.49$
H	M	E$_2$: HM	$P(E_2) = 0.7 \times 0.3 = 0.21$
M	H	E$_3$: MH	$P(E_3) = 0.3 \times 0.7 = 0.21$
M	M	E$_4$: MM	$P(E_4) = 0.3 \times 0.3 = 0.09$
		Sum	**1.00**

Our event of interest that the flying bird will get hit exactly once is:

$$A = \{E_2, E_3\} = \{HM, MH\}$$

Thus, the required probability is given by:

$$P(A) = E(E_2) + P(E_3) = 0.21 + 0.21 = 0.42.$$

Example 5.13. (MILITARY TRAINING) An army officer orders three bombers to hit a tank on the ground as shown below.

Fig. 5.35. Three bombers hitting a tank.

The probabilities that bombers A, B, and C can hit the tank are 0.6, 0.7, and 0.9 respectively. If all three bombers tried independently to hit the tank, then find the probability that:
(a) At least two bombers will hit the tank.
(b) None of bombers will hit the tank.
(c) All three bombers will hit the tank.
(d) Exactly one bomber will hit the tank.
(e) Exactly two bombers will hit the tank.
(f) At most two bombers will hit the tank.
(g) At most one bomber will hit the tank.
Solution. If three bombers A, B, and C try to hit the tank independently, then the total number of outputs (events) is given by:

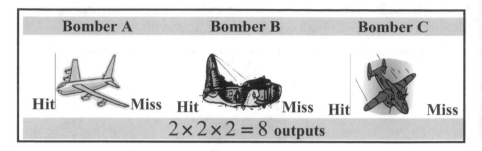

Fig. 5.36. Total number of outputs with three bombers.

Thus, if three bombers try independently, there are eight possible outcomes:

**(Hit, Hit, Hit),
(Hit, Hit, Miss), (Hit, Miss, Hit), (Miss, Hit, Hit),
(Hit, Miss, Miss), (Miss, Hit, Miss), (Miss, Miss, Hit),
and
(Miss, Miss, Miss).**

To assign the relative frequencies to each simple event in the sample space, note that the bombers A, B, and C can hit the tank with probabilities 0.6, 0.7, and 0.9 respectively. Further note that the bombers A, B, and C can miss the tank with probabilities 0.4, 0.3, and 0.1 respectively. Thus, we have the following table:

1st Bomber A	2nd Bomber B	3nd Bomber C	Event	RF = P(E_i) (Independent trials)
H	H	H	E_1: HHH	$P(E_1) = 0.6 \times 0.7 \times 0.9 = 0.378$
H	H	M	E_2: HHM	$P(E_2) = 0.6 \times 0.7 \times 0.1 = 0.042$
H	M	H	E_3: HMH	$P(E_3) = 0.6 \times 0.3 \times 0.9 = 0.162$
M	H	H	E_4: MHH	$P(E_4) = 0.4 \times 0.7 \times 0.9 = 0.252$
H	M	M	E_5: HMM	$P(E_5) = 0.6 \times 0.3 \times 0.1 = 0.018$
M	H	M	E_6: MHM	$P(E_6) = 0.4 \times 0.7 \times 0.1 = 0.028$
M	M	H	E_7: MMH	$P(E_7) = 0.4 \times 0.3 \times 0.9 = 0.108$
M	M	M	E_8: MMM	$P(E_8) = 0.4 \times 0.3 \times 0.1 = 0.012$
			Sum	**1.000**

(a) The event of interest that at least two bombers will hit the tank is:

$$A = \{\text{HHH}, \text{HHM}, \text{HMH}, \text{MHH}\} = \{E_1, E_2, E_3, E_4\}$$

Thus, the probability that at least two bombers will hit the tank is:

$$P(A) = P(E_1) + P(E_2) + P(E_3) + P(E_4)$$

$$= 0.378 + 0.042 + 0.162 + 0.252$$

$$= 0.834.$$

(b) The event of interest that none of the bombers will hit the tank is:

$$A = \{\text{MMM}\} = \{E_8\}$$

Thus, the probability that no bomber will hit the tank is given by:

$$P(A) = P(E_8) = 0.012.$$

(c) The event of interest that all three bombers will hit the tank is:

$$A = \{\text{HHH}\} = \{E_1\}$$

Thus, the probability that all three bombers will hit the tank is:

$$P(A) = P(E_1) = 0.378.$$

(d) The event of interest that exactly one bomber will hit the tank is:

$$A = \{\text{HMM}, \text{MHM}, \text{MMH}\} = \{E_5, E_6, E_7\}$$

Thus, the probability that exactly one bomber will hit the tank is:

$$P(A) = P(E_5) + P(E_6) + P(E_7)$$

$$= 0.018 + 0.028 + 0.108$$

$$= 0.154.$$

(e) The event of interest that exactly two bombers will hit the tank is:

$$A = \{HHM, HMH, MHH\} = \{E_2, E_3, E_4\}$$

Thus, the probability that exactly two bombers will hit the tank is:

$$P(A) = P(E_2) + P(E_3) + P(E_4)$$

$$= 0.042 + 0.162 + 0.252$$

$$= 0.456.$$

(f) The event of interest that at most two bombers will hit the tank is:

$$A = \{HHM, HMH, MHH, HMM, MHM, MMH, MMM\}$$

$$= \{E_2, E_3, E_4, E_5, E_6, E_7, E_8\}$$

Thus, the probability that at most two bombers will hit the tank is:

$$P(A) = P(E_2) + P(E_3) + P(E_4) + P(E_5) + P(E_6) + P(E_7) + P(E_8)$$

$$= 0.042 + 0.162 + 0.252 + 0.018 + 0.028 + 0.108 + 0.012$$

$$= 0.622 .$$

(g) The event of interest that at most one bomber will hit the tank is:

$$A = \{HMM, MHM, MMH, MMM\} = \{E_5, E_6, E_7, E_8\}$$

Thus, the probability that at most one bomber will hit the tank is:

$$P(A) = P(E_5) + P(E_6) + P(E_7) + P(E_8)$$

$$= 0.018 + 0.028 + 0.108 + 0.012$$

$$= 0.166 .$$

Example 5.14. (MARRIED COUPLES LIVE LONGER THAN A DIVORCEE OR NEVER MARRIED PERSON) In a particular society, some people **like (L)** and others **hate (H)** love-marriage (*A marriage without the consent of parents*). Three people are selected at random from the society, and their opinions about love-marriage are noted, so the opinion of each individual is known.

(a) Write down the sample space, **S**.
(b) Write out the outcomes that make up the event:
    A = "At most one person **hates** love-marriage"
(c) Write out the outcomes that make up the event:
    B = "Exactly two people **like** love-marriage"

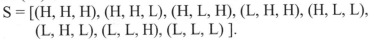

Fig. 5.37.
Ring ceremony.

Solution. (a)
    S = [(H, H, H), (H, H, L), (H, L, H), (L, H, H), (H, L, L),
       (L, H, L), (L, L, H), (L, L, L)].

(b)   A = [(H, L, L), (L, H, L), (L, L, H), (L, L, L)].

(c)   B = [(H, L, L), (L, H, L), (L, L, H)].

Example 5.15. (NEVER GET ADDICTED TO CARDS) Consider a standard deck of cards. Four independent draws are made from a well-shuffled deck using with replacement sampling. What is the probability of getting one card from each suit?

Fig. 5.38. Four suits of a deck.

Solution. We know that a standard deck of cards has four suits: 13 spades, 13 clubs, 13 hearts and 13 diamonds. Note that all four draws are independent, and are made using with replacement sampling, thus the relative frequencies (probabilities) assigned to each type of card are given below:

Card	Frequency	Relative Frequency or probability
♠ Spade	13	$P(\text{Spade}) = \frac{13}{52} = \frac{1}{4}$
♣ Club	13	$P(\text{Club}) = \frac{13}{52} = \frac{1}{4}$
♥ Heart	13	$P(\text{Heart}) = \frac{13}{52} = \frac{1}{4}$
♦ Diamond	13	$P(\text{Diamond}) = \frac{13}{52} = \frac{1}{4}$
Sum	**52**	**1**

Fig. 5.39. Distribution of four suits in a deck.

Note that the event of interest, **A**, is to draw one card of each type; a spade, a club, a heart and a diamond, that is:

A = ♠ Spade ♣ Club ♥ Heart ♦ Diamond

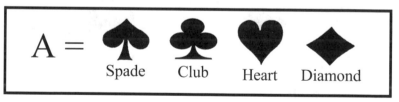

Fig. 5.40. Event of interest.

Due to the independence of draws, the required probability of getting one card from each one of the four suits is given by:

$$P(A) = P(\text{Spade}) \times P(\text{Club}) \times P(\text{Heart}) \times P(\text{Diamond})$$

$$= \frac{1}{4} \times \frac{1}{4} \times \frac{1}{4} \times \frac{1}{4} = \frac{1}{256}.$$

Example 5.16. (SOME JOBS NEED TRAVELING) At a travel and tourism company, it has been observed that:
40% of the employees like to travel by Airplane,
35% of the employees like to travel by Bus,
45% of the employees like to travel by Car,
and it is also noted that:
15% of the employees like to travel by Airplane and Bus,
5% of the employees like to travel by all three facilities,
25% of the employees like to travel by Airplane and Car, and
5% of the employees like to travel only by Bus.

Airplane

Bus   **Car**

Fig. 5.41. Probability is fun.

(a) Draw a Venn diagram with the above information.
Use it along with the probability rules to answer the following:
(b) What percent of the employees like to travel by Car only?
(c) What percent of the employees do not like a traveling job?
Solution. (a) Venn diagram:

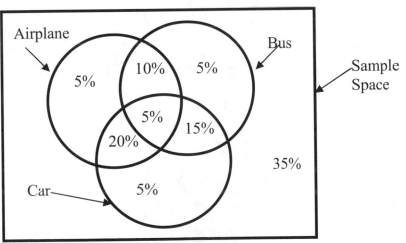

Fig. 5.42. Intersection of three events.

(b) 5% of the employees like to travel by Car only.

(c) 35% of the employees do not like a traveling job.

Example 5.17. (IDENTIFYING CARDS) Consider the experiment of drawing a card from a standard deck.   Let A = "Ace of hearts", B = "Ace of any suit", and C = "King of hearts"
(a) Are the events A and B disjoint? Explain.
(b) Are the events A and C disjoint? Explain.
(c) Are the events B and C disjoint? Explain.
Solution. We are given:

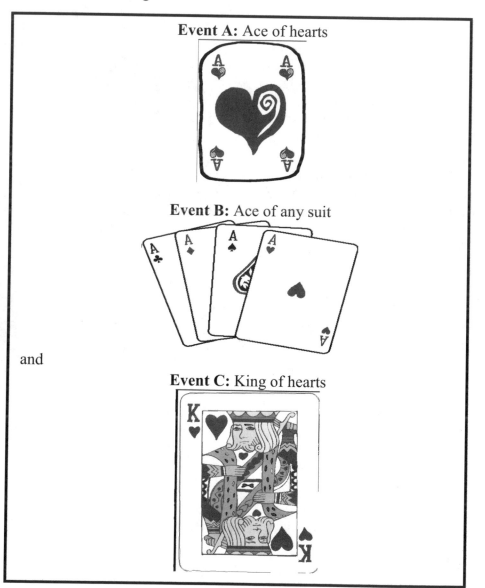

Event A: Ace of hearts

Event B: Ace of any suit

and

Event C: King of hearts

Fig. 5.43. Events of interest.

(a) Clearly the events A and B are **not** disjoint, because the ace of hearts is common to both events. The Venn diagram is as follows:

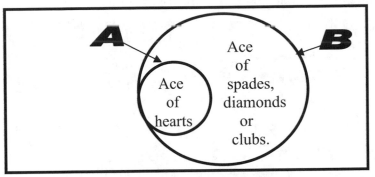

Fig. 5.44. Venn diagram.

(b) The events A and C are disjoint, because the card drawn can either be an "ace of hearts" or the "king of hearts", but not noth. The Venn diagram is as follows:

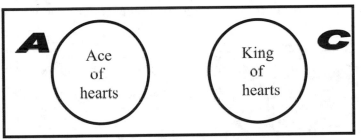

Fig. 5.45. Venn diagram.

(c) The events B and C are also disjoint because the card drawn can either be an "ace of any suit" or the "king of hearts", but not both. The Venn diagram is as follows:

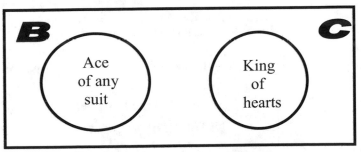

Fig. 5.46. Venn diagram.

Example 5.18. (THINK BEFORE YOU SMOKE) In an experiment studying the dependence of hypertension on smoking habits, the following data set was collected from 180 people.

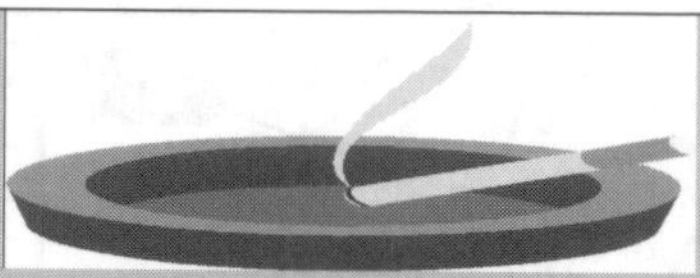

		Smoking status	
		Nonsmoker	Smoker
Hypertension status	Hypertension	30	58
	No hypertension	60	32

Fig. 5.47. Smoking is a bad habit.

(a) What is the probability that a randomly selected individual is experiencing hypertension?

(b) Given that a smoker is selected at random from this group, what is the probability that the person is experiencing hypertension?

(c) Are the events "hypertension" and "smoker" independent? Give supporting calculations.

Solution. We are given:

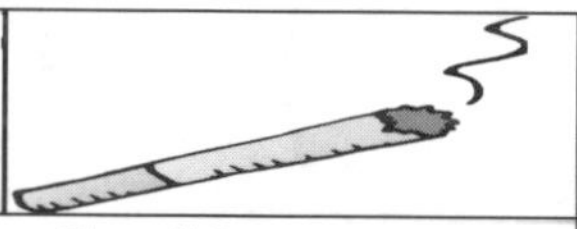

		Smoking status		
		Nonsmoker	Smoker	Total
S T A T U S	Hypertension	30	58	88
	No-hypertension	60	32	92
	Total	90	90	180

Fig. 5.48. Quit smoking.

(a) Probability that a randomly selected person has hypertension:

$$P(\text{Hypertension}) = \frac{\#\text{ of persons having hypertension}}{\text{Total}\,\#\text{ of persons}}$$

$$= \frac{88}{180} = 0.489.$$

(b) Probability that a selected smoker has hypertension:

P(Hypertension | Smoker)

$$= \frac{\text{\# of smokers having hypertension}}{\text{Total \# of smokers}}$$

$$= \frac{58}{90} = 0.644.$$

(c) Are events "hypertension" and "smoker" independent?

If the events "hypertension" and "smoker" are independent then:

P(Hypertension | Smoker) = P(Hypertension)

Now from (a) and (b), we have:

P(Hypertension | Smoker) ≠ P(Hypertension)

so the events "hypertension" and "smoker" are **not** independent.

Example 5.19. (FEEDING BIRDS IN A YARD) Assume that 2 pigeons and 3 hoopers were sitting in a yard. In the meantime, a cat entered the yard. Although the cat was very friendly with the birds, due to the fear of the cat, two birds flew from the yard. Find the probability that between these two birds, there was exactly one pigeon and one hooper.

Fig. 5.49. Always be careful.

Solution. There are two methods for such questions: (a) Ordinary method (b) Method of combinations.

(a) **Ordinary method:** Here the population is made of two pigeons, say P_1 and P_2; and three hoopers, say H_1, H_2, and H_3. Note that only two birds are flying, thus our sample space is given by:

$S =$	P_1P_2	P_1H_1	P_1H_2	P_1H_3	P_2H_1
	P_2H_2	P_2H_3	H_1H_2	H_1H_3	H_2H_3

The event of interest showing exactly one pigeon and one hooper is given by:

$A =$	P_1H_1	P_1H_2	P_1H_3
	P_2H_1	P_2H_2	P_2H_3

Obviously, by the ordinary method, the probability of the event A happening is given by:

$$P(A) = \frac{6}{10} = 0.6.$$

(b) **Method of combinations:** The total number of ways of selecting two distinct birds out of five distinct birds is given by:

$$= \binom{5}{2} = \frac{5!}{2!(5-2)!} = \frac{5!}{2! \, 3!} = \frac{5 \times 4 \times 3 \times 2 \times 1}{2 \times 1 \times 3 \times 2 \times 1} = 10.$$

The number of ways of selecting one pigeon out of two pigeons is given by:

$$= \binom{2}{1} = \frac{2!}{1!(2-1)!} = \frac{2!}{1! \, 1!} = \frac{2 \times 1}{1 \times 1} = 2.$$

The number of ways of selecting one hooper out of three hoopers is given by:

$$= \binom{3}{1} = \frac{3!}{1!(3-1)!} = \frac{3!}{1! \, 2!} = \frac{3 \times 2 \times 1}{1 \times 2 \times 1} = 3.$$

Thus the total number of ways of selecting one pigeon out of two pigeons and one hooper out of three hoopers is given by:

$$= 2 \times 3 = 6.$$

Again, the required probability of having one pigeon and one hooper by the method of combinations is given by:

$$= \frac{6}{10} = 0.6.$$

Thus, both methods (a) and (b) give the same result.

Example 5.20. (RAILWAY CROSSING) A railway engineer designed a railway crossing by making use of two black boxes (B1 and B2) and a selector S1.

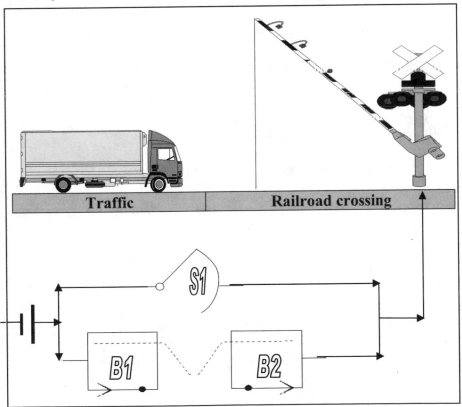

Fig. 5.50. Railway crossings.

He decided to use both black boxes in series and the selector in parallel so that the system would work if either the selector or both the black boxes worked. Both the boxes B1 and B2 and the selector S1 are assumed to work independently. Experience shows that the black box B1 works in 82% of cases, box B2 works in 75% of cases and the selector S1 works in 95% of cases. What is the chance of failure of the railway crossing?

Solution. We are given:

$$P(S1) = 0.95, \quad P(B1) = 0.82, \quad \text{and} \quad P(B2) = 0.75.$$

The railway crossing will work if either the selector S1 will work, or if both the black boxes B1 and B2 will work. Thus, the probability that the system will work is:

$$
\begin{aligned}
P(\text{work}) &= P[S1 \text{ or } (B1 \text{ and } B2)] = P[S1 \cup (B1 \cap B2)] \\
&= P[S1] + P[B1 \cap B2] - P[S1 \cap (B1 \cap B2)] \\
&= P[S1] + P[B1]P[B2] - P[S1]P[B1]P[B2] \\
&= 0.95 + 0.82 \times 0.75 - 0.95 \times 0.82 \times 0.75 \\
&= 0.98075.
\end{aligned}
$$

Thus, the chance of failure of the railway crossing will be:

$$P(\text{failure}) = 1 - P(\text{work}) = 1 - 0.98075 = 0.01925.$$

Example 5.21. (ALWAYS REMEMBER YOUR BLOOD TYPE)
Amy has alleles A and O. Bob has alleles A and B. Assume that children inherit each of a parent's two alleles with the probability of 0.5 and inherit independently from their mother and father.

(a) What is the probability that their first child has blood type A?

(b) What is the probability that their two children have blood type A?

(c) What is the probability that their two children have the same blood type?

Fig. 5.51. Amy and Bob's family.

Solution. (a) Let us make a tree diagram and use the Mendelian inheritance concept as follows:

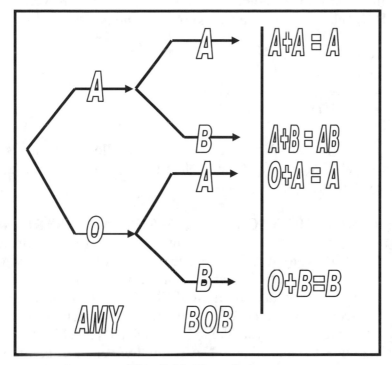

Fig. 5.52. Tree diagram.

Thus, the probability that Amy and Bob's first child will have blood type A is given by:

$$P(\text{First child has blood type A}) = \frac{2}{4} = \frac{1}{2}.$$

(b) Given both children are born independently, the probability that both children have blood type A is given by:

$P(\text{Both children have blood type A})$

$$= P(\text{First child}) \times P(\text{Second child}) = \frac{1}{2} \times \frac{1}{2} = \frac{1}{4}.$$

(c) The probability that both children have the same blood type is:

$$P(\text{same type}) = P(\text{both have A}) + P(\text{both have B}) + P(\text{both have AB})$$

$$= \frac{1}{2} \times \frac{1}{2} + \frac{1}{4} \times \frac{1}{4} + \frac{1}{4} \times \frac{1}{4} = \frac{6}{16} = \frac{3}{8}.$$

5.19 MIXING COLORS

There are three primary colors in nature: Red (R), Green (G) and Blue (B). The following are well known relations:

$$R + G + B = W$$
$$R + G = Y$$
$$R + B = M$$
$$G + B = C$$

where **W** stands for **W**hite, **Y** stands for **Y**ellow, **M** stands for **M**agenta, and **C** stands for **C**yan. Note that Y, M and C are called secondary colors.

Example 5.22. (HOW CAN WE MAKE A CARTRIDGE?) Michael designed a machine to produce color printer cartridges, which generally consist of three secondary colors: Yellow (Y), Magenta (M) and Cyan (C). Michael's machine randomly selects two colors out of three primary colors: Red (R), Green (G) and Blue (B), and makes secondary colors. What proportion of defective cartridges can be produced? Would you like to buy Michael's machine?
Solution. Since the machine selects only two colors out of three primary colors, the sample space will be as follows:

First Color	Second Color	Resultant Cartridge	Decision	
R	R	R	Defective	
R	G	Y	Good	
R	B	M	Good	
G	R	Y	Good	
G	G	G	Defective	
G	B	C	Good	
B	R	M	Good	
B	B	B	Defective	
B	G	C	Good	

Fig. 5.53. Michael's printer cartridge factory.

Thus, the probability of getting a defective cartridge will be:

$$P\left(\text{Defective cartridge}\right) = \frac{3}{9} = \frac{1}{3}.$$

No, I will not buy this machine if 33.33% cartridges are defective.

Example 5.23. (KIDS' LOTTERY STALL IS FUN) At Wonderland in Florida at a kids' lottery stall, there is a scheme to win a monkey by making magenta by randomly mixing three colors in a pot of water. Every player is to choose three packs of colors and mix them. What is the probability that you can win the monkey?

Solution. The sample space based on three primary colors, Red (R), Green (G) and Blue (B) is given below:

Possibility	Pack I	Pack II	Pack III	Resultant Color
1	R	R	R	R
2	R	R	G	Y
3	R	G	R	Y
4	G	R	R	Y
5	R	G	G	Y
6	G	R	G	Y
7	G	G	R	Y
8	R	R	B	M
9	R	B	R	M
10	B	R	R	M
11	R	B	B	M
12	B	R	B	M
13	B	B	R	M
14	G	G	G	G
15	G	G	B	C
16	G	B	G	C
17	B	G	G	C
18	G	B	B	C
19	B	G	B	C
20	B	B	G	C
21	B	B	B	B

22	R	G	B		W
23	R	B	G		W
24	G	R	B		W
25	G	B	R		W
26	B	R	G		W
27	B	G	R		W

Fig. 5.54. Color fun.

Thus, the probability of winning the monkey is given by:

$$P(\text{Monkey}) = \frac{6}{27} = \frac{2}{9}.$$

5.20 IDEA OF SIMULATING PROBABILITIES

A process that can be repeated under similar conditions is called a random process. When such a process of repetition of an experiment leads to a stable result, it is called a successful simulation. For example, when we toss a coin once, the probability of getting heads is 0.5. If we toss a coin 10 times, we may get heads three times out of 10 trials, which shows the probability of getting heads 0.3. If we toss a coin 100 times, we may get heads 45 times, so the probability of heads now becomes 0.45. If we toss a coin 1,000 times, we may get heads 480 times, so the probability of heads now becomes 0.48. If we toss a coin 10,000 times, we may get heads 489 times, so the probability of heads becomes 0.489. Thus, we can see that as the number of trials increases, the probability of heads converges to 0.5.

In the same way, if any event of interest, say A occurs x times out of a large number of n independent and identical trials, then the relative frequency x/n becomes the probability of the occurrence of the event. Such a process is called simulating the probability of an event. If you study your class notes again and again, your score will converge towards an A+ grade. You can try it. For a random process, the set of possible outcomes is known, but the exact outcome for an individual experiment or trial remains unknown or unpredictable. The meaning of simulation is to generate conditions that approximate the actual or real situation. Note that a simulation is a limitation of any random process generated with the help of random number tables or random number generators using computers or calculators. The following rules are generally adopted while doing a simulation:

Rule I. Define an event of interest, say A.
Rule II. Make a rule to simulate a single event A using a random device such as a random numbers table, calculator, or computer.
Rule III. Simulate the experiment several times say n, and note the number x of the event of interest, A happening.
Rule IV. The relative frequency will be the probability of the occurrence of the event, A.

Let us read the following example to understand it.

Example 5.24. (A LONG JOURNEY IS DIFFICULT) Consider tossing four fair coins.
(a) Find the exact probability of observing three heads.
(b) Simulate the probability of three heads using the Random Numbers **Table I** given in the Appendix.
Solution. (a) Tossing four coins:

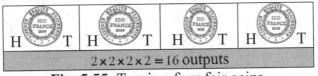

$$2 \times 2 \times 2 \times 2 = 16 \text{ outputs}$$

Fig. 5.55. Tossing four fair coins.

Thus, the sample space **S** is given by:

S =	HHHH	HHHT	HHTH	HTHH
	THHH	HHTT	HTHT	HTTH
	THTH	TTHH	THHT	TTTH
	TTHT	THTT	HTTT	TTTT

The event of interest is given by:

A =	THHH	HHHT	HHTH	HTHH

Thus, the **exact probability** of observing three heads is given by:

$$P(A) = \frac{4}{16} = 0.25.$$

(b) Simulating the probability of three heads:

Rule I. We are interested in observing three heads in an infinite number of tosses of four coins.

Rule II. Assume that if we observe any random number less than 5, then it is equivalent to observing tails, and otherwise it is heads. Let us start from the first row and the first column of the Random Numbers **Table I** given in the Appendix as:

Random No.	3	1	5	9
Outcome	T	T	H	H

Rule III. Consider we repeat the above experiment 10 times:

SIMULATED TRIALS								
Trial-1					Trial-2			
3	1	5	9		3	3	1	0
T	T	H	H		T	T	T	T
Trial-3					Trial-4			
0	0	0	0		0	0	0	9
T	T	T	T		T	T	T	H
Trial-5					Trial-6			
6	1	6	5		0	6	9	5
H	T	H	H		T	H	H	H
Trial-7					Trial-8			
2	7	9	2		9	1	6	4
T	H	H	T		H	T	H	T
Trial-9					Trial-10			
8	5	0	8		1	5	0	8
H	H	T	H		T	H	T	H

Thus, three trials (trial 5, 6 and 9) out of 10 independent and identical trials result in the event of interest A showing that the simulated probability of the occurrence of three heads is 3/10 = 0.30. If we increase the number of trials, then one can observe that this simulated probability will converge to the true probability of 0.25.

Example 5.25. (SIMULATING A DINOSAUR) Two children, Amy and Michael, are playing a game of the possibility of generating a **DINOSAUR** with the help of the Random Numbers **Table I** given in the Appendix.

Michael and Amy	DINOSAUR

Fig. 5.56. Amy and Michael's game.

Amy said to Michael, "The word **DINOSAUR** consists of 8 distinct letters **D, I, N, O, S, A, U** and **R**. Let any distinct random digit between 0 and 9 correspond to a single letter in the sequence of D, I, N, O, S, A, U and R. Thus, from the Random Numbers **Table I** given in the Appendix, if we get all 8 distinct consecutive digits between 0 and 9, then there will be a dinosaur, otherwise not."

The following example explains the appearance of a DINOSAUR.

Random digits	7	3	4	2	6	8	9	0
Result (success)	D	I	N	O	S	A	U	R

Note that the actual chance of selecting 8 distinct random digits out of 10 digits will be:

$$\text{Probability} = \frac{\binom{N}{n}}{N^n} = \frac{\binom{10}{8}}{10^8} = \frac{45}{10^8} = 0.00000045.$$

The following example explains the failure of appearance of a DINOSAUR.

Random digits	7	3	7	2	6	8	9	0
Result (failure)	D	I	D	O	S	A	U	R

Find the empirical probability of a **DINOSAUR** using a simulation.
Solution. Amy and Michael made the following 60 trials starting from the first row and first column of the Random Numbers **Table I**:

Trial	Amy's turn								Trial	Michael's turn							
1	3	1	5	9	3	3	1	0	2	0	0	0	0	0	0	0	9
3	6	1	6	5	0	6	9	5	4	2	7	9	2	9	1	6	4
5	8	5	0	8	1	5	0	8	6	1	9	9	0	1	1	3	1
7	8	0	5	8	8	6	0	7	8	8	9	0	6	5	0	5	9
9	4	6	6	6	4	2	0	0	10	6	7	4	6	8	4	2	2
11	0	3	4	4	5	3	1	7	12	3	3	5	7	6	7	4	4
13	5	5	4	4	9	7	0	6	14	0	0	9	3	0	8	5	2
15	9	5	1	5	8	6	1	3	16	6	3	3	0	1	6	2	8
17	7	4	4	3	9	6	9	3	18	0	4	6	4	2	8	7	4
19	5	0	1	2	4	0	1	2	20	0	5	7	0	1	2	2	6
21	8	7	3	4	2	7	0	6	22	0	8	8	9	4	5	5	6
23	7	1	8	3	6	4	0	0	24	1	8	9	0	5	8	1	6
25	7	3	1	9	4	0	6	0	26	0	9	8	3	1	1	3	5
27	6	8	3	7	7	8	6	8	28	7	2	8	8	7	0	5	3
29	3	2	6	9	3	9	0	9	30	0	4	3	2	2	0	7	2
31	5	4	5	4	9	9	2	0	32	8	1	0	3	6	0	5	6
33	8	8	2	3	7	4	1	5	34	6	9	7	4	6	3	7	1
35	6	9	2	1	6	8	6	6	36	2	1	3	4	8	7	4	0
37	3	5	6	2	1	1	4	2	38	9	5	5	0	0	2	4	0
39	8	7	9	4	7	9	7	9	40	7	9	4	6	0	0	7	9
41	9	4	6	6	4	3	0	5	42	2	7	7	9	1	6	7	3
43	3	0	8	9	6	6	7	6	44	1	1	3	6	6	4	3	5
45	0	3	8	0	7	1	7	8	46	1	2	2	2	3	3	9	7
47	3	1	4	3	4	3	8	0	48	8	6	1	0	1	1	0	5
49	4	2	1	2	9	8	9	2	50	0	9	3	1	0	9	7	1
51	4	4	0	3	0	4	0	3	52	6	0	0	9	3	0	3	7
53	4	2	1	0	3	4	6	3	54	1	3	4	0	2	6	0	4
55	9	5	4	8	4	9	5	6	56	3	5	9	5	6	0	7	6
57	3	8	0	6	5	6	6	3	58	1	3	4	3	9	3	5	2
59	**8**	**2**	**3**	**7**	**4**	**1**	**5**	**6**	60	1	1	6	6	9	0	2	3

At the end of the game, Amy and Michael found that there is only one (59th) trial that makes all 8 digits distinct (**82374156**) out of 60 trials, and hence according to Amy's game, the chance of the appearance of a **DINOSAUR** is 1/60 = 0.01666666. Note that if we increase the number of trials infinitely, then this empirical probability will converge to the real probability 0.00000045. Such large experiments can only be done with fast computers.

LUDI 5.1. (BE CAREFUL) Fill in the blanks:

(a) If an experiment is repeated many times, the proportion of times that an outcome will happen is called -----
    Accident   Simulation   Probability

(b) A list of all possible simple events of an experiment is called-----
    Event   Sample space   Union

(c) The ----- of events A and B is comprised of only those outcomes that are in both event A and event B .
    Union   Intersection   Complement

(d) The ----- of events A and B contains outcomes that are either in event A , or in event B , or in both A and B .
    Union   Intersection   Complement

(e) The ----- of an event A is comprised of all outcomes that are not in the event A .
    Union   Intersection   Complement

(f) If two events A and B have no outcomes in common, then A and B are said to be -----
    Dependent   Independent   Mutually exclusive

LUDI 5.2. (THREE COINS) Consider the tossing of three fair coins:

Fig. 5.57. Three coins.

(a) Give the sample space for this experiment.
(b) What is the probability of getting exactly one head?
(c) What is the probability of getting exactly one tail?
(d) What is the probability of getting at least one head?
(e) What is the probability of getting exactly three heads?
(f) Assume that the first toss is a head and is known. What is the conditional probability of getting exactly two heads?

LUDI 5.3. (VACATION TIME) Students in a certain class were asked whether they fly with different airways. Two of the airways on the questionnaire were American Airways and British Airways.

Fig. 5.58. Boarding Aircraft.

The results were used to obtain the following probabilities:

The probability that a randomly selected student has flown with American Airlines is 0.3. The probability that a randomly selected student has flown with British Airways is 0.4. The probability that a randomly selected student has flown with American Airways, but not with British Airways is 0.2.

(a) Complete the following Venn diagram by filling in each empty box with the probability for the section that the box is in:

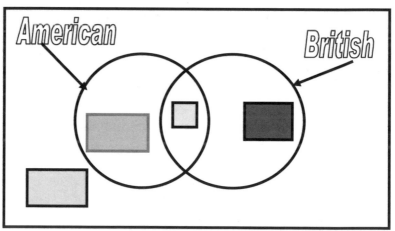

Fig. 5.59. Venn diagram.

(b) What is the probability that a student selected at random will have flown with both American and British Airways?

(c) What is the probability that a student selected at random will have flown neither with American nor with British Airways?

LUDI 5.4. (ELDERS AND KIDS NEED HELP) The results of a study relating dressing well to helping behavior are shown here:

		Dressed	
		Well	Poorly
Helping behavior	Helped	134	126
	Did Not Help	145	155

Fig. 5.60. Helping helps.

(a) What is the probability of randomly selecting a helping person?
(b) Given that the selected person is well dressed, what is the probability that the selected person is helpful?
(c) Given that the selected person is poorly dressed, what is the probability that the selected person is helpful?

LUDI 5.5. (EDUCATION LEVELS) A random sample of adults is classified according to gender and education level attained.

		Education		
		Elementary	Secondary	College
GENDER	 Male	248	238	232
	 Female	255	260	227

Fig. 5.61. Education spreads light.

If a person is selected at random from this group, what is the probability that:

(a) the person is male, given that the person has a secondary education?

(b) the person does not have a college degree, given that the person is female?

(c) Are the events "male" and "secondary education" independent? Give supportive calculations.

LUDI 5.6. (CAR LOVERS) Assume that a survey of automobile ownership was conducted for 250 families in a particular city. The results of the study showing ownership of American and foreign automobiles are given below:

|  | | Do you own an American car? | | |
		Yes	No	Total
Do you own a foreign car?	Yes	50	10	
	No	150		190
	Total			

Fig. 5.62. Car lovers.

(a) Complete the above table.

(b) What is the probability that a randomly selected family will own both American and foreign cars?

(c) What is the probability that a randomly selected family will not own either American or foreign cars?

(d) What is the probability that a randomly selected family will not own an American car, but will own a foreign car?

(e) What is the probability that a randomly selected family will own an American car, but not a foreign car?

(f) Given that a randomly selected family owns an American car, what is the probability that they will also own a foreign car?

(g) Given that a randomly selected family owns an American car, what is the probability that they will not own a foreign car?

(h) Given that a randomly selected family owns a foreign car, what is the probability that they will also own an American car?

(i) Given that a randomly selected family owns a foreign car, what is the probability that they will not own an American car?

(j) Are the events "own an American car" and "own a foreign car" independent?

LUDI 5.7. (TRUCK LOVERS) A survey of automobile ownership was conducted for 300 families in a particular city. The results of the study showing ownership of automobiles of the United States and foreign manufacturers are given below:

		Do you own a U.S. truck?		Total
		Yes	No	
Do you own a foreign truck?	Yes	120	30	
	No	110	?	
	Total			

Fig. 5.63. Truck lovers.

(a) Complete the above table.
(b) What is the probability that a randomly selected family will own both a U.S. and a foreign truck?
(c) What is the probability that a randomly selected family will not own either a U.S. or a foreign truck?
(d) What is the probability that a randomly selected family will not own a U.S. truck, but will own a foreign truck?
(e) What is the probability that a randomly selected family will own a U.S. truck, but not a foreign truck?
(f) Given that a randomly selected family owns a U.S. truck, what is the probability that they will also own a foreign truck?
(g) Given that a randomly selected family owns a U.S. truck, what is the probability that they will not own a foreign truck?
(h) Given that a randomly selected family owns a foreign truck, what is the probability that they will also own a U.S. truck?
(i) Given that a randomly selected family owns a foreign truck, what is the probability that they will not own a U.S. truck?
(j) Are the events "own a U.S. truck" and "own a foreign truck" independent?

LUDI 5.8. (REPEAT YOUR HOMEWORK) Write a short note on simulating probabilities (less than one page).
Hint: For example, if you read your class notes again and again, it will bring your grade closer to an A+.

LUDI 5.9. (COFFEE LOVERS) Based on the orders of many of its customers, a *Big-Apple-Store* manager has found that 40% of the customers are female and 50% of the customers order coffee. We also know that the events "female" and "order coffee" are independent.

(a) Complete the following Venn diagram by filling in each empty box with the probability for the section that the box is in:

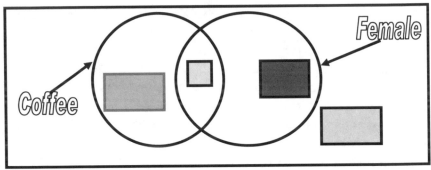

Fig. 5.64. Coffee lovers.

(b) What is the probability that a customer selected at random will be female and will not order coffee?
(c) What is the probability that a customer selected at random will be female and will also order coffee?

LUDI 5.10. (DECK) Consider the experiment of drawing a card from a standard deck. Let A = "Heart", B = "Queen" and C = "Spade".
(a) Are the events A and B disjoint? Draw a Venn diagram.
(b) Are the events A and C disjoint? Draw a Venn diagram.
(c) Are the events B and C disjoint? Draw a Venn diagram.

LUDI 5.11. (TRY IT) Toss a fair coin n times, count the number of heads x observed, compete the following table, and comment.

Trials n	10	100	1,000	2,000
x				
$p(\text{Head})$				

LUDI 5.12. (BEING A DETECTOR IS DIFFICULT) Identify the error, if any, in each statement. If there are no errors, write "no error".

	Statement
(a)	The probabilities that a shoemaker will sell 10, 20, 30, or at least 31 pairs of shoes in a given month, say March, are 0.19, 0.38, 0.29, and 0.15 respectively.
(b)	The probability that there will be snowfall tomorrow is 0.4 and the probability that there will be no snowfall is 0.52.
(c)	The probabilities that the printer will make 0, 1, 2, 3 or at least 4 mistakes/page are 0.19, 0.34, 0.25, 0.43, and -0.29, respectively.
(d)	On a single draw from a standard deck of cards, the probability of selecting a spade is ¼, the probability of selecting a red card (diamond or heart) is ½, and the probability of selecting both a spade and a red card is 1/8.
(e)	If two events are mutually exclusive, their complements will not be mutually exclusive.
(f)	If two events are independent, their complements are also independent.
(g)	If two events cannot occur together, the events are said to be mutually exclusive.

LUDI 5.13. (PRESCRIPTIONS ARE IMPORTANT). A study was conducted to assess the effectiveness of a new medication for seasickness as compared to an old medication. A group of 200 patients were randomly divided into two groups, with the first 100 assigned to receive the new medication and the other 100 assigned to receive the old medication. The data obtained is given below:

		Medication		
		New	**Old**	**Total**
Response	**Improved**	70	58	
	Did not Improve	30	42	
	Total			

Fig. 5.65. See your doctor for medication.

(a) What is the probability of selecting an improved patient?

(b) Given that the patient received the old medication, what is the probability that the patient improved?

(c) Given that the patient received the new medication, what is the probability that the patient improved?

(d) Are the events "new medication" and "improved" independent? Justify your claim.

LUDI 5.14. (THINK BEFORE YOU SPEAK) Let A and B be two events such that $P(A)= 0.5$ and $P(B)= 0.1$.

(a) If the events A and B are independent, find the probability that both events occur together, that is, $P(A \text{ and } B)$.

(b) If A and B are mutually exclusive events, find the probability that at least one of the two events occur, that is, $P(A \text{ or } B)$.

LUDI 5.15. (QUIT SMOKING) A simple random sample of size 130 was selected from a total adult male (M) population of 2060 and a simple random sample of size 120 was selected from a total adult female (F) population of 1200. The proportion of smokers among 130 males was found to be 0.60, while the total number of smokers in the overall sample was found to be 90. Complete the following table.

	Smoking Habit	Smokers (S) 	Non-smokers (NS) 	Total
G E N D E R	**Male**	78		
	Female 			
	Total	90		250

Fig. 5.66. Smoking is a bad habit.

Assume that one adult is selected out of the above 250 adults classified in the 2x2 contingency table. Find the following:

(a) $P(S)$ (b) $P(M)$ (c) $P(NS)$ (d) $P(F)$ (e) $P(S|M)$

(f) $P(S|F)$ (g) $P(F|S)$ (h) $P(M|S)$ (i) $P(S|NS)$ (j) $P(NS|M)$

(k) $P(NS|F)$ (l) $P(F|NS)$ (m) $P(M|NS)$ (n) $P(M|F)$

(o) $P(NS|S)$ (p) $P(F|M)$, (q) $P(S \cap M)$ and (r) $P(NS \cap M)$

LUDI 5.16. (SUPPORT SPORTS) The probability of a 0.6 kg ball falling in a certain 0.6×0.6 square km region of ground is 0.6, so the unit of measurement of probability will be:

(a) Kg
(b) Kg^2
(c) No units, a pure number
(d) Km^2
(e) none of these.

Fig. 5.67. Hitting a ball.

LUDI 5.17. (MILITARY TRAINING) The probability of a bomber hitting a tank is 0.8. What is the probability that the tank will be hit by at least one of two independent bombers?

(a) 0.64
(b) 0.04
(c) 0.96
(d) 1.00
(e) none of these.

Fig. 5.68. Military training.

LUDI 5.18. (ROLLING TWO DICE) Two ordinary six-sided dice are tossed.

Dice	Complete the following table					
	(1, 1)	(1, 2)	-	-	-	-
	-	-	-	-	-	-
	-	-	-	-	-	-
	-	-	-	-	-	-
	-	-	-	-	-	-
	-	-	-	-	-	(6, 6)

Fig. 5.69. Outputs of two dice.

(a) What's the probability that the number 5 appears on both dice?
(b) What's the probability that same number appears on both dice?
(c) What's the probability that the sum of both numbers is at most 4?
(d) What's the probability that the sum of both numbers is at least 5?
(e) Given that the sum of the two numbers shown is 8, what's the conditional probability that the number on the first die is less than the number on the second die?

LUDI 5.19. (FEEDING BIRDS IN A YARD) Assume that 6 parrots, 4 crows, and 5 pigeons are sitting in a yard. Due to their fear of a certain noise, four birds flew from the yard. Find the probability that among these birds there is at least one of each type.

(a) 0.4720
(b) 0.5000
(c) 0.5275
(d) 1.0000
(e) none of these.

Fig. 5.70. Feeding birds is fun.

LUDI 5.20. (LEARN TO SUCCEED INDEPENDENTLY) If A and B are independent events, then:

(a) $P(A \cap B) = P(A)P(B)$

(b) $P(A \cap B) = 0$

(c) $P(A \cap B) = P(A) + P(B)$

(d) $P(A \cap B) = P(A) / P(B)$

(e) none of these.

LUDI 5.21. (MATHEMATICS IS FUN) A problem in mathematics is given to three students: Andy, Bob and Chris. Their chances of solving the problem are 0.50, 0.75 and 0.25 respectively. What is the probability that the problem will be solved if they all try independently?

(a) 1.50
(b) 22/32
(c) 29/32
(d) 0.75
(e) none of these.

Fig. 5.71. Sitting for an exam.

LUDI 5.22.(DAY AND NIGHT ARE MUTUALLY EXCLUSIVE) If A and B are mutually exclusive events then:

(a) $P(A \cup B) = P(A)P(B)$

(b) $P(A \cup B) = 0$

(c) $P(A \cup B) = P(A) + P(B)$

(d) None of these.

LUDI 5.23. (RATE YOUR PROFESSOR) The following table shows a 2×2 contingency table of responses of 120 students regarding the teaching style of a teacher:

GENDER	Response	Good 	Poor 	Total
	Male 	60	10	
	Female 	46	4	
	Total			

Fig. 5.72. Be nice to your teacher.

Are gender and response independent?

(a) Yes, because males and females came from different 'planets'.
(b) Yes, because the calculated chi-square value 0.88 is very small at a 5% level of significance.
(c) No, because $P(\text{Gender} \mid \text{Response}) \neq P(\text{Gender})$ for every category of gender and response.
(d) Cannot apply chi-square test because one of the observed frequency is less than 5.
(e) None of these.

LUDI 5.24. (PROTECT OUR VALUABLE ANIMALS) Three cages of animals contain 3 elephants and 1 lion, 2 elephants and 2 lions, and 1 elephant and 3 lions, respectively. One animal is selected at random from each cage. What is the chance that the three selected consist of 1 elephant and 2 lions?

(a) 11/32
(b) 12/32
(c) 13/32
(d) 14/32
(e) none of these.

Cage I   Cage II   Cage III

Fig. 5.73. Kings of forests.

LUDI 5.25. (PUT BAD EVENTS IN YOUR LIFE BEHIND YOU)
Consider the following Venn diagram:

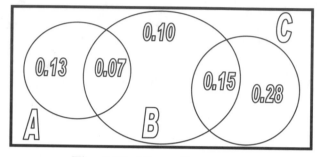

Fig. 5.74. Venn diagram.

Let $\overline{A}$, and $\overline{C}$ denote the complement events.

What is the $P(\overline{A} \cup B \cup \overline{C})$?
(a) 0.27
(b) 0.73
(c) 0.10
(d) 1.00
(e) none of these.

LUDI 5.26. (HITTING BIRDS IS NOT FUN) The probability that a
hunter can hit a flying bird is 0.8. Out of three hunters, two hunters
shoot their independent arrows from their bows to hit the flying bird.
Find the probability that the bird will get hit exactly once out of two
arrows.

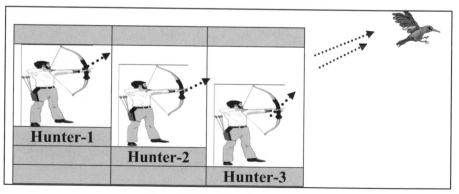

Fig. 5.75. Hunting a bird.

LUDI 5.27. (HITTING BALOONS IS FUN) Bob, Chris, and Don are at the "pop the balloon" booth at the fair. The chances of Bob, Chris, and Don hitting the balloon are 0.7, 0.8, and 0.6 respectively. If they all try to hit one balloon simultaneously and independently, what is the probability that the balloon will get hit exactly once?

Fig. 5.76. Three boys hitting a balloon.

LUDI 5.28. (PIZZA LOVERS) Eight pizza delivery vans $v_1, v_2, \ldots, v_8$ are available in a local area of St. Cloud, and the vans are divided into three groups, A, B, and C as:

$$A = \{ v_1, v_2, v_5, v_6, v_7 \}; B = \{ v_2, v_3, v_6, v_7 \} \text{ and } C = \{ v_6, v_8 \}.$$

(a) Draw a Venn diagram to show these three groups.

(b) The probabilities that each delivery van will be the one to deliver the next pizza order are:

$P(v_1) = 0.16,$
$P(v_2) = P(v_3) = P(v_4) = 0.08,$
$P(v_5) = P(v_6) = P(v_7) = P(v_8) = 0.15$

(c) Find the probabilities of the following groups:

(i) $\overline{B}$ (ii) $B \cap C$ (iii) $A \cup B$
(iv) $\overline{A} \cup B$ (v) $\overline{A \cup B}$.

Fig. 5.77. Pizza delivery

LUDI 5.29. (WHEATSTONE BRIDGE) An engineer connected an oscillating fan through a Wheatstone Bridge made of five independent relays R_i, $i = 1,2,3,4,5$ as show below:

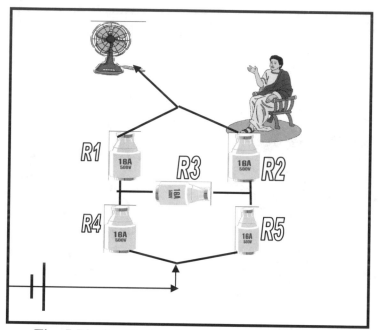

Fig. 5.78. Oscillating fans are good in the summer.

Assume that the probability of a relay being closed is θ and that the relays are open or closed independently of each other. Find the probability that the oscillating fan will work properly. If $\theta = 0.75$, then find the corresponding probability.

LUDI 5.30. (DIE DOES FAMILY PLANNING) Say that you and your spouse are making plans to have three children and you would like to have at least one boy. Assume that under natural conditions the chance of getting a boy is 0.5 and the chance of a girl is also 0.5. Roll a fair die and if it shows an odd number, then it will be a girl, and if it is an even number, it will be a boy. Simulate the probability of having (a) at least one boy, and (b) exactly one boy.

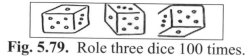

Fig. 5.79. Role three dice 100 times.

LUDI 5.31. (HOW MANY SNAKES BEFORE A DINOSAUR?)

Two children, Amy and Michael, are playing a more complex game of the possibility of generating **SNAKE** and **DINOSAUR** with the help of the Random Numbers **Table I** given in the Appendix.

| Michael and Amy | SNAKE | DINOSAUR |

Fig. 5.80. Amy and Michael's complex game.

Amy said to Michael, "The word **SNAKE** consists of 5 distinct letters: **S, N, A, K,** and **E**. Let any distinct random digit between 0 and 9 correspond to a single letter in the sequence of **S, N, A, K,** and **E**. Thus, from the Random Numbers **Table I**, if we get all 5 distinct consecutive digits between 0 and 9 (inclusive), then there will be a **SNAKE**, otherwise not."

The following example explains the appearance of a **SNAKE**.

Random digits	7	3	4	2	6
Result (success)	S	N	A	K	E

Note that the actual chance of selecting 5 distinct random digits out of 10 digits will be:

$$\text{Chance (SNAKE)} = \frac{\binom{N}{n}}{N^n} = \frac{\binom{10}{5}}{10^5} = - - - - - - -.$$

The following example explains the failure of a **SNAKE**.

Random digits	7	3	4	7	6
Result (failure)	S	N	A	S	E

In the same way, one can define the appearance (or non appearance) of a **DINOSAUR,** thus:

$$\text{Chance}(\text{DINOSAUR}) = \frac{\dbinom{N}{n}}{N^n} = \frac{\dbinom{10}{8}}{10^8} = -------$$

(I) (a) Complete the following table by assuming that Amy and Michael made 20 trials starting from the first row and first column of the Random Numbers **Table I** and **moved along the rows** as follows:

Trial	Amy's turn					Trial	Michael's turn				
1	3	7	5	6	9	2	6	1	6	0	2
3	0	1	0	3	3	4	?	?			
5						6					
7						8					
9						10					
11						12					
13						14					
15						16					
17						18					
19						20					

(b) At the end of the game, find the empirical probability of obtaining a **SNAKE** out of the 20 trials and comment.
(i) How many **SNAKEs** are generated by Amy during 20 trials?
(ii) How many **SNAKEs** are generated by Michael during 20 trials?
(iii) Who wins: Amy or Michael?

(II) (a) How many **SNAKEs** are expected before a **DINOSAUR?**
(**Hint:** Divide the exact probability (or chance) of a **SNAKE** with that of a **DINOSAUR**).
(b) Continue the game until a **DINOSAUR** appears. (**Rule:** Move along the rows as above, and if the Random Numbers **Table I** ends, then continue your game from the first row and first column by moving along the columns)
(i) How many **SNAKEs** came for Amy before a **DINOSAUR?**
(ii) How many **SNAKEs** came for Michael before a **DINOSAUR?**
(iii) Who wins: Amy or Michael?

LUDI 5.32. (A VISIT TO A FAIR) Assume that tickets for the Airplane, Helicopter, Elephant, and Horse rides at the fair are available from a local shop.

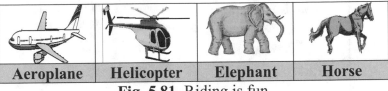

Fig. 5.81. Riding is fun.

Amy and Michael met at the fair and both had already bought two tickets independently without discussing it with each other.
(a) What is the chance that they both will have horse ride tickets?
(b) What is the chance that they both will have the same ride tickets?
(c) What is the chance that they both will have different ride tickets?
(d) What is the chance that they will have horse and helicopter ride tickets?

LUDI 5.33. (HOT AIR BALOONS LOOK CUTE) Four hot air balloons are flying as shown:

Fig. 5.82. Hot air balloons.

All the balloons land.
(a) What is the chance that two hot air balloons will land together?
(b) What is the chance that they will land one by one?
(c) What is the chance that any three will land together?
(d) What is the chance that all four will land together?
Hint: Total number of cases will be:

$$\sum_{J=1}^{4} \binom{4}{J} = \binom{4}{1} + \binom{4}{2} + \binom{4}{3} + \binom{4}{4}$$

LUDI 5.34. (RELYING ON THE LOTTERY IS RISKY) In a lottery, a person wins a toy if they draw a ticket with a toy picture on it.

Fig. 5.83. Toys.

Amy went to a fair and found that only four tickets were left. Also, only four toys teddy bear, block, rubber duckie, and the horn were left. She tried her luck to win the teddy bear by randomly drawing only two independent tickets:

(a) What is the chance that she wins the teddy bear?

(b) What is the chance that she wins the horn?

(c) What is the chance that she wins the teddy bear and the horn?

LUDI 5.35. (KITE LOVERS) Three people: Amy, Bob, and Cindy are participating in a kite competition.

Fig. 5.84. Flying kites is fun.

Assume that only two kites can compete at a time to try to cut each other's thread and that only one of them will cut the others with an equal chance. What is the probability that Bob will cut both Amy and Cindy's kites?

LUDI 5.36. (TODDLERS DO PROBABILITY) Patrick has four dice each having a picture of a puppy on them. Patrick likes to see the puppy on all four dice, but unfortunately he rolled all of these:

Fig. 5.85. Patrick likes puppies.

What is the chance that the picture of the puppy will appear on all four dice face up?

LUDI 5.37. (JUMPING ROPE IS GOOD EXERCISE) Anna and Bob participate in a rope jumping competition at the fair. They go on jumping until either they touch the rope, or until they have been jumping for 30 minutes.

Fig. 5.86. Anna and Bob are jumping.

(a) What is the probability that Anna touches the rope?
(b) What is the probability that Bob touches the rope?
(c) What is the probability that both Anna and Bob together touch the rope?
(d) What is the probability that neither of them will touch the rope?

LUDI 5.38. (REMEMBER YOUR CHILDHOOD) Tom has seven dice.

Fig. 5.87. Tom likes towers.

In how many ways can these dice be arranged to make a tower like Tom is trying to make?

LUDI 5.39. (YOUR CHILDHOOD NEVER COMES BACK) One day Michael was playing with his remote control plane. Andy and Bob did not have enough money to buy such an expensive toy, so they made slingshots to hit Michael's plane.

| Michael with a toy plane | Andy and Bob with slingshots |

Fig. 5.88. Andy, Bob and Michael's childhood story.

If both Andy and Bob tried only once independently and simultaneously and each has a 50% chance of hitting the plane, find the probability that:

(a) neither of them hit Michael's plane.
(b) exactly one of them hits Michael's plane.

LUDI 5.40. (TRY TO TOP YOUR CLASS) Andy, Bob and Chris rolled three independent and identical tops at the same time, and recorded during different phases of the game if any stopped or not.

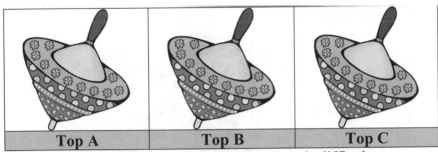

| Top A | Top B | Top C |

Fig. 5.89. Competition among tops is difficult.

(a) Construct the sample space by completing the following table:

Events	Phases of the game		
1	A is running	B is running	C is running
2	A is running	B is running	C is stopped
3			
4			
5			
6			
7			
8			

(b) What is the probability that all three tops stopped?

(c) What is the probability that all three tops are still running?

(d) Given that tops B and C are stopped, what is the probability that top A is still running?

(e) Given that top B is stopped, what is the probability that top A is still running?

(f) Given that top B is stopped, what is the probability that top A is also stopped?

(g) Given that tops B and C are stopped, what is the probability that top A is also stopped?

(h) Given that tops A and C are stopped, what is the probability that top B is also stopped?

LUDI 5.41. (SOFT DRINKS ARE GOOD FOR YOUR HEALTH)
Lisa and Michael meet each other on a patio of a drink bar. Three
types of soft drinks are available. They are grape juice, lemonade and
mango juice. Both take one type of soft drink:

Fig. 5.90. A soft drinks bar.

(a) What is the probability that both will have the same type of
drink?
(b) What is the probability that both will have different types of
drinks?
(c) What is the probability that both will have mango juice?
(d) What is the probability that both will have lemonade?
(e) What is the probability that both will have grape juice?
(f) What is the probability that one will have grape juice and the
other will have mango juice?
(g) What is the probability that one will have grape juice and the
other will have lemonade?
(h) What is the probability that Lisa orders lemonade and Michael
orders mango juice?

LUDI 5.42. (CATCH GOOD FRIENDS: A FRIEND IN NEED IS A FRIEND INDEED) Amy, Bob and Cindy independently try to catch a butterfly in a single sweep of their nets.

Fig. 5.91. Butterfly catchers.

Find the probability that:
(a) none of them will catch any butterflies.
(b) two of them will catch butterflies.
(c) all three will catch butterflies.

LUDI 5.43. (BRUSH YOUR TEETH EVERYDAY) A gumball machine in an arcade has 16 red, 14 yellow, 13 green and 12 blue candies.

Fig. 5.92. Gumball machine.

A coin of 25 cents returns four candies. If a person inserts 25 cents once:
(a) what is the probability of drawing one of each color?
(b) what is the probability of drawing two red and two blue?

LUDI 5.44. (PET LOVERS) Assume that a survey of pet owners was conducted for 740 families in a particular city. The study was conducted to find out how many animal and bird keepers are in the city.

	Cat 	Dog 	Rabbit 	Total
Parrot 	100	140	70	
Pigeon 	40	170	80	
 Duck	55	65	20	
Total				

Fig. 5.93. Animal and bird lovers.

(a) Complete the above table.
(b) What is the probability that a randomly selected family will own both a pigeon and a dog?
(c) What is the probability that a randomly selected family will own both a pigeon and a cat?
(d) What is the probability that a randomly selected family will own both a pigeon and a rabbit?
(e) What is the probability that a randomly selected family will own both a parrot and a cat?
(f) What is the probability that a randomly selected family will not own either a pigeon or a dog?
(g) What is the probability that a randomly selected family will not own a pigeon, but will own a dog?
(h) Given that a randomly selected family owns a parrot, what is the probability that they will also own a rabbit?
(i) Given that a randomly selected family owns a cat, what is the probability that they will not own a pigeon?
(j) Are the events "own a pigeon" and "own a cat" independent?
(k) Are the events "own a parrot" and "own a rabbit" independent?

LUDI 5.45. (TOSSING TWO DICE) A white and a colored die are tossed. The possible outcomes are shown in the following figure:

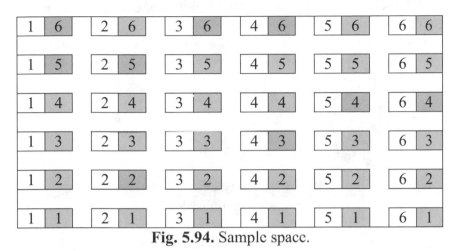

Fig. 5.94. Sample space.

(a) Identify the events A = [Sum = 6], B = [Sum =7], C = [Sum is even], D = [Same number on each die].

(b) Assuming both dice are "fair," assign probability to each simple event.

(c) Find P(A), P(B), P(C) and P(D).

(d) Find $P(A \cap B)$, $P(A \cap C)$ and $P(B \cap C)$.

(e) Find $P(A \cup B)$, $P(A \cup C)$ and $P(B \cup C)$.

(f) Find $P(A \cup B \cup C)$ and $P(A \cap B \cap C)$.

LUDI 5.46. (A GOOD MATCH IS RARE) Match the given probability of an event *A* with the given appropriate verbal description (more than one matching may apply).

Probability			Match	
(a)	0.95	(i)	No chance of occurrence
(b)	0.03	(ii)	Very likely to occur
(c)	4.0	(iii)	As much chance of happening as not
(d)	-0.23	(iv)	Very little chance of occurrence
(e)	0.41	(v)	May occur but by no means sure
(f)	0.50	(vi)	An incorrect assignment
(g)	0.00	(vii)	Sure to occur
(h)	1.00	(viii)	None of the above applies

LUDI 5.47. (WHICH CAME FIRST: THE CHICKEN OR THE EGG?) At a bus stop shop, while traveling from NY to California, three types of eggs are available: golden eggs, runny eggs, and scrambled eggs.

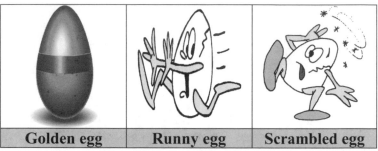

| Golden egg | Runny egg | Scrambled egg |

Fig. 5.95. Breakfast lovers.

Three passengers, Gina, Renee, and Stephanie, go there for breakfast. Find the probability that:
(a) all three will order golden eggs.
(b) all three will order scrambled eggs.
(c) all three will order runny eggs.
(d) Gina will order a golden egg, Renee will order a runny egg, and Stephanie will order a scrambled egg.
(e) none of the three will order golden eggs.

LUDI 5.48. (PRACTICE FORMULAE) (a) Let A and B be two events such that $P(A \cup B) = \frac{5}{6}$, $P(A \cap B) = \frac{1}{3}$, and $P(\bar{B}) = \frac{1}{2}$. Determine $P(A)$ and $P(B)$. Are the events A and B independent?

(b) Let A and B be two events such that $P(A \cup B) = \frac{3}{4}$, $P(A \cap B) = \frac{1}{4}$, and $P(\bar{A}) = \frac{2}{3}$. Find $P(A)$, $P(B)$, and $(A \cap \bar{B})$.

(c) For the two events A and B, let $P(A) = P(A \mid B) = \frac{1}{4}$ and $P(B \mid A) = \frac{1}{2}$. Are the events A and B independent?

(d) For the two events A and B, if $P(A \mid B) = P(B \mid A)$, show that $P(A) = P(B)$. When will the events A and B be independent?

(e) For the two events A and B, if $P(A) = P(B)$, show that $P(A \mid B) = P(B \mid A)$. When will the events A and B be independent?

LUDI 5.49. (SOME JOBS NEED TRAVELING) At a travel and tourism company, it has been observed that:

50% of the employees like to travel by airplane,
35% of the employees like to travel by bus,
45% of the employees like to travel by car, and it is also noted that:
17% of the employees like to travel by airplane and bus,
7% of the employees like to travel by all three facilities,
20% of the employees like to travel by airplane and car, and
6% of the employees like to travel only by bus.

Airplane

Bus   Car

Fig. 5.96. Probability is fun.

(a) Draw a Venn diagram with the above information.
Use it along with the probability rules to answer the following questions:
(b) What percentage of the employees like to travel by car only?
(c) What percentage of the employees do not like to travel?
(d) What percentage of employees like to travel by airplane only?
(e) What percentage of employees like to travel by both car and bus?
(f) What percentage of employees like to travel either by car or by bus?
(g) What percentage of employees like to travel by bus only?

LUDI 5.50. (ALWAYS REMEMBER YOUR BLOOD TYPE)

Angie has alleles A and O. Bob has alleles B and O. They have three kids.

(a) What is the probability that all three kids have blood type O?
(b)What is the probability that all three kids have the same blood type?
(c) What is the probability that all the three kids have blood type B?
(d) What is the probability that the first child will have either blood type O or B?

Fig. 5.97. Angie and Bob's family.

LUDI 5.51. (GOOD GRADES ARE ASSETS) In a senior class of statistics, it has been observed that:

55% of the students have A grades,
25% of the students have B grades,
45% of the students have C grades,

 and also noted that:

15% of the students have A and B grades,
 5% of the students have all three A, B, and C grades,
25% of the students have A and C grades,
 5% of the students have only B grades.

Fig. 5.98. Grades.

(a) Complete the following Venn diagram to present the above information.

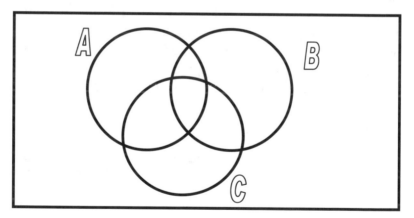

Fig. 5.99. Venn diagram.

Use it along with the probability rules to answer the following questions:

(b) What percent of the students have only A grades?
(c) What percent of the students have a D, F, or an I grade?

LUDI 5.52. (NEVER GIVE FALSE STATEMENTS IN COURT)
Once Amy and Bob were very angry due to some unavoidable circumstances, so they went to court to get an answer to their argument.

Fig. 5.100. Courts are made for justice.

Amy speaks the truth in 75% of cases and Bob in 80% of cases. What is the probability that they will contradict each other in court?
Choice: (a) 6/20 (b) 5/20 (c) 7/20 (d) 8/20 (e) None of these.

LUDI 5.53. (MOBILE POPULATIONS) Angie bought two cages of birds: cage I contains 5 parrots and 1 pigeon, and cage II contains 6 parrots.

Fig. 5.101. Birds like to fly.

One day, Angie forgot to lock both cages and two birds flew from cage I to cage II. Then two birds flew back from cage II to cage I. Assume all the birds have an equal chance of flying.

(i) Find the probability that the pigeon is still in cage I.
Choice: (a) 1/6 (b) 2/4 (c) 3/4 (d) 2/6 (e) None of these.

(ii) Find the probability that the pigeon is now in cage II.
Choice: (a) 1/3 (b) 2/4 (c) 1/4 (d) 3/6 (e) None of these.

LUDI 5.54. (FLOWER LOVERS) In a flower shop, there are two lotus plants, three zinnia, and four rose plants available based on a lottery draw as Amy enters the shop. Fortunately, there are only nine tickets left and none of them are losers.

Fig. 5.102. Do not pluck the flowers.

Amy bought four lottery tickets. What is the probability that she wins exactly two roses?
Choice: (a) 9/21 (b) 8/21 (c) 10/21 (d) 11/21 (e) none of these.

LUDI 5.55. (CHERRY LOVERS)   A basket of 100 cherries contains 10 rotten cherries.

| Cherries | Chandra with a bowl |

Fig. 5.103. Chandra likes cherries.

Chandra randomly picks 5 cherries and puts them in her bowl. What is the chance that her bowl has at least one rotten cherry?

Choice: (a) 1 (b) 0 (c) $1-\left(\dfrac{^{10}C_0 \; ^{90}C_5}{^{100}C_5}\right)$ (d) $1-\left(\dfrac{^{10}C_5 \; ^{90}C_0}{^{100}C_5}\right)$

(e) none of these.

LUDI 5.56. (HOME COOKED FOOD IS THE BEST) Beth goes to a *punjabi* restaurant and orders a breakfast with an omelette made of four eggs. *Sardar-Ji* (owner) had one dozen eggs in the refrigerator, but did not realize that two of them were rotten. He randomly picked up four eggs and served the breakfast to Beth.

| Beth | Punjabi Restaurant |

Fig. 5.104. Too much fast food may not be good.

(i) What is the probability that Beth's breakfast has no rotten eggs?
Choice: (a) 0/12 (b) 4/12 (c) 210/495 (d) 2/4 (e) None.
(ii) What is the probability that Beth's breakfast has exactly one rotten egg?
Choice: (a) 1/12 (b) 4/12 (c) 240/495 (d) 1/4 (e) None.
(iii) What is the probability that Beth's breakfast has both rotten eggs?
Choice: (a) 1/12 (b) 4/12 (c) 45/495 (d) 1/4 (e) None.

LUDI 5.57. (MUSIC LOVERS) In a competition among three types of musical instruments, it has been observed:

| Accordion Player | Banjo Player | Conga Player |

Fig. 5.105. Music is must for life.

40% of the listeners liked music by the accordion player,
35% of the listeners liked music by the banjo player,
50% of the listeners liked music by the conga player,
and it is also noted that:
15% of the listeners liked both the accordion and the banjo player,
5% of the listeners liked all three music players,
25% of the listeners liked the accordion and the conga player,
5% of the listeners liked only the banjo player.
What percent of the listeners did not like the three music players?
Choice: (a) 0% (b) 100% (c) 30% (d) 50% (e) None.

LUDI 5.58. (NEWS IS IMPORTANT)
Three newspapers "American Express," "British News," and "Canadian Express" are available at a local stand. A survey shows that 20% read American Express, 16% read British News, 14% read Canadian Express, 8% read both American Express and British News, 5% read American Express and Canadian Express, 4% read British Express and Canadian Express and 2% read all three newspapers.

Fig. 5.106. A stand.

The shopkeeper has a system to keep a record of the phone numbers of all his customers. Assume the shopkeeper randomly selected a phone number and talked to the customer about these three newspapers. Complete the following Venn diagram:

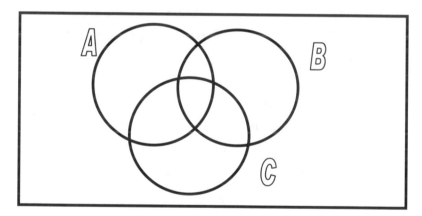

Fig. 5.107. Venn diagram.

(a) What is the probability that a randomly selected customer reads none of these three newspapers?
(b) What is the probability that a randomly selected customer reads all of these three newspapers?
(c) What is the probability that a randomly selected customer reads only American Express?
(d) What is the probability that a randomly selected customer reads only British News?
(e) What is the probability that a randomly selected customer reads only Canadian Express?

LUDI 5.59. (SAFETY SAVES) An engineer designed a special type of antenna to stay in touch with an airplane during its flight time. The antenna is made of two independent relay systems: System 1 and System 2. One system is based on a standard Wheatstone bridge whereas the second system is specially designed for more safety.

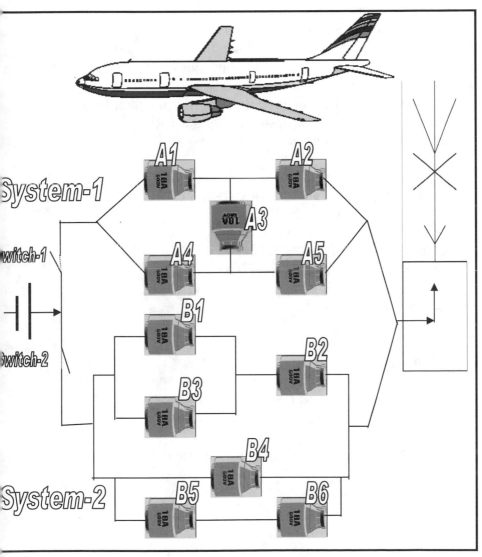

Fig. 5.108. Neighborhood watch.

System 1, the Wheatstone bridge, is made of five independent relays, A_i, $i = 1, 2, 3, 4, 5$. Let P be the probability that any one of these five relays works well independently.

System 2 is made of six independent relays, B_i, $i = 1, 2, 3, 4, 5, 6$, as shown in the figure. Let T be the probability that any of these six relays works well independently.

(a) Find the probability in terms of P and T that the antenna works well. Derive its value if $P = 0.7$ and $T = 0.8$.

(b) Consider the situation that due to some fault in system 1, it started making some noise, so the operator had to shut it off. Find the probability that the antenna will still work.

(c) Consider another situation that due to some fault in system 2, it started to make some noise, so the operator had to shut it off. Find the probability that the antenna will still work.

LUDI 5.60. (ALWAYS REMEMBER YOUR BLOOD TYPE)

Amy has alleles A and B. Bob has alleles B and O. Assume that children inherit each of a parent's two alleles with a probability of 0.5 and that they inherit independently from their mother and father.

(a) Make a tree diagram to display the sample space.

(b) What is the probability that their first child has blood type A?

Fig. 5.109. Amy and Bob's family.

(c) What is the probability that their two children have blood type A?

(d) What is the probability that their two children have the same blood type?

(e) What is the probability that their first child will have blood type O?

6. DISCRETE DISTRIBUTIONS

6.1 INTRODUCTION

In this chapter, we discuss discrete random variable and two discrete distributions: binomial distribution and Poisson distribution.

6.2 DISCRETE RANDOM VARIABLE

A variable X is said to be a discrete random variable if it assumes a non-negative integer value x corresponding to the outcome of an experiment, chance, or random event.

For example, the number of defects, X, in a randomly selected piece of furniture can be 0, 1, 2 etc., and we say:

$$x = 0, 1, 2, 3, ...\infty$$

Fig. 6.1. Bedroom.

In other words, $x = 0$ means there is no defect, $x = 1$ means there is one defect, $x = 2$ means there are two defects and so on. Note that uppercase X is a random variable, whereas lowercase x is a particular value of X.

6.3 DISCRETE PROBABILITY DISTRIBUTION

The discrete probability distribution of a quantitative discrete random variable, X, is a **formula**, **table**, or **graph** that provides all possible values of X along with $P(x)$ or $P(X = x)$, that is, the probability associated with each value x of X. Note the use of lowercase x and uppercase X in the formula $P(X = x)$.

(**a**) **Formula:** In mathematical notation, a random variable X is said to follow a discrete probability distribution if:

$$0 \le P(x) \le 1 \text{ and } \sum_x P(x) = 1$$

where $\sum_x$ denotes the sum over all possible values of X. Note that $P(x)$ is also called the probability mass function (p.m.f) of X.

(b) Table: In a table, a random variable X is said to follow the discrete probability distribution mass function if it takes the form:

Probability mass function (p.m.f)	
X	$P(X = x)$
x_1	$P(X = x_1)$
x_2	$P(X = x_2)$
.	.
.	.
Sum	**1.00**

In the above table, the random variable X can take values x_1, x_2, ... with probabilities $P(X = x_1)$, $P(X = x_2)$,...., respectively, such that the sum of all the probabilities is one.

(c) Graph: In a graph, a random variable X is said to follow the discrete probability distribution mass function if it takes the shape:

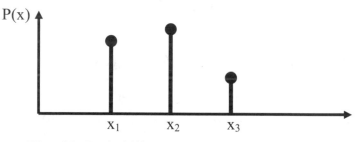

Fig. 6.2. Probability mass function of X.

The above graph is also called a **stick graph**. It shows that the distribution function of a discrete random variable X attaches some mass $P(x)$ at each value x of X and that the sum of the mass attached to all possible values is one. It is named the probability mass function (p.m.f).

6.3.1 MEAN VALUE OF A DISCRETE RANDOM VARIABLE

Let X be a discrete random variable with a probability mass function $P(X = x)$, then:

$$\text{Mean of X} = \mu = E(X) = \sum_x \{x\, P(X = x)\}$$

mu

Expected value of X

6.3.2 VARIANCE OF A DISCRETE RANDOM VARIABLE

Let X be a discrete random variable with a probability mass function $P(X = x)$, then its population variance will be given either by:

(i)   Variance of X $= \sigma^2 = \sum_x \{P(X = x)(x - \mu)^2\}$

or by

(ii)   Variance of X $\to \sigma^2 = \left[\sum_x \{x^2 P(X = x)\}\right] - \mu^2$

Sigma-Square

Note that both formulae in (i) and (ii) will give the same result. Formula (i) is by the definition of variance whereas formula (ii) is called the computing formula. Note the use of uppercase X and lowercase x.

6.3.3 STANDARD DEVIATION OF A DISCRETE RANDOM VARIABLE

The population standard deviation of a discrete random variable, X, is defined as a positive square root of its population variance, that is:

$$\sigma = \sqrt{\sigma^2}$$

and is denoted by the lowercase Greek letter sigma (σ).

Example 6.1. (GAMBLING IS RISKY) Toss two fair coins and let X be the number of heads observed.
(a) Find the probability mass function (p.m.f) of the discrete random variable X .
(b) How many heads are you guessing on average?
(c) Find the mean or expected value of X .
(d) Does your guess match the expected value?
(e) Find the variance of X .
(f) Derive the standard deviation of X .

Fig. 6.3. Coin.

Solution. (a) While tossing two fair coins, we have:

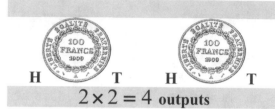

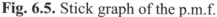

Fig. 6.4. Total number of outputs with two coins.

Thus, tossing two fair coins there will be four possible outcomes:

(Head, Head), (Head, Tail), (Tail, Head), and (Tail, Tail).

Here, the discrete random variable X is the number of heads observed and it can take the values $x = 0$, 1 or 2. Thus the required probability mass function (p.m.f) can be a table, graph, or a formula as follows:

Table (p.m.f):

No. of heads X	Events	Tally	Frequency f	RF or $P(X = x)$
0	TT	I	1	0.25
1	HT, TH	II	2	0.50
2	HH	I	1	0.25
		Sum	**4**	**1.00**

Graph (p.m.f):

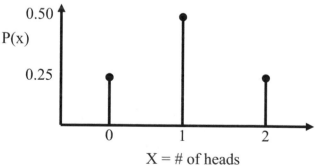

Fig. 6.5. Stick graph of the p.m.f.

From the above table or graph, we can also present probability mass function (p.m.f) with a **formula** as follows:

$$P(X = 0) = 0.25, \quad P(X = 1) = 0.50 \quad \text{and} \quad P(X = 2) = 0.25$$

such that:

$$P(X = 0) + P(X = 1) + P(X = 2) = 0.25 + 0.50 + 0.25 = 1.$$

In other words, 25% of the mass is hanging on the point X = 0, 50% of the mass is hanging on the point X = 1, and 25% of the mass is hanging on the point X = 2. Note that the area under a single point for a discrete random variable is non-zero. Thus, the area under each point is the mass assigned to each discrete value of the random variable, and the total mass is one.

(b) On the average, in two trials with a fair coin, our guess is to observe one head.

(c) Mean or expected number of heads:

$$\text{Mean} = E(X) = \sum_x \{xP(X = x)\} = \sum_{x=0}^{2} \{xP(X = x)\}$$

$$= \{0 \times P(X = 0)\} + \{1 \times P(X = 1)\} + \{2 \times P(X = 2)\}$$

$$= \{0 \times 0.25\} + \{1 \times 0.50\} + \{2 \times 0.25\}$$

$$= 0 + 0.5 + 0.5$$

$$= 1 \text{ (Heads)}.$$

(d) Yes! Our guess was fantastic!

(e) We shall find variance of X with two methods:

Method I (formula by definition): We have:

$$\text{Variance} = \sigma^2 = \sum_{x=0}^{2} \{P(X = x)(x - \mu)^2\}$$

$$= \{P(X = 0) \times (0 - 1)^2\} + \{P(X = 1) \times (1 - 1)^2\} + \{P(X = 2) \times (2 - 1)^2\}$$

$$\begin{array}{lll}
= \{0.25 \times 1\} & + \quad \{0.5 \times 0\} & + \quad \{0.25 \times 1\} \\
= 0.25 & + \quad 0.00 & + \quad 0.25
\end{array}$$

$$= 0.50 .$$

Method II (computing formula): We have:

$$\text{Variance} = \sigma^2 = \left[\sum_{x=0}^{2} \left\{ x^2 P(X=x) \right\} \right] - \mu^2$$

$$= \left[\left\{ 0^2 \times P(X=0) \right\} + \left\{ 1^2 \times P(X=1) \right\} + \left\{ 2^2 \times P(X=2) \right\} \right] - (1)^2$$

$$= \left[\left\{ 0 \times 0.25 \right\} + \left\{ 1 \times 0.50 \right\} + \left\{ 4 \times 0.25 \right\} \right] - 1$$

$$= \left[0 + 0.50 + 1 \right] - 1$$

$$= 0.50.$$

Thus, both methods give the same variance.

(f) Standard deviation is given by:

$$\sigma = \sqrt{\sigma^2} = \sqrt{0.50} = 0.707.$$

Example 6.2. (LOST AND FOUND) In a girls' college, the distribution of the girls' ages, X (years) is given below:

Girls of different age group						
X (years)	21	22	23	24	25	26
$P(X=x)$	0.05	0.20	?	0.25	0.15	0.05

Fig. 6.6. Some people do not like to disclose their age.

(a) Find $P(X=23)$, the proportion of 23 year old girls.
(b) Find the average age.
(c) Find the variance.
(d) Find the standard deviation.
Solution. (a) We know that the sum of all the probabilities should be equal to one, that is:

$$\sum_{x=21}^{26} P(X=x) = 1$$

which implies:

$$P(X=21) + P(X=22) + P(X=23) + P(X=24) + P(X=25) + P(X=26) = 1$$

or

$$0.05 + 0.20 + P(X = 23) + 0.25 + 0.15 + 0.05 = 1$$

or

$$0.70 + P(X = 23) = 1$$

or

$$P(X = 23) = 1 - 0.70 = 0.30 .$$

Thus, the complete probability mass function is given by:

X	21	22	23	24	25	26
$P(X = x)$	0.05	0.20	0.30	0.25	0.15	0.05

(b) The mean of the random variable X (or average age) is given by:

$$\mu = E(X) = \sum_{x=21}^{26} \{xP(X = x)\}$$

$$= \{21 \times P(X = 21)\} + \{22 \times P(X = 22)\} + \{23 \times P(X = 23)\}$$
$$+ \{24 \times P(X = 24)\} + \{25 \times P(X = 25)\} + \{26 \times P(X = 26)\}$$
$$= \{21 \times 0.05\} + \{22 \times 0.20\} + \{23 \times 0.30\}$$
$$+ \{24 \times 0.25\} + \{25 \times 0.15\} + \{26 \times 0.05\}$$
$$= 1.05 + 4.40 + 6.90 + 6.00 + 3.75 + 1.30$$
$$= 23.40 \text{ years.}$$

(c) **Variance : Method I (formula by definition):** The variance is given by:

$$\sigma^2 = \sum_{x=21}^{26} \{P(X = x)(x - \mu)^2\}$$

$$= \{P(X = 21) \times (21 - 23.40)^2\} + \{P(X = 22) \times (22 - 23.40)^2\}$$
$$+ \{P(X = 23) \times (23 - 23.40)^2\} + \{P(X = 24) \times (24 - 23.40)^2\}$$
$$+ \{P(X = 25) \times (25 - 23.40)^2\} + \{P(X = 26) \times (26 - 23.40)^2\}$$
$$= \{0.05 \times (-2.4)^2\} + \{0.20 \times (-1.4)^2\} + \{0.30 \times (-0.4)^2\}$$
$$+ \{0.25 \times (0.6)^2\} + \{0.15 \times (1.6)^2\} + \{0.05 \times (2.6)^2\}$$
$$= 0.288 + 0.392 + 0.048$$
$$+ 0.09 + 0.384 + 0.338$$
$$= 1.54 \text{ years}^2.$$

Method II (computing formula): The variance is given by:

$$\sigma^2 = \left[\sum_{x=21}^{26} \left\{ x^2 P(X = x) \right\} \right] - \mu^2$$

$$= \left[\left\{ 21^2 \times P(X = 21) \right\} + \left\{ 22^2 \times P(X = 22) \right\} + \left\{ 23^2 \times P(X = 23) \right\} \right.$$

$$\left. + \left\{ 24^2 \times P(X = 24) \right\} + \left\{ 25^2 \times P(X = 25) \right\} + \left\{ 26^2 \times P(X = 26) \right\} \right] - (23.40)^2$$

$$= \left[\{441 \times 0.05\} + \{484 \times 0.20\} + \{529 \times 0.30\} + \{576 \times 0.25\} \right.$$

$$\left. + \{625 \times 0.15\} + \{676 \times 0.05\} \right] - (547.56)$$

$$= 549.1 - 547.56 = 1.54 \text{ years}^2.$$

(d) Standard deviation: The population standard deviation is given by:

$$\sigma = \sqrt{\sigma^2} = \sqrt{1.54 \text{ years}^2} = 1.24 \text{ years}.$$

Example 6.3. (STUDY HOURS HELP) Suppose the numbers of hours per week students spend studying for their statistics courses is:

Number of hours (X)	2	3	4	5	6	7	8	9
Proportion of students	0.05	0.15	0.25	0.30	0.10	0.07	0.05	0.03

Fig. 6.7. A girl studying at home.

(a) Is this a density function or a mass function? Why?

(b) What is the mode of this distribution?

(c) What proportion of students study more than 4 hours? (X>4)

(d) What proportion of students study 4 or more hours? ($X \geq 4$)

(e) What proportion of students study less than 2 hours? (X<2)

(f) What proportion of students spend 2 or less hours? ($X \leq 2$)

(g) What proportion of students study at least 4, but no more than 6 hours? ($4 \leq X \leq 6$)

(h) What proportion of students study between 4 and 6 hours?

(i)What proportion of students study exactly 4 hours? (X=4)

(j) What proportion of students study more than 9 hours? (X>9)

(k) What proportion of students study at least 6 hours? ($X \geq 6$)

Solution.
(a) It is a mass function, because the number of hours (x) is a discrete variable. Also, the sum of all proportions is equal to one.
(b) Mode = 5, because the maximum is the 30% of students who spend 5 hours studying.
(c) $0.30 + 0.10 + 0.07 + 0.05 + 0.03 = 0.55$.
(d) $0.25 + 0.30 + 0.10 + 0.07 + 0.05 + 0.03 = 0.80$.
(e) zero.
(f) 0.05.
(g) $0.25 + 0.30 + 0.10 = 0.65$.
(h) 0.30.
(i) 0.25.
(j) zero.
(k) $0.10 + 0.07 + 0.05 + 0.03 = 0.25$.

Example 6.4. (ALWAYS WATCH THE NEWS) Let X be the number of televisions in randomly selected city households. The mass function of X is given in the following table.

Value of X	x_i	0	1	2	3	4
$P(X = x)$	p_i	0.05	0.30	0.40	0.20	?

Fig. 6.8. A girl watching TV after doing her homework.

(a) Complete the mass function of X and sketch its stick graph.
(b) What is the probability that a randomly selected household has at least one TV?
(c) What is the probability that a randomly selected household has exactly two TVs?
(d) What is the expected number of televisions per household?
(e) Find the variance of X using the formula by definition.
(f)   Find the variance of X using the computing formula.
Solution. (a) Since the sum of all proportions needs to be one, the complete mass function of X is given by:

Value of X	x_i	0	1	2	3	4
$P(X = x)$	p_i	0.05	0.30	0.40	0.20	0.05

Please note the use of new symbols x_i and p_i where the index i changes from 1 to 5. The stick graph is given by:

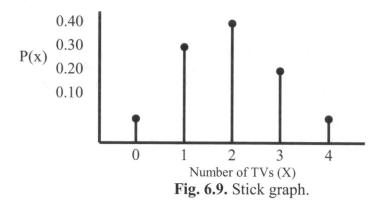

Fig. 6.9. Stick graph.

(b) Probability of at least one TV is given by:

$$P(\text{At least one TV}) = P(X = 1) + P(X = 2) + P(X = 3) + P(X = 4)$$
$$= p_2 + p_3 + p_4 + p_5$$
$$= 0.30 + 0.40 + 0.20 + 0.05$$
$$= 0.95.$$

(c) Probability of exactly two TVs is given by:

$$P(\text{Exactly two TVs}) = P(X = 2) = p_3 = 0.40.$$

(d) Expected number of TV's is given by:

$$\text{Mean} = \mu = E(X) = \sum_{i=1}^{5} p_i x_i = p_1 x_1 + p_2 x_2 + p_3 x_3 + p_4 x_4 + p_5 x_5$$
$$= (0.05 \times 0) + (0.30 \times 1) + (0.40 \times 2) + (0.20 \times 3) + (0.05 \times 4)$$
$$= 0.00 + 0.30 + 0.80 + 0.60 + 0.20 = 1.9 \text{ TVs.}$$

(e) Variance of X (using the formula by definition):

$$\sigma^2 = \sum_{i=1}^{5} \left[p_i (x_i - \mu)^2 \right]$$

$$= p_1 (x_1 - \mu)^2 + p_2 (x_2 - \mu)^2 + p_3 (x_3 - \mu)^2 + p_4 (x_4 - \mu)^2 + p_5 (x_5 - \mu)^2$$
$$= 0.05(0 - 1.9)^2 + 0.30(1 - 1.9)^2 + 0.40(2 - 1.9)^2 + 0.20(3 - 1.9)^2 + 0.05(4 - 1.9)^2$$
$$= (0.05 \times 3.61) + (0.30 \times 0.81) + (0.40 \times 0.01) + (0.20 \times 1.21) + (0.05 \times 4.41)$$
$$= 0.1805 + 0.2430 + 0.0040 + 0.2420 + 0.2205 = 0.89 \ (\text{TVs})^2.$$

(f) Variance of X (using the computing formula):

$$\sigma^2 = \left[\sum_{i=1}^{5} (p_i x_i^2) \right] - (\mu^2) = \left[p_1 x_1^2 + p_2 x_2^2 + p_3 x_3^2 + p_4 x_4^2 + p_5 x_5^2 \right] - (\mu^2)$$

$$= \left[(0.05 \times 0^2) + (0.30 \times 1^2) + (0.40 \times 2^2) + (0.20 \times 3^2) + (0.05 \times 4^2) \right] - (1.9^2)$$
$$= \left[(0.05 \times 0) + (0.30 \times 1) + (0.40 \times 4) + (0.20 \times 9) + (0.05 \times 16) \right] - (3.61)$$
$$= \left[0.00 + 0.30 + 1.60 + 1.80 + 0.80 \right] - (3.61)$$
$$= 4.50 - 3.61 = 0.89 \ (\text{TVs})^2. \text{ (Same result as above).}$$

Two discrete distributions frequently used are:

(a) Binomial distribution
(b) Poisson distribution.

We shall discuss these distributions in the rest of this chapter.

6.4 BINOMIAL DISTRIBUTION

Consider two players Amy and Bob who both like to play with a fair coin.

Fig. 6.10. A coin.

Amy says to Bob, "If I toss a fair coin and heads appears, then I win."

There are two cases:

If heads (H) appears, then it will be a success (S) for Amy.

If tails (T) appears, then it will be a failure (F) for Amy.

Note carefully in this case we do not care about Bob, only Amy. We said either Amy would have a "success" or a "failure".

What is the probability that Amy will get a heads?

Obviously:

$$P(\text{H}) = \frac{1}{2} = P(\text{success}) = p$$

and

$$P(\text{T}) = \frac{1}{2} = P(\text{failure}) = q \ (= 1 - p)$$

Note that:

$$P(\text{success}) + P(\text{failure}) = 1 \ \text{or} \ p + q = 1$$

6.4.1 BINOMIAL EXPERIMENT

A binomial experiment, of a discrete random variable X, is one that has the following five characteristics:

(i) The experiment consists of n identical trials.

(ii) Each trial results in one of the two outcomes:
(a) success (b) failure
(iii) The:
$$P(\text{success}) = p$$
for a single trial and remains the same from trial to trial, and the:
$$P(\text{failure}) = q = 1 - p$$
Note that:
$$p + q = 1$$

(iv) The trials are independent.

(v) The probability of x successes out of n trails is given by:

$$P(X = x) = \binom{n}{x} p^x q^{(n-x)}, \quad x = 0,\ 1,\ 2...,\ n$$

where

$$\binom{n}{x} = \frac{n!}{x!(n - x)!}$$

We say $X \sim Bin(n, p)$, that is, the random variable X follows a binomial distribution with parameters n and p.

Recall that: $n! = n(n-1)(n-2)....2.1$ and $0! = 1$.

Also, remember that: $2! = 2 \times 1$, $3! = 3 \times 2 \times 1$ and so on.

6.4.2 MEAN AND VARIANCE OF THE BINOMIAL VARIABLE

Let a random variable X follow binomial distribution with the probability of success p on each trial. Let n be the independent and identical trials performed. If $X \sim Bin(n,\ p)$, then the probability of x successes is given in the following table.

$X = x$	$P(X = x)$
0	$P(X = 0) = \binom{n}{0} p^0 q^{(n-0)}$
1	$P(X = 1) = \binom{n}{1} p^1 q^{(n-1)}$
2	$P(X = 2) = \binom{n}{2} p^2 q^{(n-2)}$
•	•
•	•
n	$P(X = n) = \binom{n}{n} p^n q^{(n-n)}$
Sum	**1.00**

It can be shown that the mean **or** the expected value of X is given by:

$$\text{Mean} = E(X) = \mu = \sum_{x=0}^{n} xP(X = x) = np$$

and the variance of X is given by:

$$\text{Variance} = \sigma^2 = \sum_{x=0}^{n} \left\{ P(X = x)(x - \mu)^2 \right\} = npq$$

The proofs of these results are beyond the scope of this book. Note that the variance of a binomial variable is always less than its mean value.

Example 6.5. (COMBINATIONS) Compute the following combinations.

(a) $\binom{7}{0}$ (b) $\binom{7}{1}$ (c) $\binom{7}{7}$ (d) $\binom{7}{3}$ (e) $\binom{7}{9}$

Solution.

(a) $\binom{7}{0} = \dfrac{7!}{0!(7-0)!} = \dfrac{7!}{0! \ 7!} = \dfrac{7 \times 6 \times 5 \times 4 \times 3 \times 2 \times 1}{1 \times 7 \times 6 \times 5 \times 4 \times 3 \times 2 \times 1} = 1.$

(b) $\dbinom{7}{1} = \dfrac{7!}{1!\,(7-1)!} = \dfrac{7!}{1!\ 6!} = \dfrac{7 \times 6 \times 5 \times 4 \times 3 \times 2 \times 1}{1 \times 6 \times 5 \times 4 \times 3 \times 2 \times 1} = 7.$

(c) $\dbinom{7}{7} = \dfrac{7!}{7!\,(7-7)!} = \dfrac{7!}{7!\ 0!} = \dfrac{7 \times 6 \times 5 \times 4 \times 3 \times 2 \times 1}{7 \times 6 \times 5 \times 4 \times 3 \times 2 \times 1 \times 1} = 1.$

(d) $\dbinom{7}{3} = \dfrac{7!}{3!\,(7-3)!} = \dfrac{7!}{3!\ 4!} = \dfrac{7 \times 6 \times 5 \times 4 \times 3 \times 2 \times 1}{3 \times 2 \times 1 \times 4 \times 3 \times 2 \times 1} = \dfrac{7 \times 5}{1} = 35.$

(e) $\dbinom{7}{9} =$ Error in the question, we cannot choose 9 distinct objects out of 7 o

Example 6.6. (BINOMIAL DISTRIBUTION) Suppose that the random variable X has binomial distribution with $n = 6$ and $p = 0.3$. Find each of the following quantities.

(a) $P(X = 0)$
(b) $P(X = 2)$
(c) $P(X > 4)$
(d) $P(X \le 3)$
(e) $P(X = 6)$
(f) $P(X = 1.2)$
(g) Mean of X
(h) Variance of X
(i) Standard deviation.

Fig. 6.11. Use your own calculator during an examination.

Solution. We are given $X \sim \text{Bin}(n,\ p) = \text{Bin}(6,\ 0.3)$, that is, X follows a binomial distribution with $n = 6$ and $p = 0.3$. Obviously, $q = 1 - p = 1 - 0.3 = 0.7$.

(a)

$$P(X = 0) = \dbinom{6}{0}(0.3)^0 (0.7)^{6-0} = \dfrac{6!}{0!\,(6-0)!}(0.3)^0 (0.7)^6 = \dfrac{6!}{0! \times 6!}(0.3)^0 (0.7)^6$$

$$= \dfrac{6 \times 5 \times 4 \times 3 \times 2 \times 1}{1 \times 6 \times 5 \times 4 \times 3 \times 2 \times 1} \times 1 \times 0.7 \times 0.7 \times 0.7 \times 0.7 \times 0.7 \times 0.7 = 0.117649$$

$$\big(\text{Something non - zero}\big)^0 = 1 \quad \text{C.T.M}$$

(b)

$$P(X = 2) = \binom{6}{2}(0.3)^2 (0.7)^{6-2}$$

$$= \frac{6!}{2!(6-2)!}(0.3)^2 (0.7)^4 = \frac{6!}{2! \times 4!}(0.3)^2 (0.7)^4$$

$$= \frac{6 \times 5 \times 4 \times 3 \times 2 \times 1}{2 \times 1 \times 4 \times 3 \times 2 \times 1} \times 0.3 \times 0.3 \times 0.7 \times 0.7 \times 0.7 \times 0.7$$

$$= 0.324135.$$

(c)

$$P(X > 4) = P(X = 5) + P(X = 6)$$

$$= \binom{6}{5}(0.3)^5 (0.7)^{6-5} + \binom{6}{6}(0.3)^6 (0.7)^{6-6}$$

$$= \frac{6!}{5!(6-5)!}(0.3)^5 (0.7)^1 + \frac{6!}{6!(6-6)!}(0.3)^6 (0.7)^0$$

$$= \frac{6!}{5! \times 1!}(0.3)^5 (0.7)^1 + \frac{6!}{6! \times 0!}(0.3)^6 (0.7)^0$$

$$= \frac{6 \times 5 \times 4 \times 3 \times 2 \times 1}{5 \times 4 \times 3 \times 2 \times 1 \times 1} \times 0.3 \times 0.3 \times 0.3 \times 0.3 \times 0.3 \times 0.7$$

$$+ \frac{6 \times 5 \times 4 \times 3 \times 2 \times 1}{6 \times 5 \times 4 \times 3 \times 2 \times 1 \times 1} \times 0.3 \times 0.3 \times 0.3 \times 0.3 \times 0.3 \times 0.3 \times 1$$

$$= 0.010206 + 0.000729 = 0.010935.$$

(d)

$$P(X \leq 3) = P(X = 0) + P(X = 1) + P(X = 2) + P(X = 3)$$

$$= \binom{6}{0}(0.3)^0 (0.7)^{6-0} + \binom{6}{1}(0.3)^1 (0.7)^{6-1}$$

$$+ \binom{6}{2}(0.3)^2 (0.7)^{6-2} + \binom{6}{3}(0.3)^3 (0.7)^{6-3}$$

$$= \frac{6!}{0!(6-0)!}(0.3)^0 (0.7)^{6-0} + \frac{6!}{1!(6-1)!}(0.3)^1 (0.7)^{6-1}$$

$$+ \frac{6!}{2!(6-2)!}(0.3)^2 (0.7)^{6-2} + \frac{6!}{3!(6-3)!}(0.3)^3 (0.7)^{6-3}$$

$$= \frac{6!}{0!\times6!}(0.3)^0(0.7)^6 + \frac{6!}{1!\times5!}(0.3)^1(0.7)^5$$

$$+ \frac{6!}{2!\times4!}(0.3)^2(0.7)^4 + \frac{6!}{3!\times3!}(0.3)^3(0.7)^3$$

$$= \frac{6\times5\times4\times3\times2\times1}{1\times6\times5\times4\times3\times2\times1}\times1\times0.7\times0.7\times0.7\times0.7\times0.7\times0.7$$

$$+ \frac{6\times5\times4\times3\times2\times1}{1\times5\times4\times3\times2\times1}\times0.3\times0.7\times0.7\times0.7\times0.7\times0.7$$

$$+ \frac{6\times5\times4\times3\times2\times1}{2\times1\times4\times3\times2\times1}\times0.3\times0.3\times0.7\times0.7\times0.7\times0.7$$

$$+ \frac{6\times5\times4\times3\times2\times1}{3\times2\times1\times3\times2\times1}\times0.3\times0.3\times0.3\times0.7\times0.7\times0.7$$

$$= 0.117649 + 0.302526 + 0.324135 + 0.185220 = 0.929530.$$

(e)

$$P(X = 6) = \binom{6}{6}(0.3)^6(0.7)^{6-6}$$

$$= \frac{6!}{6!(6-6)!}(0.3)^6(0.7)^0 = \frac{6!}{6!\times0!}(0.3)^6(0.7)^0 = 0.000729.$$

(f) $P(X = 1.2) = 0$, because X is a discrete variable and cannot take any fractional value.

(g) Mean of X :

$$\text{Mean of X} = \mu = E(X) = np = 6\times0.3 = 1.8.$$

(h) Variance of X :

$$V(X) = \sigma^2 = npq = 6\times0.3\times0.7 = 1.26.$$

(i) Standard deviation:

$$\sigma = \sqrt{\sigma^2} = \sqrt{1.26} = 1.12.$$

Example 6.7. (HANDLING LARGE NUMBERS) Compute the following combinations:

(a) $\begin{pmatrix} 100 \\ 0 \end{pmatrix}$ (b) $\begin{pmatrix} 100 \\ 1 \end{pmatrix}$ (c) $\begin{pmatrix} 100 \\ 100 \end{pmatrix}$ (d) $\begin{pmatrix} 100 \\ 5 \end{pmatrix}$

Solution.

(a) $\begin{pmatrix} 100 \\ 0 \end{pmatrix} = \dfrac{100!}{0!(100-0)!} = \dfrac{100!}{0! \times 100!} = \dfrac{100!}{1 \times 100!} = 1$.

(b) $\begin{pmatrix} 100 \\ 1 \end{pmatrix} = \dfrac{100!}{1!(100-1)!} = \dfrac{100!}{1! \times 99!} = \dfrac{100(100-1)!}{1 \times 99!} = \dfrac{100 \times 99!}{99!} = 100$.

(c) $\begin{pmatrix} 100 \\ 100 \end{pmatrix} = \dfrac{100!}{100!(100-100)!} = \dfrac{100!}{100! \times 0!} = \dfrac{100!}{100! \times 0!} = \dfrac{100!}{100! \times 1} = 1$.

(d) $\begin{pmatrix} 100 \\ 5 \end{pmatrix} = \dfrac{100!}{5!(100-5)!} = \dfrac{100!}{5! \times 95!}$

$$= \frac{100(100-1)(100-2)(100-3)(100-4)(100-5)!}{5! \times 95!}$$

$$= \frac{100 \times 99 \times 98 \times 97 \times 96 \times 95!}{5 \times 4 \times 3 \times 2 \times 1 \times 95!}$$

$$= \frac{100 \times 99 \times 98 \times 97 \times 96}{5 \times 4 \times 3 \times 2 \times 1}$$

$$= 75{,}287{,}520.$$

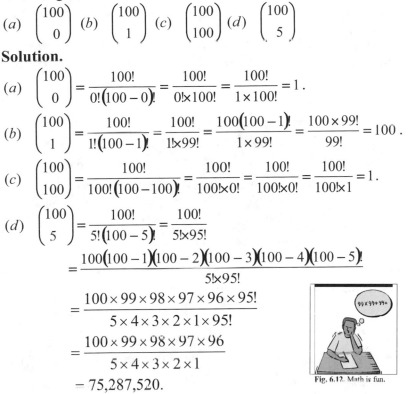

Fig. 6.12. Math is fun.

Example 6.8. (HUNTING BIRDS IS NOT FUN) Consider five birds flying in the sky. A hunter can hit a flying bird with a probability of 0.1 as shown below.

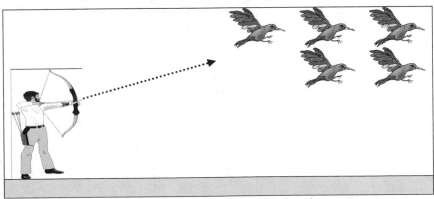

Fig. 6.13. Hitting birds in the sky.

A hunter tries independently five times to hit a bird. Given that the hunter can hit only one bird with a given arrow, find the probability that at the end of five trials:
(a) None of the birds will get hit.
(b) One bird will get hit.
(c) Two birds will get hit.
(d) Three birds will get hit.
(e) Four birds will get hit.
(f) All five birds will get hit.

Solution. Given the number of hits: $X \sim Bin(5, 0.1)$. Here $n = 5$ and $p = 0.1$, which implies $q = 1 - p = 1 - 0.1 = 0.9$. Also, we know that the probability of x successes (hits) out of n trials is given by:

$$P(X = x) = \binom{n}{x} p^x q^{(n-x)}, \quad x = 0, 1, 2, ..., n$$

(a) None of the birds will get hit $(x = 0)$: The probability that no birds will get hit is given by:

$$P(X = 0) = \binom{5}{0}(0.1)^0 (0.9)^{(5-0)} = \frac{5!}{0!(5-0)!}(0.1)^0 (0.9)^5$$

$$= \frac{5!}{0! \ 5!} \times 1 \times (0.9)^5$$

$$= \frac{5 \times 4 \times 3 \times 2 \times 1}{1 \times 5 \times 4 \times 3 \times 2 \times 1} \times 1 \times 0.9 \times 0.9 \times 0.9 \times 0.9 \times 0.9$$

$$= 0.59049.$$

(b) One bird will get hit $(x = 1)$: The probability of hitting one bird is given by:

$$P(X = 1) = \binom{5}{1}(0.1)^1 (0.9)^{(5-1)} = \frac{5!}{1!(5-1)!}(0.1)^1 (0.9)^4$$

$$= \frac{5!}{1! \ 4!} \times (0.1) \times (0.9)^4$$

$$= \frac{5 \times 4 \times 3 \times 2 \times 1}{1 \times 4 \times 3 \times 2 \times 1} \times 0.1 \times 0.9 \times 0.9 \times 0.9 \times 0.9$$

$$= 0.32805.$$

(c) Two birds will get hit $(x = 2)$: The probability of hitting two birds is given by:

$$P(X=2)=\binom{5}{2}(0.1)^2(0.9)^{(5-2)}=\frac{5!}{2!(5-2)!}(0.1)^2(0.9)^3$$

$$=\frac{5!}{2!\ 3!}\times(0.1)^2\times(0.9)^3$$

$$=\frac{5\times4\times3\times2\times1}{2\times1\times3\times2\times1}\times0.1\times0.1\times0.9\times0.9\times0.9$$

$$=0.0729.$$

(d) Three birds will get hit $(x=3)$: The probability of hitting three birds is given by:

$$P(X=3)=\binom{5}{3}(0.1)^3(0.9)^{(5-3)}=\frac{5!}{3!(5-3)!}(0.1)^3(0.9)^2$$

$$=\frac{5!}{3!\ 2!}\times(0.1)^3\times(0.9)^2$$

$$=\frac{5\times4\times3\times2\times1}{3\times2\times1\times2\times1}\times0.1\times0.1\times0.1\times0.9\times0.9$$

$$=0.0081.$$

(e) Four birds will get hit $(x=4)$: The probability of hitting four birds is given by:

$$P(X-4)=\binom{5}{4}(0.1)^4(0.9)^{(5-4)}=\frac{5!}{4!(5-4)!}(0.1)^4(0.9)^1$$

$$=\frac{5!}{4!\ 1!}\times(0.1)^4\times(0.9)^1$$

$$=\frac{5\times4\times3\times2\times1}{4\times3\times2\times1\times1}\times0.1\times0.1\times0.1\times0.1\times0.9$$

$$=0.00045.$$

(f) All five birds will get hit $(x=5)$: The probability of hitting all five birds is given by:

$$P(X=5)=\binom{5}{5}(0.1)^5(0.9)^{(5-5)}=\frac{5!}{5!(5-5)!}(0.1)^5(0.9)^0$$

$$=\frac{5!}{5!\ 0!}\times(0.1)^5\times(0.9)^0$$

$$=\frac{5\times4\times3\times2\times1}{5\times4\times3\times2\times1\times1}\times0.1\times0.1\times0.1\times0.1\times0.1\times1$$

$$=0.00001.$$

Thus, we have the following table:

Number of birds that get hit	None	 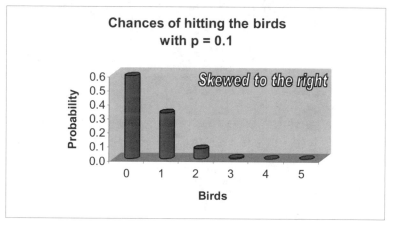				
$X = x$	0	1	2	3	4	5
$P(X = x)$	0.59045	0.32805	0.07290	0.00810	0.00045	0.00001

Fig. 6.14. Table showing chance of hits.

A bar chart showing the chances of hitting the birds is given below:

Chances of hitting the birds with p = 0.1

Skewed to the right

Probability — 0.6, 0.5, 0.4, 0.3, 0.2, 0.1, 0.0

Birds — 0, 1, 2, 3, 4, 5

Fig. 6.15. A bar chart showing chance of hits.

The above bar chart shows that if the probability of success is less than 0.5 (here p = 0.1), then the binomial distribution is skewed to the right.

Example 6.9. (MILITARY TRAINING) A soldier is trying to hit five bombers flying in the sky. A soldier can hit a flying bomber with a probability of 0.9. The soldier tries independently five times to hit the flying bombers. Given that the soldier can hit only one bomber at a time, find the probability that at the end of five trials:
(a) None of the bombers will get hit.
(b) One bomber will get hit.
(c) Two bombers will get hit.
(d) Three bombers will get hit.
(e) Four bombers will get hit.
(f) All five bombers will get hit.

Fig. 6.16. An officer.

Fig. 6.17. A soldier hitting bombers.

Solution. Given the number of hits: $X \sim Bin(5, 0.9)$. Here $n = 5$ and $p = 0.9$ which implies $q = 1 - p = 1 - 0.9 = 0.1$. Also we know that the probability of x successes (hits) out of n trials is given by:

$$P(X = x) = \binom{n}{x} p^x q^{(n-x)}, \quad x = 0, 1, 2, ..., n$$

(a) None of the bombers will get hit $(x = 0)$: The probability that no bomber will get hit is given by:

$$P(X = 0) = \binom{5}{0}(0.9)^0 (0.1)^{(5-0)} = \frac{5!}{0!(5-0)!}(0.9)^0 (0.1)^5$$

$$= \frac{5!}{0! \ 5!} \times 1 \times (0.1)^5$$

$$= \frac{5 \times 4 \times 3 \times 2 \times 1}{1 \times 5 \times 4 \times 3 \times 2 \times 1} \times 1 \times 0.1 \times 0.1 \times 0.1 \times 0.1 \times 0.1$$

$$= 0.00001.$$

(b) One bomber will get hit $(x = 1)$: The probability of hitting one bomber is given by:

$$P(X=1)=\binom{5}{1}(0.9)^1(0.1)^{(5-1)}=\frac{5!}{1!(5-1)!}(0.9)^1(0.1)^4$$

$$=\frac{5!}{1!\ 4!}\times(0.9)\times(0.1)^4$$

$$=\frac{5\times4\times3\times2\times1}{1\times4\times3\times2\times1}\times0.9\times0.1\times0.1\times0.1\times0.1$$

$$=0.00045\ .$$

(c) Two bombers will get hit $(x=2)$: The probability of hitting two bombers is given by:

$$P(X=2)=\binom{5}{2}(0.9)^2(0.1)^{(5-2)}=\frac{5!}{2!(5-2)!}(0.9)^2(0.1)^3$$

$$=\frac{5!}{2!\ 3!}\times(0.9)^2\times(0.1)^3$$

$$=\frac{5\times4\times3\times2\times1}{2\times1\times3\times2\times1}\times0.9\times0.9\times0.1\times0.1\times0.1$$

$$=0.00810\ .$$

(d) Three bombers will get hit $(x=3)$: The probability of hitting three bombers is given by:

$$P(X=3)=\binom{5}{3}(0.9)^3(0.1)^{(5-3)}=\frac{5!}{3!(5-3)!}(0.9)^3(0.1)^2$$

$$=\frac{5!}{3!\ 2!}\times(0.9)^3\times(0.1)^2$$

$$=\frac{5\times4\times3\times2\times1}{3\times2\times1\times2\times1}\times0.9\times0.9\times0.9\times0.1\times0.1$$

$$=0.07290\ .$$

(e) Four bombers will get hit $(x=4)$: The probability of hitting four bombers is given by:

$$P(X=4)=\binom{5}{4}(0.9)^4(0.1)^{(5-4)}=\frac{5!}{4!(5-4)!}(0.9)^4(0.1)^1$$

$$=\frac{5!}{4!\ 1!}\times(0.9)^4\times(0.1)^1$$

$$=\frac{5\times4\times3\times2\times1}{4\times3\times2\times1\times1}\times0.9\times0.9\times0.9\times0.9\times0.1$$

$$=0.32805.$$

(f) All five bombers will get hit $(x = 5)$: The probability of hitting all five bombers is given by:

$$P(X = 5) = \binom{5}{5}(0.9)^5 (0.1)^{(5-5)} = \frac{5!}{5!(5-5)!}(0.9)^5 (0.1)^0$$

$$= \frac{5!}{5!\ 0!} \times (0.9)^5 \times (0.1)^0$$

$$= \frac{5 \times 4 \times 3 \times 2 \times 1}{5 \times 4 \times 3 \times 2 \times 1 \times 1} \times 0.9 \times 0.9 \times 0.9 \times 0.9 \times 0.9 \times 1$$

$$= 0.59049 .$$

Thus, we have the following table:

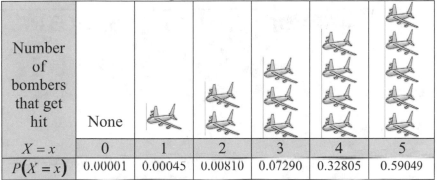

Number of bombers that get hit	None					
$X = x$	0	1	2	3	4	5
$P(X = x)$	0.00001	0.00045	0.00810	0.07290	0.32805	0.59049

Fig. 6.18. Probabilities of hitting bombers.

A bar chart showing the chance of hitting bombers is given below:

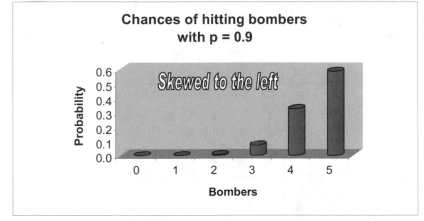

Fig. 6.19. A bar chart showing the chance of hits.

The above bar chart shows that if the probability of success is more than 0.5 (here p = 0.9), the binomial distribution is skewed to the left.

Example 6.10. (HITTING BALOONS IS FUN) At a festival, if anyone can hit all five balloons with a toy gun in five trials, then they win a scooter. A boy is trying and his probability of hitting a balloon is 0.5. Given that the boy can hit only one balloon with a given trial, find the probability that at the end of five trials:

(a) None of the balloons will get hit.

(b) One balloon will get hit.

(c) Two balloons will get hit.

(d) Three balloons will get hit.

(e) Four balloons will get hit.

(f) All five balloons will get hit.

Fig. 6.20. A boy hitting balloons.

Solution. Given the number of hits: $X \sim Bin(5, 0.5)$. Here $n = 5$ and $p = 0.5$ which implies $q = 1 - p = 1 - 0.5 = 0.5$. Also we know that the probability of x successes (hits) out of n trials is given by:

$$P(X = x) = \binom{n}{x} p^x q^{(n-x)}, \quad x = 0, 1, 2, ..., n$$

(a) None of the balloons will get hit $(x = 0)$: The probability that no balloons will get hit is given by:

$$P(X=0)=\binom{5}{0}(0.5)^0(0.5)^{(5-0)}=\frac{5!}{0!(5-0)!}(0.5)^0(0.5)^5$$

$$=\frac{5!}{0!\ 5!}\times 1\times(0.5)^5$$

$$=\frac{5\times 4\times 3\times 2\times 1}{1\times 5\times 4\times 3\times 2\times 1}\times 1\times 0.5\times 0.5\times 0.5\times 0.5\times 0.5$$

$$=0.03125.$$

(b) One balloon will get hit $(x=1)$: The probability of hitting one balloon is given by:

$$P(X=1)=\binom{5}{1}(0.5)^1(0.5)^{(5-1)}=\frac{5!}{1!(5-1)!}(0.5)^1(0.5)^4$$

$$=\frac{5!}{1!\ 4!}\times(0.5)\times(0.5)^4$$

$$=\frac{5\times 4\times 3\times 2\times 1}{1\times 4\times 3\times 2\times 1}\times 0.5\times 0.5\times 0.5\times 0.5\times 0.5$$

$$=0.15625.$$

(c) Two balloons will get hit $(x=2)$: The probability of hitting two balloons is given by:

$$P(X=2)=\binom{5}{2}(0.5)^2(0.5)^{(5-2)}=\frac{5!}{2!(5-2)!}(0.5)^2(0.5)^3$$

$$=\frac{5!}{2!\ 3!}\times(0.5)^2\times(0.5)^3$$

$$=\frac{5\times 4\times 3\times 2\times 1}{2\times 1\times 3\times 2\times 1}\times 0.5\times 0.5\times 0.5\times 0.5\times 0.5$$

$$=0.31250.$$

(d) Three balloons will get hit $(x=3)$: The probability of hitting three balloons is given by:

$$P(X=3)=\binom{5}{3}(0.5)^3(0.5)^{(5-3)}=\frac{5!}{3!(5-3)!}(0.5)^3(0.5)^2$$

$$=\frac{5!}{3!\ 2!}\times(0.5)^3\times(0.5)^2$$

$$= \frac{5 \times 4 \times 3 \times 2 \times 1}{3 \times 2 \times 1 \times 2 \times 1} \times 0.5 \times 0.5 \times 0.5 \times 0.5 \times 0.5$$
$$= 0.31250.$$

(e) Four balloons will get hit $(x = 4)$: The probability of hitting four balloons is given by:

$$P(X = 4) = \binom{5}{4}(0.5)^4 (0.5)^{(5-4)} = \frac{5!}{4!(5-4)!}(0.5)^4 (0.5)^1$$

$$= \frac{5!}{4! \ 1!} \times (0.5)^4 \times (0.5)^1$$

$$= \frac{5 \times 4 \times 3 \times 2 \times 1}{4 \times 3 \times 2 \times 1 \times 1} \times 0.5 \times 0.5 \times 0.5 \times 0.5 \times 0.5$$

$$= 0.15625.$$

(f) All five balloons will get hit $(x = 5)$: The probability of hitting all five balloons is given by:

$$P(X = 5) = \binom{5}{5}(0.5)^5 (0.5)^{(5-5)} = \frac{5!}{5!(5-5)!}(0.5)^5 (0.5)^0$$

$$= \frac{5!}{5! \ 0!} \times (0.5)^5 \times (0.5)^0$$

$$= \frac{5 \times 4 \times 3 \times 2 \times 1}{5 \times 4 \times 3 \times 2 \times 1 \times 1} \times 0.5 \times 0.5 \times 0.5 \times 0.5 \times 0.5 \times 1$$

$$= 0.03125 .$$

Thus, we have the following table:

Number of balloons that get hit	None					
$X = x$	0	1	2	3	4	5
$P(X = x)$	0.03125	0.15625	0.31250	0.31250	0.15625	0.03125

Fig. 6.21. Probabilities of hitting balloons.

A bar chart showing the chances of hitting the balloons is given below:

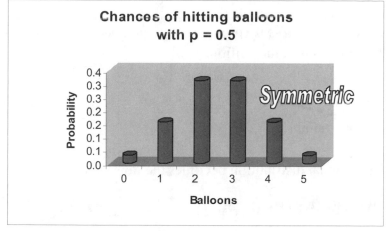

Fig. 6.22. A bar chart showing chance of hits.

The above bar chart shows that if the probability of success is equal to 0.5, then the binomial distribution is symmetric.

Thus, from the above three examples related to hitting birds, bombers and balloons, we are at the following conclusion:

In a binomial distribution:
(a) If $p < 0.5$, then the distribution is skewed to the left.
(b) If $p = 0.5$, then the distribution is bell-shaped and symmetric.
(c) If $p > 0.5$, then the distribution is skewed to the right.

Example 6.11. (EASY TO REMEMBER) If $X \sim \text{Bin}(100,\ 0.75)$, find $E(X)$, $V(X)$, and σ.

Solution. We are given:
$$X \sim \text{Bin}(100,\ 0.75) \sim \text{Bin}(n,\ p),$$
which implies:
$$n = 100 \text{ and } p = 0.75, \text{ thus } q = 1 - p = 1 - 0.75 = 0.25.$$
Thus,
$$E(X) = \text{mean} = np = 100 \times 0.75 = 75,$$
$$V(X) = npq = 100 \times 0.75 \times 0.25 = 18.75,$$
and the standard deviation $= \sigma = \sqrt{V(X)} = \sqrt{18.75} = 4.33$.

Note that if the sample size is large relative to the population size, that is, if $\frac{n}{N} \geq 0.05$, then the resulting experiment is **not** binomial.

Further note that if $p \to 0$ (probability of success is small) and $n \to \infty$ (number of trials becomes infinite),

and

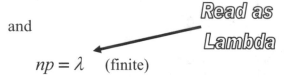

$np = \lambda$ (finite)

then the binomial distribution becomes the Poisson distribution.

6.5 POISSON DISTRIBUTION

A random variable X is said to follow the Poisson distribution if the probability of x successes out of an infinite number of trials is given by:

$$P(X = x) = \frac{e^{-\lambda} \lambda^x}{x!}, \quad x = 0, 1, 2,..., \infty$$

where

$$\text{mean} = \lambda = np, \quad e \approx 2.7182, \quad \text{and} \quad x! = x(x-1)(x-2)....2.1$$

Also, note that:

$$\text{variance} = \lambda$$

Thus, it is most important to note that for this distribution the mean and the variance are equal. The following table lists a few useful values of $e^{-\lambda}$ for different values of λ:

Values of $e^{-\lambda}$							
λ	0.00	0.25	0.50	0.75	1.00	1.25	1.50
$e^{-\lambda}$	1.00000	0.77880	0.60653	0.47237	0.36788	0.28650	0.22313
λ	1.75	2.00	2.25	2.50	2.75	3.00	3.25
$e^{-\lambda}$	0.17377	0.13534	0.10539	0.08208	0.06393	0.04978	0.03877

Example 6.12. (ALWAYS OBEY THE TRAFFIC RULES) In the daily life of a big city, there are an infinite number of traffic vehicles passing through a particular intersection. Although efforts have been made to control the traffic by using traffic lights, traffic police, traffic rules, and traffic signs, some accidents still happen. Based on the past 100 years, it has been found that the chance of an accident during a period of one month is 0.00025. It is predicted that during the next month, there will be 10,000 vehicles passing through the intersection as shown in **Figure 6.23**.

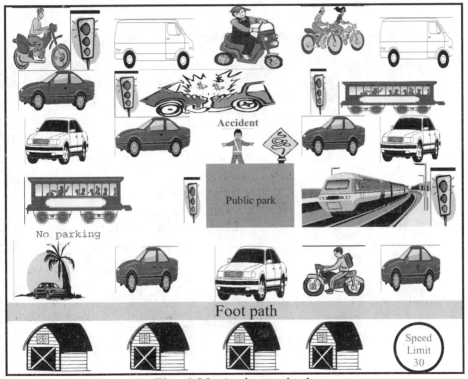

Fig. 6.23. A city outlook.

Find the probability that:
(a) there will be no accident during the month.
(b) there will be exactly one accident during the month.
(c) there will be exactly two accidents during the month.
(d) there will be at least one accident during the month.
(e) there will be at least two accidents during the month.
(f) there will be at most two accidents during the month.

Solution. We are given:

$$p = 0.00025 \text{ and } n = 10,000$$

so that

$$\lambda = np = 10000 \times 0.00025 = 2.5.$$

We know that the probability of x successes is given by:

$$P(X = x) = \frac{e^{-\lambda} \lambda^x}{x!}, \quad x = 0, 1, 2, \dots, \infty$$

Note that unfortunately, in this example a success means an accident.

Thus we have:

(a) the probability of no accidents $(x = 0)$:

$$P(X = 0) = \frac{e^{-\lambda} \lambda^0}{0!} = \frac{e^{-2.5}(2.5)^0}{0!} = e^{-2.5} = 0.08208.$$

(b) the probability of exactly one accident $(x = 1)$:

$$P(X = 1) = \frac{e^{-\lambda} \lambda^1}{1!} = \frac{e^{-2.5}(2.5)^1}{1!} = e^{-2.5} \times 2.5 = 0.20521.$$

(c) the probability of exactly two accidents $(x = 2)$:

$$P(X = 2) = \frac{e^{-\lambda} \lambda^2}{2!} = \frac{e^{-2.5}(2.5)^2}{2 \times 1} = \frac{e^{-2.5} \times 2.5 \times 2.5}{2} = 0.25652.$$

(d) the probability of at least one accident $(x \geq 1)$:   Note that the total probability is one. Thus, the probability that there will be at least one accident can be written as:

$$P(\text{At least one accident}) = P(x \geq 1)$$
$$= P(X = 1) + P(X = 2) + P(X = 3) + \dots\dots$$
$$= 1 - P(X < 1)$$

$$= 1 - \{P(X = 0)\}$$
$$= 1 - \{0.08208\}$$
$$= 0.91792.$$

(e) the probability of at least two accidents $(x \geq 2)$: Note that the total probability is one, thus the probability that there will be at least two accidents can be written as:

$$P(\text{At least two accidents}) = P(X \geq 2)$$
$$= P(X = 2) + P(X = 3) + \ldots\ldots$$
$$= 1 - P(X < 2)$$
$$= 1 - \{P(X = 0) + P(X = 1)\}$$
$$= 1 - \{0.08208 + 0.20521\}$$
$$= 0.71271.$$

(f) the probability of at most two accidents $(x \leq 2)$: The probability that there will be at most two accidents will be:

$$P(\text{At most two accidents}) = P(X \leq 2)$$
$$= P(X = 0) + P(X = 1) + P(X = 2)$$
$$= 0.08208 + 0.20521 + 0.25652$$
$$= 0.54381.$$

LUDI 6.1. (MISSING PETS) The number of parrots X, owned by a university's women has the probability mass function:

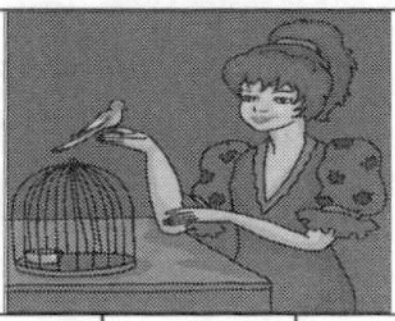

Number of parrots (X)	0	1	2	3	4
Proportion of women	0.65	0.20		0.03	0.01

Fig. 6.24. Bird lovers.

(a) Complete the above distribution.
(b) What proportion of university women own 1 parrot or less?
(c) What proportion of university women own 3 or less parrots?
(d) What proportion of university women own 3 parrots?
(e) What proportion of university women own less than 3 parrots?
(f) What proportion of university women own more than 3 parrots?
(g) What proportion of university women own 3 or more parrots?
(h) What proportion of university women own 4 parrots?
(i) What proportion of university women own 5 parrots?
(j) What proportion of university women own no parrots?

LUDI 6.2. (CARRY IMPORTANT PHONE NUMBERS) Let X be a random variable representing the number of calls received in one hour by a 911 emergency service. A portion of the probability distribution of X is given below:

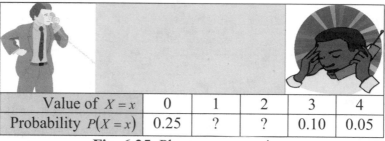

Value of $X = x$	0	1	2	3	4
Probability $P(X = x)$	0.25	?	?	0.10	0.05

Fig. 6.25. Phone conversation.

(a) Suppose the probability that $X = 1$ and the probability that $X = 2$ are the same. What are these probabilities?
(b) What is the expected number of 911 calls in one hour?
(**Hint**: Find mean)
(c) Find the standard deviation of the number of calls.
(**Hint**: Use the computing formula to find the variance)
(d) What is the probability of expecting more than three calls?
(e) What is probability of expecting at most three calls?

LUDI 6.3. (LEARNING COMBINATIONS) Calculate:

(a) $\binom{10}{3}$ (b) $\binom{12}{7}$ (c) $\binom{12}{8}$

LUDI 6.4. (THE LOTTERY IS EITHER A SUCCESS OR A FAILURE) Suppose the random variable X follows the binomial distribution with $n = 8$ and $p = 0.25$.

(a) What is the expected value of the random variable X?
(b) What is the variance of X?
(c) What is the standard deviation of X?
(d) What is $P(X = 3)$?
(e) What is $P(X \le 6)$?
(f) What is $P(X > 6)$?

Fig. 6.26. Never lose your balance in success or failure.

LUDI 6.5. (HELP FIND MISSING KIDS) Let the variable X represent the number of children per family in a small city. The distribution of X is:

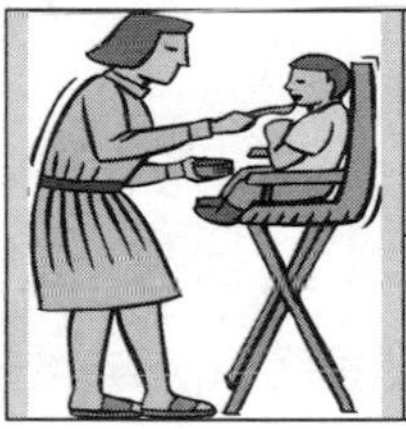

Value of X ($X = x$)	x_i	0	1	2	3	4
P($X = x$)	p_i	0.40	?	0.20	0.10	0.05

Fig. 6.27. Mother feeding her child.

(a) Complete the mass function of X and sketch it.
(b) What is the probability that a randomly selected household has at least two children?
(c) What is the probability that a randomly selected household has exactly two children?
(d) What is the expected number of children per household?
$$\text{Mean} = \mu = E(X) = \sum_i p_i x_i$$

(e) Find the variance of X by using the formula by definition.
$$\sigma^2 = \sum_i \left[p_i (x_i - \mu)^2 \right]$$

(f) Find the variance of X by using the computing formula.
$$\sigma^2 = \left[\sum_i (p_i x_i^2) \right] - (\mu^2)$$

(g) Find the standard deviation.

LUDI 6.6. (PRACTICE FORMULAE) There is a 70% chance of rain on a particular day. Three people leave from home independently.

(a) Complete the following table to find the probability that none, one, two, or all three will need an umbrella.

RAINY DAY				
x	0	1	2	3
$p(x)$				

Fig. 6.28. A rainy day.

(b) Find the standard deviation of X.
Hint: Given: $X \sim B(n, p)$ with $n = 3$ and $p = 0.7$.

LUDI 6.7. (MORE PRACTICE WITH FACTORIALS) A barber opened a new shop. He can have at most three appointments, say X, in each hour. The probability of having an appointment in a particular hour on a particular day has the probability mass function (p.m.f) given by the formula:

$$p(x) = \frac{1}{84}\binom{5}{x}\binom{4}{3-x} \quad \text{for } x = 0, 1, 2, 3$$

(a) Compute the numerical probabilities and fill in the following table.

HAIR CUT				
x	0	1	2	3
$p(x)$				

Fig. 6.29. A barbershop.

(b) Calculate the mean and standard deviation of the random variable X.

LUDI 6.8. (SPORT LOVERS) A basketball player gets three chances, $n = 3$, in a game to jump and score, say X. His probability to score, X, has been found to follow the probability mass function:

$$p(x) = \frac{12}{25}\left(\frac{1}{x+1}\right) \quad \text{for} \quad x = 0, 1, 2, 3$$

(a) Compute the numerical probabilities and fill in the following table.

SPORT LOVERS				

x	0	1	2	3
$p(x)$				

Fig. 6.30. A basketball player.

(b) Calculate the mean and standard deviation of the random variable X and interpret.

LUDI 6.9. (BE A SUCCESSFUL PERSON) In each case, find the probability of x successes in n Bernoulli trials with the probability of success p for each trial.

(a) $x = 2$, $n = 4$, $p = 1/3$.
(b) $x = 3$, $n = 6$, $p = 0.25$.
(c) $x = 2$, $n = 6$, $p = 0.75$.
(d) $x = 2$, $n = 2$, $p = 0.75$.
(e) $x = 2$, $n = 5$, $p - 0.5$.
(f) $x = 8$, $n = 10$, $p = 0.4$.

Fig. 6.31. Remember Mom's lunch box on every success in life.

LUDI 6.10. (LEARNING COMBINATIONS) Calculate:

(a) $\binom{10}{8}$ (b) $\binom{8}{6}$ (c) $\binom{6}{4}$

LUDI 6.11. (LIFE IS MADE OF SUCCESSES AND FAILURES)

For the binomial distribution with $n = 4$ and $p = 0.25$, find the probability of:

(a) no success.
(b) exactly one success.
(c) exactly two successes.
(d) exactly three successes.
(e) exactly four successes.
(f) three or more successes.
(g) at the most three successes.
(h) two or more failures.

Fig. 6.32. Remember your teachers on every success in life.

LUDI 6.12. (THE CHANCE TO WIN A GAME IS LESS) Use the formula for the Poisson distribution to find the following probabilities:

(a) $P(x = 2)$ with $\lambda = 2$.
(b) $P(x \leq 2)$ with $\lambda = 2$.
(c) $P(x \geq 1)$ with $\lambda = 2$.
(d) $P(x = 2)$ with $\lambda = 2.5$.
(e) $P(x \leq 2)$ with $\lambda = 2.5$.
(f) $P(x \geq 1)$ with $\lambda = 2.5$.
(g) $P(x = 3)$ with $\lambda = 2.5$.
(h) $P(x \leq 3)$ with $\lambda = 2.5$.
(i) $P(x > 3)$ with $\lambda = 2.5$.
(j) $P(2 < x \leq 5)$ with $\lambda = 2.5$.
(k) $P(2 \leq x \leq 5)$ with $\lambda = 2.5$.

Fig. 6.33. Beware of any poison.

LUDI 6.13. (IDEAL PEOPLE ARE ABSOLUTELY RIGHT)

Verify that the probability mass function (p.m.f) of a discrete random variable X is given by:

$$p(x) = \frac{6 - |x - 7|}{36}, \text{ for } x = 2, 3, 4, \ldots, 12$$

and derive the variance of X.

Fig. 6.34. Math is fun.

LUDI 6.14. (BEACH VISITORS) The number of swimmers the shark can bite on the beach follows the Poisson distribution with a mean of 0.002.

Fig. 6.35. A beach outlook.

(a) What is the probability that none of the swimmers will be bitten?
(b) What is the probability that one of the swimmers will be bitten?
(c) What is the probability that at least one of the swimmers will be bitten?
(d) What is the probability that at most one of the swimmers will be bitten?

LUDI 6.15. (WRONG NUMBERS) The probability of getting a wrong number during a day is $3.25e^{-3.25}$

(a) What is the probability that there will be no wrong numbers called?
(b) What is the probability that there will be more than one wrong call?
(c) What is the probability that there will be more than two wrong calls?

Fig. 6.36. On phones.

LUDI 6.16. (CUTE RESULT: 0!=1 and 1!=1) Verify if the probability mass function (p.m.f) of a discrete random variable X is given by:

$$p(x) = \frac{\binom{2}{x}\binom{4}{3-x}}{\binom{6}{3}}, \text{ for } x = 0,1,2$$

and derive the variance of X.

Fig. 6.37. Math is fun.

LUDI 6.17. (ALWAYS WATCH YOUR LUGGAGE) A thief tries to open three, $n = 3$, locks. Assume that on each trial, the probability of opening a lock is 0.6, that is, the probability of success $p = 0.6$. Let the random variable X denote the number of locks opened by the thief.

(a) Complete the following table:

BEAWARE OF THIEVES				
Number of locks opened (x)	0	1	2	3
Find $p(X = x)$				

Fig. 6.38. A thief carrying a key.

Hint: Use binomial distribution: $p(X = x) = \binom{n}{x} p^x q^{(n-x)}$, and find:

$p(X = 0) = - - - - - - - -$
$p(X = 1) = - - - - - - - -$
$p(X = 2) = - - - - - - - -$
$p(X = 3) = - - - - - - - -$
(b) Find the expected value of X.
(c) Find the standard deviation of X.

LUDI 6.18. (FOUR QUARTERS MAKE A DOLLAR) A fair coin is tossed four times. List the elements of the sample space that are presumed to be equally likely and the corresponding values of x of the random variable X, the total number of heads.
(a) Construct a probability mass function table, and sketch it with a stick graph.
(b) Find the expected value of the number of heads.
(c) Find the variance of the number of heads.
(d) Find the standard deviation. Is it a parameter or a statistic?

LUDI 6.19. (READ A FIRST TIME TYPED MANUSCRIPT VERY CAREFULLY) A typist knows from previous experience that even though he is an experienced typist, he does make mistakes. Let X be the random variable representing the number of mistakes made by the typist on a randomly selected page from all those he has typed. The probability mass function of X is given by:

Number of Mistakes (X = x)	x_i	0	1	2	3	4	5
P(X = x)	p_i	0.50	0.10		0.15	0.08	0.01

Fig. 6.39. A typist.

(a) Complete the probability mass function of X and sketch it.
(b) What is the probability that a randomly selected page has at least two mistakes?
(c) What is the probability that a randomly selected page has exactly two mistakes?
(d) What is the expected number of mistakes per page?
(e) Find the variance of X using the formula by definition.

$$\sigma^2 = \sum_i \left[p_i (x_i - \mu)^2 \right]$$

(f) Find the variance of X using the computing formula.

$$\sigma^2 = \left[\sum_i \left(p_i x_i^2 \right) \right] - \left(\mu^2 \right)$$

(g) Find the standard deviation.

LUDI 6.20. (BE SMART)
(a) A random variable X follows a binomial distribution with a mean of 20 and a variance of 16. Find the values of n, p, and q. State if you observe anything wrong.
(b) A random variable X follows a binomial distribution with a mean of 40 and a standard deviation of 5. Find the values of n, p and q. Comment on these values.
(c) Comment on the statement: A random variable X follows a binomial distribution with a mean of 16 and a variance of 20.
(d) Comment on the statement: A random variable X follows the Poisson distribution with a mean of 6.25 and a variance of 16.

LUDI 6.21. (KIDS LEARN ALL TRADES) A kid tries to hit a target at archery camp. The probability of an arrow hitting the bull's-eye is 0.70.

Fig. 6.40. Kids scout camp.

The kid makes three tries. Find the probability that there will be:
(a) no successful hits.
(b) one successful hit.
(c) two successful hits.
(d) all three successful hits.
(e) more than one successful hit.

LUDI 6.22. (HUNTING IS NOT FUN) Consider four birds flying in the sky. A hunter can hit a flying bird, in a single trial, with a probability of 0.2 and makes four independent trials, as shown below.

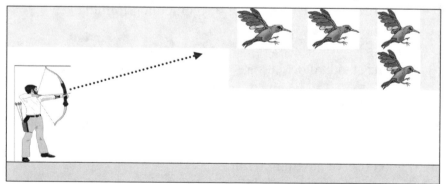

Fig. 6.41. Hitting birds in the sky.

Hint: Use binomial distribution: $p(X = x) = \binom{n}{x} p^x q^{(n-x)}$, and find:

(a) $p(X = 0)$, $p(X = 1)$, $p(X = 2)$, $p(X = 3)$, and $p(X = 4)$.
(b) What is the probability of at least three hits?
(c) What is the probability of at most three hits?
(d) Find the standard deviation of X.

LUDI 6.23. (SAFE DRIVING IS VERY IMPORTANT) An engineer designed a subway for trains on a circular track and also designed a bike path that intersects the train track.

Fig. 6.42. Avoid accidents.

Although the cyclists are very careful in crossing the train track, there is a probability that a train and a cyclist may cross the intersection at the same time, leading to an accident. Experience shows that the probability of an accident happening is 0.00020. If 10,000 cyclists cross the train track, find the probability that there will be:
(a) no accidents.
(b) one accident.
(c) at least two accidents.
(d) at the most four accidents.
(e) more than three but less than seven accidents.
(f) either less than three or more than five accidents.
(g) either seven or thirteen accidents.

LUDI 6.24. (POWER OF THE PRESS)   The probability of no typos on a page of a newspaper is $e^{-2.5}$.

(a) What is the probability that a page contains exactly one typo?
(b) What is the probability that a page contains exactly two typos?
(c) What is the probability that a page contains more than two typos?

Fig. 6.43. Update yourself.

LUDI 6.25. (MILITARY TRAINING) An army officer plans to hit a target with five independent and identical missiles. The chance that a missile will hit the target is 0.85.

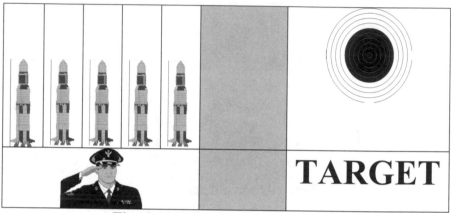

Fig. 6.44. Scientists in the military.

(I) Find the probability that:
(a) None of the missiles will hit the target.
(b) One missile will hit the target.
(c) Two missiles will hit the target.
(d) Three missiles will hit the target
(e) Four missiles will hit the target.
(f) All five missiles will hit the target.
(g) Sketch the number of hits versus the probability of hits.
(h) At least three missiles will hit the target.
(i) At most three missiles will hit the target.
(j) The third and fifth missiles will hit the target.
(k) The third or the fifth missiles will hit the target.
(l) Find the average number of hits.
(m) Find the variance of the number of hits.
(II) Repeat the above LUDI if the chance of hitting the target is 0.15, and write a short paragraph on the differences you observed.
(III) Repeat the above LUDI if the chance of hitting the target is 0.50, and write a short paragraph on the differences you observed.

7. CONTINUOUS DISTRIBUTIONS

7.1 INTRODUCTION

In this chapter we discuss continuous random variables and two continuous distributions: uniform distribution and normal distribution.

7.2 CONTINUOUS RANDOM VARIABLE

A variable, X, is said to be a continuous random variable if it assumes all possible real values x on a line or an interval of a line in response to the outcome of an experiment.

For example, the GPA of a student, X, can be any value between 0 and 4, that is:

$$x = 0.00, 0.001, ... , 4.00$$

Note that there are an infinite number of values for GPA between $x = 0.00$ and $x = 4.00$. The use of an uppercase X is a random variable and a lowercase x is its particular value.

7.3 PROBABILITY DENSITY FUNCTION

A probability density function (p.d.f) of a continuous random variable, X, is a non-negative function or curve that describes the overall shape of its distribution. The total area under the entire curve is unity (one) and the proportions are measured as area under the probability density function.

Consider a seller selling 1,000 pens in a year.

Fig.7.1. Pen.

At the end of the year, consider the following frequency distribution table obtained from a list of prices of 1,000 pens in the store.

Class Boundary Price ($)	Class width CW = UB-LB	Frequency (f)	Relative Frequency (RF)	Height of the bar H = RF/CW
2 to less than 4	4 – 2 = 2	70	0.070	0.070/2 = 0.035
4 to less than 6	6 – 4 = 2	132	0.132	0.132/2 = 0.066
6 to less than 8	8 – 6 = 2	190	0.190	0.190/2 = 0.095
8 to less than 10	10 – 8 = 2	216	0.216	0.216/2 = 0.108
10 to less than 12	12 – 10 = 2	190	0.190	0.190/2 = 0.095
12 to less than 14	14 –12 = 2	132	0.132	0.132/2 = 0.066
14 to less than 16	16 – 14 = 2	70	0.070	0.070/2 = 0.035
	Sum	**1,000**	**1.000**	

where **UB** and **LB** stand for upper and lower class boundaries.

Thus, there were 70 pens having a price between $2 and $4, 132 pens having a price between $4 and $6, and so on. From the above table, we can make the following histogram:

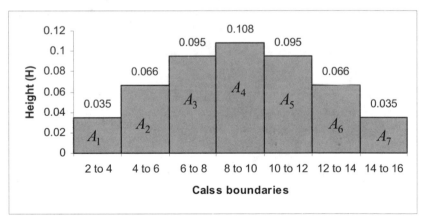

Fig. 7.2. Histogram.

Total Area = Sum of the areas of all the rectangles

$$= A_1 + A_2 + A_3 + A_4 + A_5 + A_6 + A_7$$
$$= (0.035 \times 2) + (0.066 \times 2) + (0.095 \times 2) + (0.108 \times 2) + (0.095 \times 2)$$
$$+ (0.066 \times 2) + (0.035 \times 2)$$
$$= 0.070 + 0.132 + 0.190 + 0.216 + 0.190 + 0.132 + 0.070$$
$$= 1.000$$

Such a graph is called a probability density function (p.d.f).

Two well-known examples of continuous distributions are:

(a) Uniform distribution   (b) Normal distribution.

7.4 UNIFORM DISTRIBUTION

A uniform distribution of a continuous random variable X is a flat, rectangular shaped curve in a given interval of a line from a to b.

We say:

$$X \sim U(a, b)$$

that is, X follows a uniform distribution between a and b.

Fig.7.3. A good sprinkler has a uniform distribution.

The general form of a uniform probability density function (p.d.f) is given by:

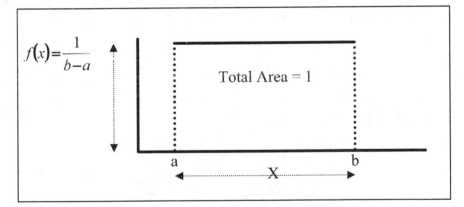

$$f(x) = \frac{1}{b-a}$$

Total Area = 1

a

b

X

Fig. 7.4. Uniform distribution.

In the above graph, we have:

       Minimum value of X = a

       Maximum value of X = b

       Length of the rectangle = Range of X $= (b - a)$

Note that:

       Total area under the rectangle = 1.

So:

$$\text{Height of the rectangle} = f(x) = \frac{1}{(b-a)}$$

Thus, we can state the following:

If the values of a random variable X are uniformly distributed between any two real numbers a and b, then the probability density function (p.d.f) of X is given by:

$$f(x) = \begin{cases} \dfrac{1}{(b-a)} & \text{if } a \leq x \leq b \\ 0 & \text{otherwise} \end{cases}$$

Note the following results:

(a) The mean and median values of a random variable X having a uniform distribution are the same and are given by:

$$\text{Mean} = \text{Medain} = \frac{a+b}{2}$$

(b) There is no mode value, that is, all values have equal frequency or relative frequency.

(c) The range of the data values is given by:

$$\text{Range} = (\text{Maximum value}) - (\text{Minimum value}) = b - a$$

(d) The variance of a random variable X having a uniform distribution is given by:

$$\text{Variance} = V(X) = \sigma^2 = \frac{(b-a)^2}{12}$$

Read as "Sigma square"

(e) The area under any single point for a continuous random variable is always zero, because a single point has no length and no width.

(f) The area between any two points c and d is given by:

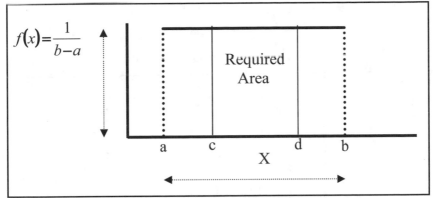

$$f(x) = \frac{1}{b-a}$$

Required Area

a   c   d   b

X

Fig. 7.5. Uniform distribution.

Required Area = length × width = length × height

$$= (d - c) \times f(x) = (d - c) \times \frac{1}{(b-a)}$$

Example 7.1. (FARMERS NEED UNIFORM DISTRIBUTION)
The variable X is $U(19, 29)$.

(a) Write $f(x)$ and sketch it.

(b) Find the minimum and maximum values of X.

(c) Find the mean and median values.

(d) Find the proportion of values below 19.

(e) Find the proportion of values above 22.

(f) Find the proportion of values between 22 and 25.

(g) Find the proportion of values equal to 22.

(h) Find the variance of X.

(i) Find the range of X.

Fig. 7.6. Farmers.

Solution. Here $a = 19$ and $b = 29$.

(a)

$$f(x) = \begin{cases} \dfrac{1}{b-a} & a \le x \le b \\ 0 & \text{otherwise} \end{cases} = \begin{cases} \dfrac{1}{29-19} & 19 \le x \le 29 \\ 0 & \text{otherwise} \end{cases}$$

$$= \begin{cases} \dfrac{1}{10} & 19 \leq x \leq 29 \\ 0 & \text{otherwise} \end{cases}$$

Thus we have:

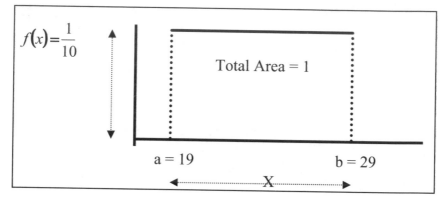

Fig. 7.7. Uniform distribution.

(b) Minimum value: $a = 19$ and Maximum value: $b = 29$.

(c) Mean = Median = $\dfrac{a+b}{2} = \dfrac{19+29}{2} = \dfrac{48}{2} = 24$.

(d) Proportion of values below 19 or $(X < 19) = 0$.

(e) Proportion of values above 22:

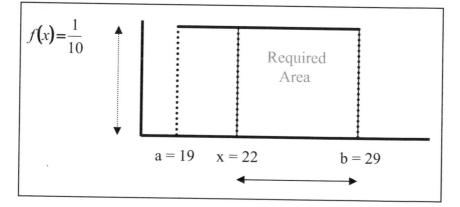

Fig. 7.8. Uniform distribution.

Required area or proportion = length × width

$$= (29 - 22) \times \text{height}$$

$$= 7 \times \frac{1}{10}$$

$$= 0.7 \quad \text{or} \quad 70\%$$

(f) Proportion of values between 22 and 25:

Required area or proportion = length × width

$$= (25 - 22) \times \text{height}$$

$$= 3 \times \frac{1}{10}$$

$$= 0.3 \quad \text{or} \quad 30\%$$

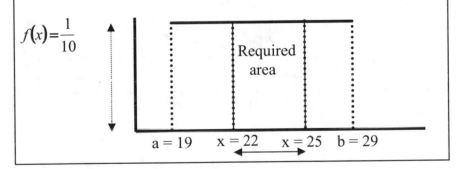

Fig. 7.9. Uniform distribution.

(g) Proportion of values equal to 22:   Because it is a continuous distribution, the area under a single point is zero. Thus, the proportion of values equal to 22 is zero.

(h) The variance of X is given by:

$$\text{Variance} = V(X) = \sigma^2 = \frac{(b-a)^2}{12} = \frac{(29-19)^2}{12} = \frac{10^2}{12} = \frac{100}{12} = 8.33.$$

(i) The range of X is given by:

$$\text{Range} = b - a = 29 - 19 = 10.$$

Example 7.2. (FARMERS NEED UNIFORM DISTRIBUTION)
Five cities are located on a straight road five miles long. The road administrators decided to decorate the side of the road with maple trees.

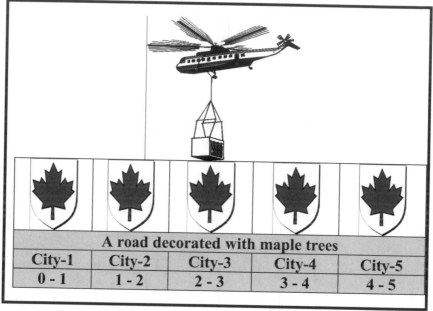

A road decorated with maple trees				
City-1	City-2	City-3	City-4	City-5
0 - 1	1 - 2	2 - 3	3 - 4	4 - 5

Fig. 7.10. Distribution of maple trees.

Assume that there is no gap from city to city and all the cities are of equal size (the road length of each city is one mile). It was decided to grow 1,000 maple trees in the five cities along the road. A helicopter is used to uniformly distribute all 1,000 seeds on the road among these five cities. Assuming that all the seeds will grow, then:

(a) How many maple trees are expected to grow in each city?
(b) Construct a frequency distribution table.
(c) Construct an appropriate graph to present it.
(d) If possible, find the total area under the graph.

Solution. (a) We are given that the distribution of trees among the five cities is uniform. Therefore, we expect an equal number of seeds distributed among them, and hence the expected number of maple trees in each city:

$$= \frac{\text{Total no. of seeds}}{\text{No. of cities}} = \frac{1,000}{5} = 200 \text{ trees}.$$

(b) A frequency distribution table is given below:

Cities	Class Boundaries	No. of seeds (f)	Relative Frequency
City 1	0.00 to less than 1.00	200	0.2
City 2	1.00 to less than 2.00	200	0.2
City 3	2.00 to less than 3.00	200	0.2
City 4	3.00 to less than 4.00	200	0.2
City 5	4.00 to less than 5.00	200	0.2
Sum		1,000	1.0

(c) If the cities are joined together with no gap, then we shall make a histogram.

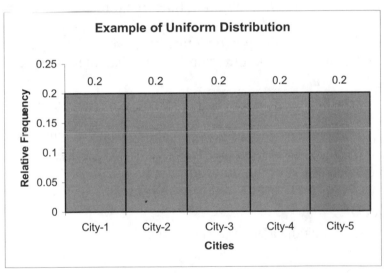

Fig. 7.11. Uniform distribution.

(d) If we assume that each city represents one mile, then the relative frequency gives the proportion of maple trees distributed per mile, and the area of each rectangle is 0.2. Hence, the total area under all the five rectangles is one.

Example 7.3. (ALWAYS CARE FOR PUBLIC PROPERTY) Consider the following density function giving the waiting time (in minutes) of a train in a subway:

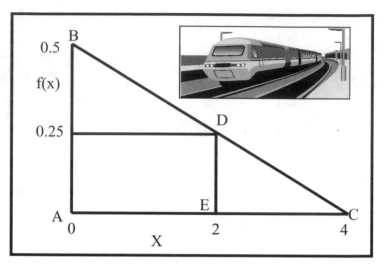

Fig.7.12. Waiting time of a train.

(a) Are you sure that the above picture gives you a density function?
(b) What is the maximum time you have to wait after entering the subway?
(c) What is the minimum time you have to wait after entering the subway?
(d) What proportion of times do you have to wait more than two minutes?
(e) The value of $X = 2$ minutes is called $- - - - - - -$

(i) Q_1 (ii) Q_2 (iii) Q_3 (iv) Mean

Solution.

(a) We know that the area of a triangle is:

$$= \frac{1}{2} \times \text{Height} \times \text{Base} = \frac{1}{2} \times \frac{1}{2} \times 4 = 1.$$

Since the total area under the big triangle ABC = 1, it is a density function.

(b) Maximum time = 4 minutes.

(c) Minimum time = 0 minutes.

(d) Area of the small triangle DEC $= \frac{1}{2} \times 0.25 \times 2 = 0.25$.

That is, 25% of the time you may have to wait more than 2 minutes if you enter the subway randomly several times.

(e) The value of $X = 2$ is called the third quartile Q_3 because 75% of the area lies below it.

7.5 NORMAL DISTRIBUTION

A normal distribution of a continuous random variable X is a symmetric and bell shaped curve. It has two tails: the right tail and left tail. The mean, median and mode are identical. Its mean is denoted by a Greek letter *mu* (μ) and the spread is denoted by a Greek lowercase letter *sigma* (σ). The total area under the normal curve is one, and 50% of the area lies around the central value. If a random variable X follows normal distribution, we say:

$$X \sim N(\mu, \quad \sigma)$$

X follows a normal distribution with a mean mu and standard deviation sigma

Fig. 7.13. Bell.

A pictorial representation of a normal distribution is given below:

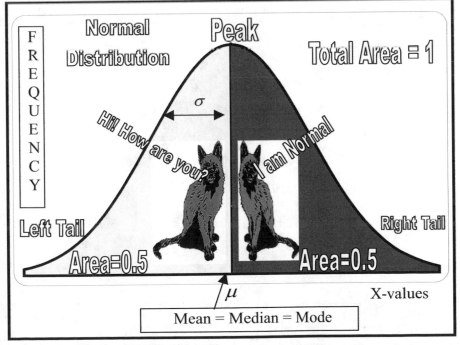

Fig. 7.14. Normal distribution.

7.5.1 POPULATION Z-SCORE

The population Z-score or standard Z-score or standard normal variate for an observation tells us how far the observed value is from the population mean in standard deviation units. Note that the value of X goes from negative infinity to positive infinity, that is $-\infty < X < +\infty$, and hence the value of the Z-score also goes from negative infinity to positive infinity, that is $-\infty < Z < +\infty$.

It is defined as:

$$Z = \frac{X - \mu}{\sigma}$$

Note that if $X = \mu$, then $Z = \frac{X - \mu}{\sigma} = \frac{\mu - \mu}{\sigma} = 0$. This means that at the mean value of the X variable, the value of Z is always zero. Further, note that if X is more than μ, then Z is positive; and if X is less than μ, then Z is negative. Also, note that the variance of the standard normal variate Z is always one. The variable X may have any unit of measurement like *kg, $, lbs* etc., but the Z-score is a unit free number between $-\infty$ and $+\infty$.

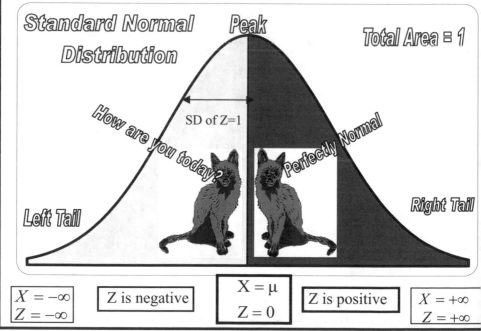

Fig. 7.15. Standard normal distribution.

Thus, we have the following result:

If $X \sim N(\mu, \sigma)$ then $Z \sim N(0, 1)$. In other words, if any variable X follows a normal distribution with a mean mu (μ) and standard deviation sigma (σ), then the Z-score follows a standard normal distribution with a mean zero (0) and a standard deviation one (1).

The following **Table II** (also cited as **Table II** in the Appendix) gives the area under the standard normal curve to the left side of the Z value as marked on the horizontal axis in the right corner of the table.

Table II. Area under the standard normal curve.

Area to the left side of the Z value

In this table:
  Minimum value of the Z = -3.49
  Maximum value of the Z = +3.49

Z	0.00	0.01	0.02	0.03	0.04	0.05	0.06	0.07	0.08	0.09
-3.4	0.0003	0.0003	0.0003	0.0003	0.0003	0.0003	0.0003	0.0003	0.0003	0.0002
-3.3	0.0005	0.0005	0.0005	0.0004	0.0004	0.0004	0.0004	0.0004	0.0004	0.0003
-3.2	0.0007	0.0007	0.0006	0.0006	0.0006	0.0006	0.0006	0.0005	0.0005	0.0005
-3.1	0.0010	0.0009	0.0009	0.0009	0.0008	0.0008	0.0008	0.0008	0.0007	0.0007
-3.0	0.0013	0.0013	0.0013	0.0012	0.0012	0.0011	0.0011	0.0011	0.0010	0.0010
-2.9	0.0019	0.0018	0.0018	0.0017	0.0016	0.0016	0.0015	0.0015	0.0014	0.0014
-2.8	0.0026	0.0025	0.0024	0.0023	0.0023	0.0022	0.0021	0.0021	0.0020	0.0019
-2.7	0.0035	0.0034	0.0033	0.0032	0.0031	0.0030	0.0029	0.0028	0.0027	0.0026
-2.6	0.0047	0.0045	0.0044	0.0043	0.0041	0.0040	0.0039	0.0038	0.0037	0.0036
-2.5	0.0062	0.0060	0.0059	0.0057	0.0055	0.0054	0.0052	0.0051	0.0049	0.0048
-2.4	0.0082	0.0080	0.0078	0.0075	0.0073	0.0071	0.0069	0.0068	0.0066	0.0064
-2.3	0.0107	0.0104	0.0102	0.0099	0.0096	0.0094	0.0091	0.0089	0.0087	0.0084
-2.2	0.0139	0.0136	0.0132	0.0129	0.0125	0.0122	0.0119	0.0116	0.0113	0.0110
-2.1	0.0179	0.0174	0.0170	0.0166	0.0162	0.0158	0.0154	0.0150	0.0146	0.0143
-2.0	0.0228	0.0222	0.0217	0.0212	0.0207	0.0202	0.0197	0.0192	0.0188	0.0183
-1.9	0.0287	0.0281	0.0274	0.0268	0.0262	0.0256	0.0250	0.0244	0.0239	0.0233
-1.8	0.0359	0.0351	0.0344	0.0336	0.0329	0.0322	0.0314	0.0307	0.0301	0.0294
-1.7	0.0446	0.0436	0.0427	0.0418	0.0409	0.0401	0.0392	0.0384	0.0375	0.0367
-1.6	0.0548	0.0537	0.0526	0.0516	0.0505	0.0495	0.0485	0.0475	0.0465	0.0455
-1.5	0.0668	0.0655	0.0643	0.0630	0.0618	0.0606	0.0594	0.0582	0.0571	0.0559
-1.4	0.0808	0.0793	0.0778	0.0764	0.0749	0.0735	0.0721	0.0708	0.0694	0.0681
-1.3	0.0968	0.0951	0.0934	0.0918	0.0901	0.0885	0.0869	0.0853	0.0838	0.0823
-1.2	0.1151	0.1131	0.1112	0.1093	0.1075	0.1056	0.1038	0.1020	0.1003	0.0985
-1.1	0.1357	0.1335	0.1314	0.1292	0.1271	0.1251	0.1230	0.1210	0.1190	0.1170
-1.0	0.1587	0.1562	0.1539	0.1515	0.1492	0.1469	0.1446	0.1423	0.1401	0.1379

	.00	.01	.02	.03	.04	.05	.06	.07	.08	.09
-0.9	0.1841	0.1814	0.1788	0.1762	0.1736	0.1711	0.1685	0.1660	0.1635	0.1611
-0.8	0.2119	0.2090	0.2061	0.2033	0.2005	0.1977	0.1949	0.1922	0.1894	0.1867
-0.7	0.2420	0.2389	0.2358	0.2327	0.2296	0.2266	0.2236	0.2206	0.2177	0.2148
-0.6	0.2743	0.2709	0.2676	0.2643	0.2611	0.2578	0.2546	0.2514	0.2483	0.2451
-0.5	0.3085	0.3050	0.3015	0.2981	0.2946	0.2912	0.2877	0.2843	0.2810	0.2776
-0.4	0.3446	0.3409	0.3372	0.3336	0.3300	0.3264	0.3228	0.3192	0.3156	0.3121
-0.3	0.3821	0.3783	0.3745	0.3707	0.3669	0.3632	0.3594	0.3557	0.3520	0.3483
-0.2	0.4207	0.4168	0.4129	0.4090	0.4052	0.4013	0.3974	0.3936	0.3897	0.3859
-0.1	0.4602	0.4562	0.4522	0.4483	0.4443	0.4404	0.4364	0.4325	0.4286	0.4247
-0.0	0.5000	0.4960	0.4920	0.4880	0.4840	0.4801	0.4761	0.4721	0.4681	0.4641
				- - - - - Be careful - - - - -						
+0.0	0.5000	0.5040	0.5080	0.5120	0.5160	0.5199	0.5239	0.5279	0.5319	0.5359
+0.1	0.5398	0.5438	0.5478	0.5517	0.5557	0.5596	0.5636	0.5675	0.5714	0.5753
+0.2	0.5793	0.5832	0.5871	0.5910	0.5948	0.5987	0.6026	0.6064	0.6103	0.6141
+0.3	0.6179	0.6217	0.6255	0.6293	0.6331	0.6368	0.6406	0.6443	0.6480	0.6517
+0.4	0.6554	0.6591	0.6628	0.6664	0.6700	0.6736	0.6772	0.6808	0.6844	0.6879
+0.5	0.6915	0.6950	0.6985	0.7019	0.7054	0.7088	0.7123	0.7157	0.7190	0.7224
+0.6	0.7257	0.7291	0.7324	0.7357	0.7389	0.7422	0.7454	0.7486	0.7517	0.7549
+0.7	0.7580	0.7611	0.7642	0.7673	0.7704	0.7734	0.7764	0.7794	0.7823	0.7852
+0.8	0.7881	0.7910	0.7939	0.7967	0.7995	0.8023	0.8051	0.8078	0.8106	0.8133
+0.9	0.8159	0.8186	0.8212	0.8238	0.8264	0.8289	0.8315	0.8340	0.8365	0.8389
+1.0	0.8413	0.8438	0.8461	0.8485	0.8508	0.8531	0.8554	0.8577	0.8599	0.8621
+1.1	0.8643	0.8665	0.8686	0.8708	0.8729	0.8749	0.8770	0.8790	0.8810	0.8830
+1.2	0.8849	0.8869	0.8888	0.8907	0.8925	0.8944	0.8962	0.8980	0.8997	0.9015
+1.3	0.9032	0.9049	0.9066	0.9082	0.9099	0.9115	0.9131	0.9147	0.9162	0.9177
+1.4	0.9192	0.9207	0.9222	0.9236	0.9251	0.9265	0.9279	0.9292	0.9306	0.9319
+1.5	0.9332	0.9345	0.9357	0.9370	0.9382	0.9394	0.9406	0.9418	0.9429	0.9441
+1.6	0.9452	0.9463	0.9474	0.9484	0.9495	0.9505	0.9515	0.9525	0.9535	0.9545
+1.7	0.9554	0.9564	0.9573	0.9582	0.9591	0.9599	0.9608	0.9616	0.9625	0.9633
+1.8	0.9641	0.9649	0.9656	0.9664	0.9671	0.9678	0.9686	0.9693	0.9699	0.9706
+1.9	0.9713	0.9719	0.9726	0.9732	0.9738	0.9744	0.9750	0.9756	0.9761	0.9767
+2.0	0.9772	0.9778	0.9783	0.9788	0.9793	0.9798	0.9803	0.9808	0.9812	0.9817
+2.1	0.9821	0.9826	0.9830	0.9834	0.9838	0.9842	0.9846	0.9850	0.9854	0.9857
+2.2	0.9861	0.9864	0.9868	0.9871	0.9875	0.9878	0.9881	0.9884	0.9887	0.9890
+2.3	0.9893	0.9896	0.9898	0.9901	0.9904	0.9906	0.9909	0.9911	0.9913	0.9916
+2.4	0.9918	0.9920	0.9922	0.9925	0.9927	0.9929	0.9931	0.9932	0.9934	0.9936
+2.5	0.9938	0.9940	0.9941	0.9943	0.9945	0.9946	0.9948	0.9949	0.9951	0.9952
+2.6	0.9953	0.9955	0.9956	0.9957	0.9959	0.9960	0.9961	0.9962	0.9963	0.9964
+2.7	0.9965	0.9966	0.9967	0.9968	0.9969	0.9970	0.9971	0.9972	0.9973	0.9974
+2.8	0.9974	0.9975	0.9976	0.9977	0.9977	0.9978	0.9979	0.9979	0.9980	0.9981
+2.9	0.9981	0.9982	0.9982	0.9983	0.9984	0.9984	0.9985	0.9985	0.9986	0.9986
+3.0	0.9987	0.9987	0.9987	0.9988	0.9988	0.9989	0.9989	0.9989	0.9990	0.9990
+3.1	0.9990	0.9991	0.9991	0.9991	0.9992	0.9992	0.9992	0.9992	0.9993	0.9993
+3.2	0.9993	0.9993	0.9994	0.9994	0.9994	0.9994	0.9994	0.9995	0.9995	0.9995
+3.3	0.9995	0.9995	0.9995	0.9996	0.9996	0.9996	0.9996	0.9996	0.9996	0.9997
+3.4	0.9997	0.9997	0.9997	0.9997	0.9997	0.9997	0.9997	0.9997	0.9997	0.9998

Source: Generated in Excel using the $NORMSDIST(x)$.

Let us now learn how to use **Table II**.

Example 7.4. (THE Z-TABLE IS THE BACKBONE OF ALL STATISTICS COURSES) Obtain the area under the standard normal curve:
(a) to the left of $Z = -1.96$.
(b) to the left of $Z = -0.55$.
(c) to the right of $Z = +1.96$.
(d) to the right of $Z = +1.83$.
Solution. (a) Area to the left of $Z = -1.96$:

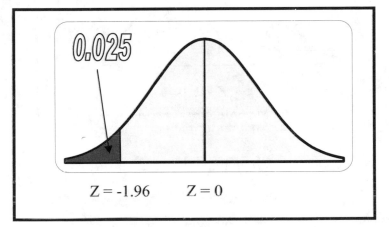

Z = -1.96   Z = 0

Fig. 7.16. Area under the normal curve.

Thus

$$A(Z < -1.96) = A(\text{to the left of } Z\text{-}1.96) = 0.025 .$$

(b) Area to the left of $Z = -0.55$:

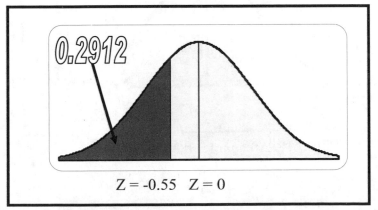

Z = -0.55   Z = 0

Fig. 7.17. Area under the normal curve.

Thus

$$A(Z < -0.55) = A(\text{to the left of } Z = -0.55) = 0.2912.$$

(c) Area to the right of $Z = +1.96$:

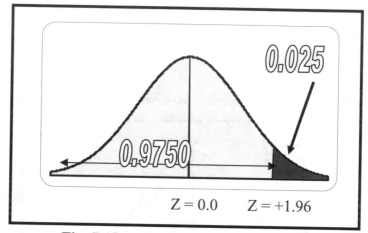

Fig. 7.18. Area under the normal curve.

$$\text{Required Area} = 1 - A(\text{to the left of } Z = +1.96)$$
$$= 1 - 0.9750 = 0.0250.$$

(d) Area to the right of $Z = +1.83$:

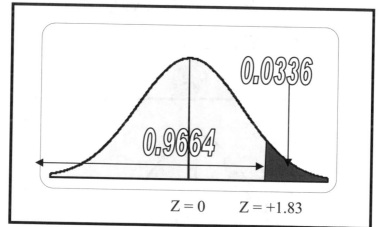

Fig. 7.19. Area under the normal curve.

$$A(\text{to the right of } Z = +1.83) = 1 - A(\text{to the left of } Z = +1.83)$$
$$= 1 - 0.9664 = 0.0336.$$

Example 7.5. (MORE PRACTICE WITH THE Z-TABLE) Find
the area under the standard normal curve:
(a) between $Z = 0$ and $Z = +1.90$.
(b) between $Z = 0$ and $Z = +1.75$.
(c) between $Z = +1.27$ and $Z = +2.37$.
(d) between $Z = -2.78$ and $Z = -1.53$.
(e) between $Z = -1.67$ and $Z = +2.34$.
(f) Area under the single point $Z = +2.16$.
Solution (a) Area between $Z = 0$ and $Z = +1.90$:

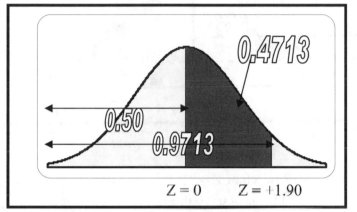

Fig. 7.20. Area under the normal curve.

Required area $= A$(to the left of $Z = +1.90$) $- A$(to the left of $Z = 0.00$)
$\qquad = 0.9713 - 0.5000 = 0.4713$.

(b) Area between $Z = 0$ and $Z = +1.75$:

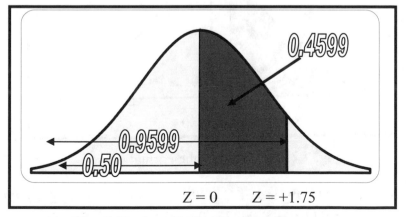

Fig. 7.21. Area under the normal curve.

Required area = A(to the left of $Z = +1.75$) - A(to the left of $Z = 0.00$)
$$= 0.9599 - 0.5000 = 0.4599.$$

(c) Area between $Z = +1.27$ and $Z = +2.37$:

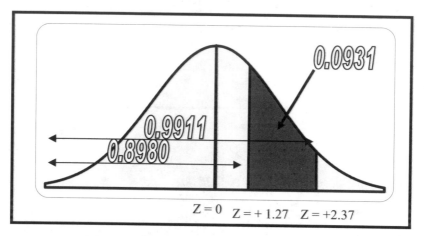

Fig. 7.22. Area under the normal curve.

Required area = A(to the left of $Z = +2.37$) - A(to the left of $Z = +1.27$)
$$= 0.9911 - 0.8980 = 0.0931.$$

(d) Area between $Z = -2.78$ and $Z = -1.53$:

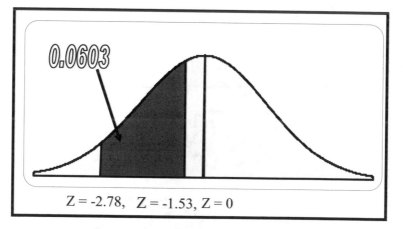

Fig. 7.23. Area under the normal curve.

Required area = A(to the left of $Z = -1.53$) - A(to the left of $Z = -2.78$)
$$= 0.0630 - 0.0027 = 0.0603.$$

(c) Area between $Z = -1.67$ and $Z = +2.34$:

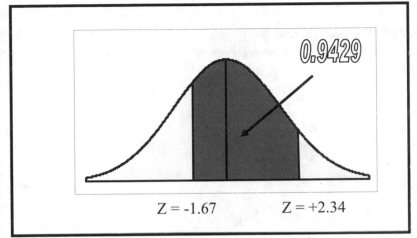

Fig. 7.24. Area under the normal curve.

Required area $= A\left(\text{to the left of } Z = +\, 2.34\right) - A\left(\text{to the left of } Z = -1.67\right)$
$$= 0.9904 - 0.0475 = 0.9429.$$

(f) Area under the single point $Z = +2.16$:

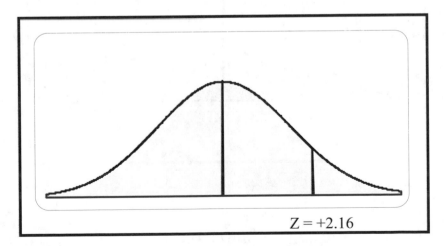

Fig. 7.25. Area under the normal curve.

Remember that the area under a single point for a continuous variable is always zero, because a single point has no length and no width.

Example 7.6. (USE OF Z-SCORE) Find the following areas under a normal distribution curve with a mean of 20 and a standard deviation of 4.

(a) Area to the left of $X = 7.76$.
(b) Area to the right of $X = 7.76$.
(c) Area between $X = 7.76$ and $X = 12$.
(d) Area between $X = 23$ and $X = 25$.
(e) Area to the left of $X = 23$.
(f) Area to the right of $X = 25$.
Solution. We are given:

$$\mu = 20 \text{ and } \sigma = 4.$$

(a) Area to the left of $X = 7.76$:

If $X = 7.76$ then $Z = \dfrac{X - \mu}{\sigma} = \dfrac{7.76 - 20}{4} = -3.06$.

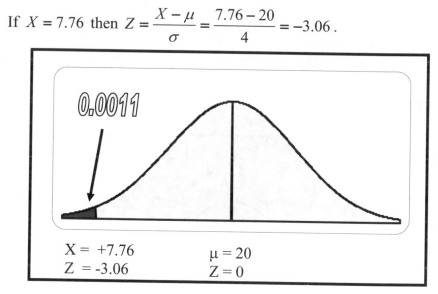

X = +7.76   μ = 20
Z = -3.06   Z = 0

Fig. 7.26. Area under normal curve.

Thus:

$A(\text{to the left of } X = 7.76) = A(\text{to the left of } Z = -3.06) = 0.0011$.

(b) Area to the right of $X = 7.76$:

If $X = 7.76$ then $Z = \dfrac{X - \mu}{\sigma} = \dfrac{7.76 - 20}{4} = -3.06$.

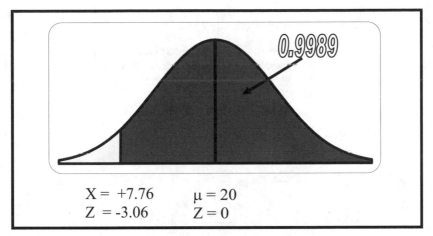

$$X = +7.76 \qquad \mu = 20$$
$$Z = -3.06 \qquad Z = 0$$

Fig. 7.27. Area under normal curve.

Thus:

$A(\text{to the right of } X = +7.76) = 1 - A(\text{to the left of } Z = -3.06)$
$$= 1 - 0.0011 = 0.9989.$$

(c) Area between $X = 7.76$ and $X = 12$:

If $X = 7.76$ then $Z = \dfrac{X - \mu}{\sigma} = \dfrac{7.76 - 20}{4} = -3.06$.

If $X = 12$ then $Z = \dfrac{X - \mu}{\sigma} = \dfrac{12 - 20}{4} = -2.00$.

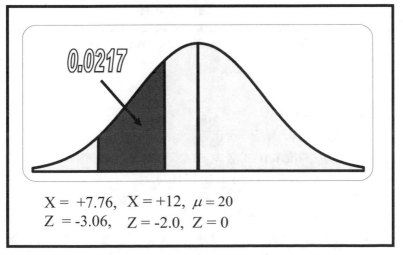

$$X = +7.76, \quad X = +12, \quad \mu = 20$$
$$Z = -3.06, \quad Z = -2.0, \quad Z = 0$$

Fig. 7.28. Area under normal curve.

Thus:

$A\big(\text{Between } X = +7.76 \text{ and } X = +12\big)$

$\qquad = A\big(\text{Between } Z = -3.06 \text{ and } Z = -2\big)$

$\qquad = A\big(\text{to the left of } Z = -2\big) - A\big(\text{to the left of } Z = -3.06\big)$

$\qquad = 0.0228 - 0.0011 = 0.0217.$

(d) Area between $X = 23$ and $X = 25$:

If $X = 23$, then $Z = \dfrac{X - \mu}{\sigma} = \dfrac{23 - 20}{4} = +0.75$.

If $X = 25$, then $Z = \dfrac{X - \mu}{\sigma} = \dfrac{25 - 20}{4} = +1.25$.

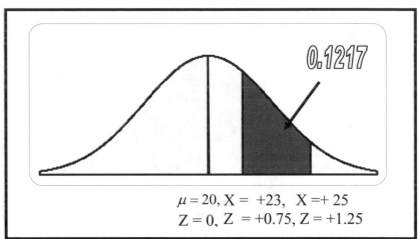

$$\mu = 20, X = +23, \ X = +25$$
$$Z = 0, \ Z = +0.75, Z = +1.25$$

Fig. 7.29. Area under normal curve.

Thus:

$A\big(\text{Between } X = +23 \text{ and } X = +25\big)$

$\qquad = A\big(\text{Between } Z = +0.75 \text{ and } Z = +1.25\big)$

$\qquad = A\big(\text{to the left of } Z = +1.25\big) - A\big(\text{to the left of } Z = +0.75\big)$

$\qquad = 0.8944 - 0.7734 = 0.1210.$

(e) Area to the left of $X = 23$:

If $X = 23$, then $Z = \dfrac{X - \mu}{\sigma} = \dfrac{23 - 20}{4} = +0.75$.

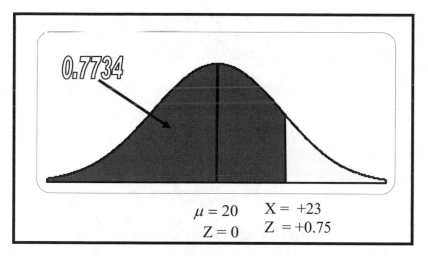

$$\mu = 20 \qquad X = +23$$
$$Z = 0 \qquad Z = +0.75$$

Fig. 7.30. Area under normal curve.

Thus:

$$A(\text{to the left of } X = +23) = A(\text{to the left of } Z = +0.75) = 0.7734.$$

(f) Area to the right of $X = 25$:

If $X = 25$, then $Z = \dfrac{X - \mu}{\sigma} = \dfrac{25 - 20}{4} = +1.25$.

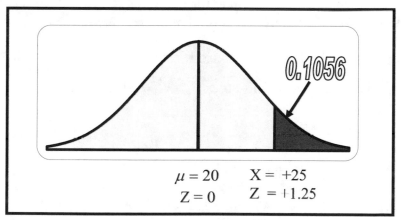

$$\mu = 20 \qquad X = +25$$
$$Z = 0 \qquad Z = +1.25$$

Fig. 7.31. Area under normal curve.

Thus

$$A(\text{to the right of } X = +25) = A(\text{to the right of } Z = +1.25)$$
$$= 1 - A(\text{to the left of } Z = +1.25)$$
$$= 1 - 0.8944 = 0.1056.$$

Example 7.7. (STANDARDIZE YOUR SCORE)   In the population of students who took the graduate record exam in a particular year, the analytical ability scores were normally distributed with a mean of 520 and a variance of 100.

(a) What proportion of students had scores between 510 and 540?
or $(510 < X < 540)$?
(b) What proportion of these students scored higher than 530?
or $(X > 530)$?
(c) What proportion of these students scored less than 500?
or $(X < 500)$?
(d) What proportion of students scored higher than 520?
or $(X > 520)$?
(e) What proportion of students scored exactly 540?
or $(X = 540)$?
(f) What proportion of students scored either higher than 534 or less than 504?
or $(X > 534$ or $X < 504)$?
(g) What score is the 90.15th percentile?
or (Given area to the left of a certain point $= 0.9015$, find X =?)
(h) What score is the 75th percentile?
or (Given area to the left of a certain point $= 0.75$, find X =?)

Solution. We are given:
$$\text{variance} = \sigma^2 = 100 \text{ so } \sigma = \sqrt{\sigma^2} = \sqrt{100} = 10$$
and
$$\text{mean } \mu = 520$$
so that
$$X \sim N(\mu, \sigma) = N(520, 10).$$

(a) Area between X = 510 and X = 540 (510 < X < 540):

If $X = 510$, then $Z = \dfrac{X - \mu}{\sigma} = \dfrac{510 - 520}{10} = \dfrac{-10}{10} = -1$.

If $X = 540$, then $Z = \dfrac{X - \mu}{\sigma} = \dfrac{540 - 520}{10} = \dfrac{+20}{10} = +2$.

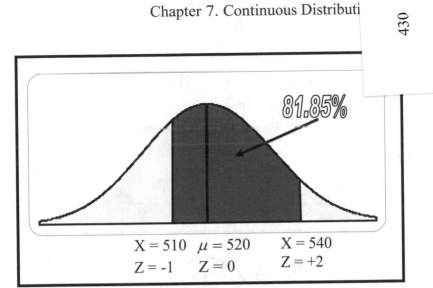

Fig. 7.32. Area under a normal curve.

Area between X = 510 and X = 540

$$= A(\text{between } Z = -1 \text{ and } Z = +2)$$
$$= A(\text{to the left of } Z = +2) - A(\text{to the left of } Z = -1)$$
$$= 0.9772 - 0.1587 = 0.8185 \quad \text{or} \quad 81.85\%$$

(b) More than 530 (X > 530):

If $X = 530$, then $Z = \dfrac{X - \mu}{\sigma} = \dfrac{530 - 520}{10} = \dfrac{10}{10} = +1$.

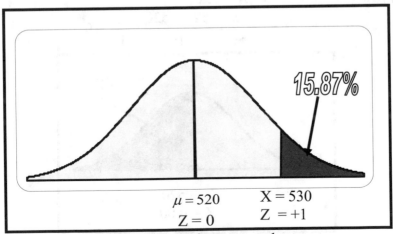

Fig. 7.33. Area under a normal curve.

Area to the right of $X = 530$

$$= A(\text{to the right of } Z = +1)$$
$$= 1 - A(\text{to the left of } Z = +1)$$
$$= 1 - 0.8413 = 0.1587 \text{ or } 15.87\%$$

(c) Less than 500 (X < 500):

If $X = 500$, then $Z = \dfrac{X - \mu}{\sigma} = \dfrac{500 - 520}{10} = -\dfrac{20}{10} = -2$.

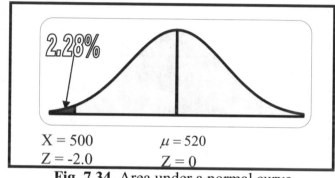

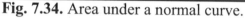

$$X = 500 \qquad\qquad \mu = 520$$
$$Z = -2.0 \qquad\qquad Z = 0$$

Fig. 7.34. Area under a normal curve.

Area to the left of $X = 500$

$$= A(\text{to the left of } Z = -2)$$
$$= 0.0228 \text{ or } 2.28\%$$

(d) More than X = 520 (X > 520):

If $X = 520$, then $Z = \dfrac{X - \mu}{\sigma} = \dfrac{520 - 520}{10} = \dfrac{0}{10} = 0$.

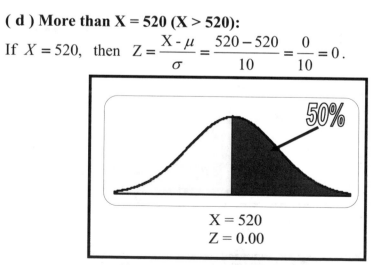

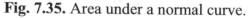

$$X = 520$$
$$Z = 0.00$$

Fig. 7.35. Area under a normal curve.

Area to the right of X = 520

$$= 1 - A(\text{to the left of } Z = 0.0).$$
$$= 1 - 0.50 = 0.50 \quad \text{or} \quad 50\%$$

(e) Exactly X = 540:

We know that the area under a single point of a **continuous distribution** is always zero. So the proportion of such values or score is zero.

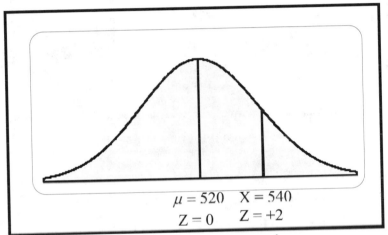

$\mu = 520 \quad X = 540$
$Z = 0 \qquad Z = +2$

Fig. 7.36. Area under a normal curve.

(f) Either X > 534 or X < 504:

If $X = 504$, then $Z = \dfrac{X - \mu}{\sigma} = \dfrac{504 - 520}{10} = -\dfrac{16}{10} = -1.6$.

$$A(\text{to the left of } X = 504) = A(\text{to the left of } Z = -1.6)$$
$$= 0.0548.$$

If $X = 534$, then $Z = \dfrac{X - \mu}{\sigma} = \dfrac{534 - 520}{10} = \dfrac{14}{10} = +1.4$.

A(to the right of $X = 534$)

$$= 1 - A(\text{to the left of } X = 534)$$
$$= 1 - A(\text{to the left of } Z = +1.4)$$
$$= 1 - 0.9192 = 0.0808.$$

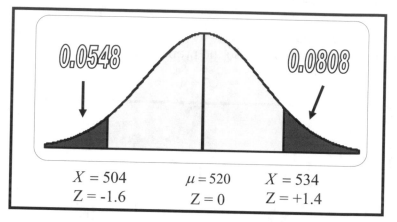

$X = 504$	$\mu = 520$	$X = 534$
$Z = -1.6$	$Z = 0$	$Z = +1.4$

Fig. 7.37. Area under a normal curve.

A(either $X < 504$ or $X > 534$)

$$= A(\text{to the left of } X = 504)$$
$$+ A(\text{to the right of } X = 534)$$
$$= 0.0548 + 0.0808$$
$$= 0.1356 \quad \text{or} \quad 13.56\%$$

(g) What score is the 90.15[th] percentile? OR (Given the area to the left of a certain point = 0.9015, find X = ?):

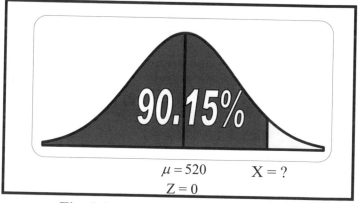

$\mu = 520$	$X = ?$
$Z = 0$	

Fig. 7.38. Area under a normal curve.

We know that:

$$Z = \frac{X - \mu}{\sigma} \text{ or } Z\sigma = X - \mu \text{ or } X = \mu + Z\sigma .$$

Using **Table II**, the 90.15% of the area is to the left of $Z = +1.29$, therefore:

$$X = \mu + Z\sigma = 520 + 1.29 \times 10 = 520 + 12.9 = 532.9 .$$

Thus, 90.15% of the students had scores below 532.9.

(h) What score is the 75th percentile? or (Given area to the left of a certain point = 0.75, find X = ?):

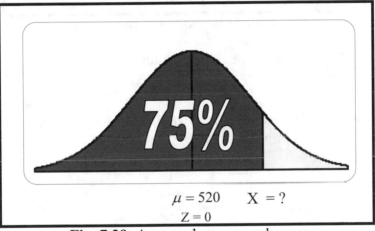

$\mu = 520$ X = ?

$Z = 0$

Fig. 7.39. Area under a normal curve.

We know that:

$$Z = \frac{X - \mu}{\sigma} \text{ or } Z\sigma = X - \mu \text{ or } X = \mu + Z\sigma .$$

Using **Table II**, approximately 75% of the area is to the left of $Z = +0.674$ therefore:

$$X = \mu + Z\sigma = 520 + 0.674 \times 10 = 520 + 6.74 = 526.74 .$$

Thus, 75% of the students had scores below 526.74.

Example 7.8. (QUARTILES) The variable $Z \sim N(0, 1)$. Find Q_1 and Q_3.

Solution. We are given:

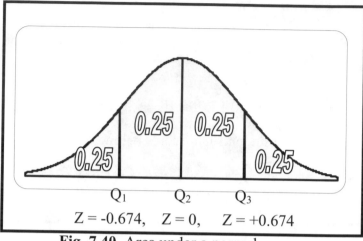

Fig. 7.40. Area under a normal curve.

Using the **Table II**, we have approximately:
the first quartile $Q_1 = -0.674$ and the third quartile $Q_3 = +0.674$.

Example 7.9. (READING GRAPHS) Consider that we are given the following two normal graphs:

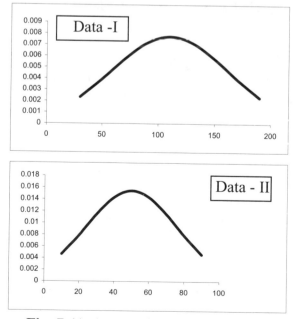

Fig. 7.41. Area under a normal curve.

(a) What are the expected mean values of data I and data II?
(b) Which data is expected to have more variation and why?
Solution. (a) Data I and II are expected to have mean values around 110 and 50, respectively.
(b) Data I has a wider peak than data II, thus data I is expected to have more variation than data II. In other words, the value of the standard deviation σ will be more for data I than for data II.

7.5.2 DETECTION OF OUTLIERS WITH Z-SCORE

The population Z-score has been found to be useful in detecting the outliers in a given data set. We know that the Z-score is defined as:

$$Z = \frac{X - \mu}{\sigma}$$

(a) If the value of the Z-score lies between -2 and +2 for the given value of X, then this value cannot be treated as an outlier.
(b) If the value of the Z-score lies either between –3 and –2 or between +2 and +3, then that value of X is a mild outlier.
(c) If the value of Z is either less than –3, or more than +3, then that value of X can be treated as an extreme outlier.

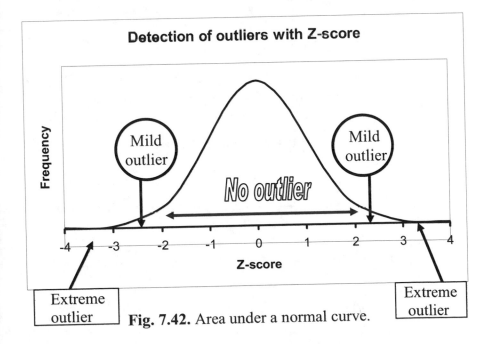

Fig. 7.42. Area under a normal curve.

Example 7.10. (ALWAYS WATCH FOR STRANGERS) Let $X \sim N(100, 16)$ represent the IQ of 10th grade students. Amy is a 10th grade student and has an IQ of 140.

(a) What proportion of students have an IQ less than Amy's?

(b) Can Amy be considered an extraordinarily intelligent student?

Solution. We are given $X \sim N(100, 16)$, that is, $\mu = 100$ and $\sigma = 16$. The value of Amy's score is given by $X = 140$. Thus, the value of the Z-score is given by:

$$Z = \frac{X - \mu}{\sigma} = \frac{140 - 100}{16} = \frac{40}{16} = 2.5 .$$

(a) Using **Table II**, the area to the left of $Z = 2.5$ is 0.9938. Thus, 99.38% of the students have IQs less than Amy's.

(b) The value of the Z-score $Z = 2.5$ lies between +2 and +3.0 and hence can be treated as an outlier for the rest of the students. The score lies on the right tail of the curve, so Amy can be considered one of the extraordinarily intelligent students.

7.5.3 HOW TO GENERATE A NORMAL DISTRIBUTION?

The probability density function (p.d.f) of a normal random variable X is given by:

$$f(x) = \frac{1}{\sqrt{2\pi}\,\sigma} \exp\left\{ -\frac{1}{2}\left(\frac{X - \mu}{\sigma} \right)^2 \right\}$$

where X is a random variable that can take any value between negative infinity and positive infinity, that is, $-\infty < x < +\infty$,

$\mu = \int xf(x)dx \approx \dfrac{\sum\limits_{I=1}^{N} X_I}{N}$ denotes the population mean,

$\sigma^2 = \int (X - \mu)^2 f(x)dx \approx \dfrac{\sum\limits_{I=1}^{N}(X_I - \mu)^2}{N}$ denotes the population variance,

$\pi = \dfrac{22}{7}$ and $\exp = e^{power} \approx 2.718^{power}$. If we define a standard score as:

$$Z = \frac{(X - \mu)}{\sigma},$$

then $Z \sim N(0, 1)$, that is, Z follows a standard normal variate with a probability density function (p.d.f):

$$f(Z) = \frac{1}{\sqrt{2\pi}} \exp\left\{-\frac{Z^2}{2}\right\} \approx 0.4e^{\left(-z^2/2\right)}, \quad -\infty < Z < +\infty$$

Example 7.11. (STANDARDIZE YOUR PRODUCT) Consider a population of $N = 7$ values, say:

X: 30, 40, 50, 60, 70, 80, and 90.

Plot a normal probability density function (p.d.f).
Solution. We are given:

$N = 7$, $X_1 = 30$, $X_2 = 40$, $X_3 = 50$, $X_4 = 60$, $X_5 = 70$, $X_6 = 80$, and $X_7 = 90$.

If we wish to make a normal probability density function (p.d.f), then we have to go through the following steps:

(i) Compute the population mean:

$$\mu = \frac{\sum\limits_{I=1}^{N} X_I}{N} = \frac{30 + 40 + 50 + 60 + 70 + 80 + 90}{7} = 60.$$

(ii) Compute the population variance:

$$\sigma^2 = \frac{\sum\limits_{I=1}^{N}(X_I - \mu)^2}{N} = \frac{\sum\limits_{I=1}^{7}(X_I - \mu)^2}{7}$$

$$= \frac{(X_1 - \mu)^2 + (X_2 - \mu)^2 + \dots\dots\dots + (X_7 - \mu)^2}{7}$$

$$= \frac{(30-60)^2 + (40-60)^2 + (50-60)^2 + (60-60)^2 + (70-60)^2 + (80-60)^2 + (90-60)^2}{7}$$

$$= \frac{(-30)^2 + (-20)^2 + (-10)^2 + (0)^2 + (10)^2 + (20)^2 + (30)^2}{7}$$

$$= \frac{900 + 400 + 100 + 0 + 100 + 400 + 900}{7} = \frac{2800}{7} = 400.$$

(iii) Compute the population standard deviation:

$$\sigma = \sqrt{\sigma^2} = \sqrt{400} = 20.$$

(iv) Compute the following table:

X_I	$X_I - \mu$	$Z = \dfrac{(X_I - \mu)}{\sigma}$	$f(Z) \approx 0.4e^{-\frac{z^2}{2}}$	Relative Frequency
30	30-60 = -30	$-\dfrac{30}{20} = -1.5$	$0.4e^{\frac{-(-1.5)^2}{2}} = 0.130$	$\dfrac{0.130}{1.852} = 0.070$
40	40-60 = - 20	$-\dfrac{20}{20} = -1.0$	$0.4e^{\frac{-(-1)^2}{2}} = 0.243$	$\dfrac{0.243}{1.852} = 0.131$
50	50-60 = - 10	$-\dfrac{10}{20} = -0.5$	$0.4e^{\frac{-(-0.5)^2}{2}} = 0.353$	$\dfrac{0.353}{1.852} = 0.191$
60	60-60 = 0	$-\dfrac{0}{20} = 0.0$	$0.4e^{\frac{-(0)^2}{2}} = 0.400$	$\dfrac{0.400}{1.852} = 0.216$
70	70-60 = +10	$+\dfrac{10}{20} = +0.5$	$0.4e^{\frac{-(+0.5)^2}{2}} = 0.353$	$\dfrac{0.353}{1.852} = 0.191$
80	80-70 = +20	$+\dfrac{20}{20} = +1.0$	$0.4e^{\frac{-(+1)^2}{2}} = 0.243$	$\dfrac{0.243}{1.852} = 0.131$
90	90-60 = +30	$+\dfrac{30}{20} = +1.5$	$0.4e^{\frac{-(1.5)^2}{2}} = 0.130$	$\dfrac{0.130}{1.852} = 0.070$
Sum	**0**	**0**	**1.852**	**1.000**

Then we have the following probability curve:

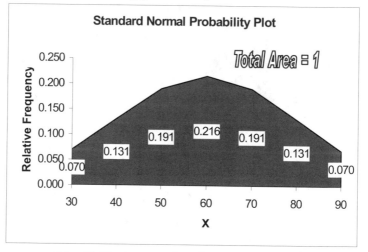

Fig. 7.43. Area under the normal curve.

The corresponding Z-values can also be shown on the horizontal axis instead of X values. The curve will become smoother as we increase the number of data values.

LUDI 7.1. (A ONE WORD DECISION IS DIFFICULT) True/ False statements (state only one):

Q	Statement	True	False
a	In a normal distribution, reducing the value of the variance makes the density wider and flatter.		
b	The mean, mode, and median of a standard normal variable are always zero.		
c	The variance and standard deviation of a standard normal variable is always one.		
d	The standard score can be used to detect outliers.		
e	The Z-score at the mean value of the X variable is always zero.		
f	If for a given value of X, the Z-score is negative, then it means that the value of X is below the mean value of X.		
g	A uniform density function is a bell shaped curve.		

LUDI 7.2. (THINK ABOUT THE FUTURE) Use **Table II** from the Appendix to find the following Z-scores:

(a) $Z_{0.05}$ = - - - - - - - -

(b) $Z_{0.025}$ = - - - - - - - -

(c) $Z_{0.01}$ = - - - - - - - -

(d) $Z_{0.005}$ = - - - - - - - -

LUDI 7.3. (STANDARDIZED PRODUCTS ARE EXPENSIVE)
If a variable $Z \sim N(0, 1)$, then find the following and sketch:

(a) proportion of values less than $Z = -1.8$ or $(Z < -1.8)$.

(b) proportion of values less than $Z = 2.2$ or $(Z < 2.2)$.

(c) proportion of values more than $Z = -1.5$ or $(Z > -1.5)$.

(d) proportion of values more than $Z = 1.5$ or $(Z > 1.5)$.

(e) proportion of values between $Z = 1.6$ and $Z = 2.8$ or $(1.6 < Z < 2.8)$.

(f) proportion of values between $Z = -1.2$ and $Z = -2.4$ or $(-2.4 < Z < -1.2)$.

(g) proportion of values between $Z = -2.85$ and $Z = 2.46$
or $(-2.85 < Z < 2.46)$.

(h) proportion of values to the left of $Z = 0$ or $(Z \leq 0)$.

(i) proportion of values to the right of $Z = 0$ or $(Z \geq 0)$.

(j) proportion of values to the left of $Z = -4$ or $(Z \leq -4)$.

(k) proportion of values to the right of $Z = 4$ or $(Z \geq 4)$.

(l) proportion of values to the left of $Z = 4$ or $(Z \leq 4)$.

(m) proportion of values to the right of $Z = -4$ or $(Z \geq -4)$.

(n) proportion of values to the left of $Z = -2.85$
or to the right of $Z = 2.46$ or $(Z < -2.85) \cup (Z > 2.46)$.

(o) proportion of values under a single point $Z = 2.46$.

LUDI 7.4. (TRY TO STANDARDIZE YOUR PRODUCT)

If a variable $X \sim N(32, 6)$, then find the following areas and sketch:

(a) the value of X lying between 20 and 50.

(b) the value of X lying between 15 and 25.

(c) the value of X lying between 45 and 55.

(d) the value of X lying to the left of 40.

(e) the value of X lying to the right of 20.

(f) the value of X lying to the left of 42.

(g) the value of X lying to the right of 42.

(h) the value of X lying between 42 and 48.

(i) the value of X lying between 36 and 42.

LUDI 7.5. (WHERE DO I STAND IN CLASS?) For a midterm,
scores follow a normal distribution $N(500, 100)$. On the final, scores follow a normal distribution $N(18, 6)$. John scored 680 on the midterm and Michael scored 27 on the final exam:

(a) What is John's standard Z-score?

(b) What is Michael's standard Z-score?

(c) Whose standard Z-score is higher: Michael or John's?

(d) What proportion of students lies between Michael and John?

LUDI 7.6. (LEARN TO READ THE Z-TABLE) Find the Z-score
in each of the following cases:

(a) $A(Z < z) = 0.1736$.

(b) $A(Z > z) = 0.1003$.

(c) $A(-z < Z < +z) = 0.9522$.

LUDI 7.7. (READING AREA FROM THE Z-TABLE) Use the standard normal **Table II** to find the following areas:

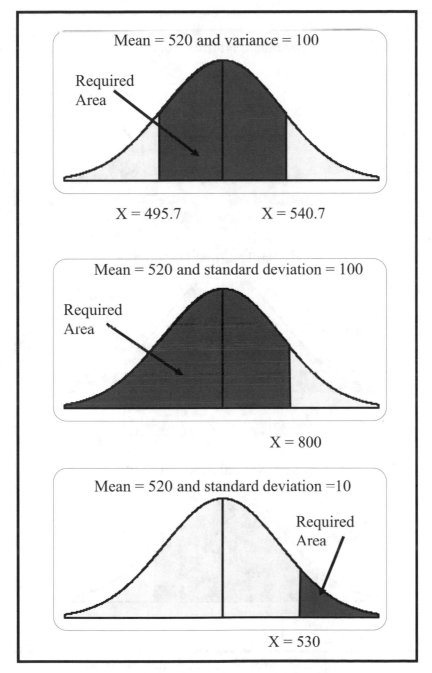

Fig. 7.44. Area under a normal curve.

LUDI 7.8. (READING Z-VALUES FROM TABLE II) Find the value of Z for the following areas:

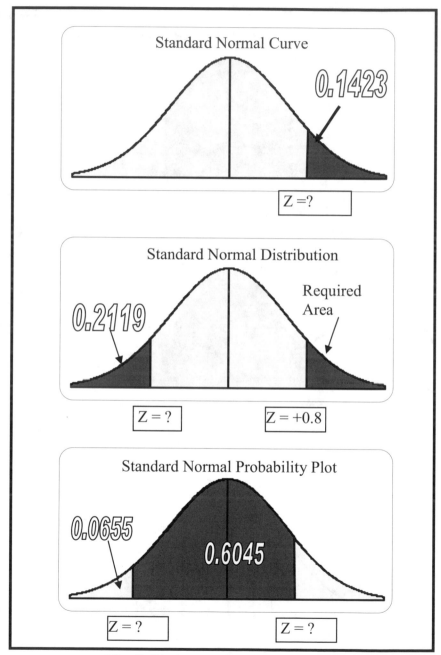

Fig. 7.45. Area under a normal curve.

LUDI 7.9. (KEEP NEAT AND CLEAN TRAINS) Consider the following density function giving the waiting time (in minutes) of a train in a subway:

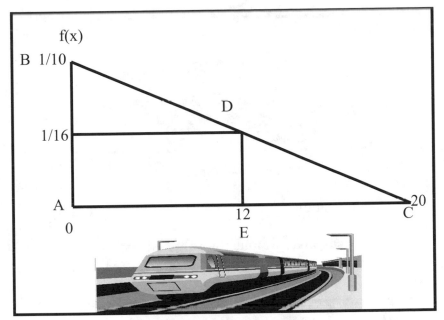

Fig. 7.46. Area under a curve.

(a) Are you sure that the above picture gives you a density function? Justify your answer.
(b) What is the maximum time you have to wait after entering the subway?
(c) What is the minimum time you have to wait after entering the subway?
(d) What proportion of times do you have to wait more than 12 minutes?

LUDI 7.10. (LOOKING AT THE SCORES OF A CLASS) In the population of students who took STAT 193 in a particular year, the analytical ability scores were normally distributed with a mean score of 600 and a standard deviation of 20.
(a) Sketch the normal distribution plot.
(b) Jessica has a score of 650. Is Jessica an outlier?
(c) What proportion of students have a higher score than Jessica's score?

LUDI 7.11. (ALWAYS TAKE CARE OF PUBLIC PROPERTY)
Consider the following density function giving the waiting time of a train in a subway in minutes as:

Fig. 7.47. Train.

(a) Are you sure that this gives you a density function? Justify your answer.
(b) What is the maximum waiting time after entering the subway?
(c) What is the minimum waiting time after entering the subway?

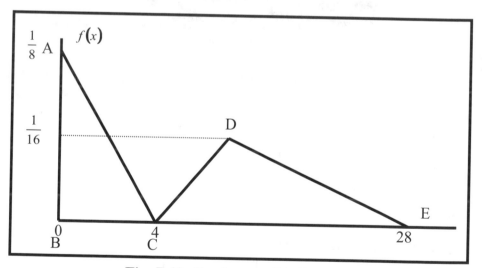

Fig. 7.48. Continuous distribution.

(d) What proportion of the time do you have to wait more than 4 minutes?
(e) The value of X = 4 minutes is called - - - - - - -
   (*i*) Q_1 (*ii*) Q_2 (*iii*) Q_3 (*iv*) None

LUDI 7.12. (THREE MEASURES OF MIDDLE VALUES) Let X represent the income (in $) of the employees at a university campus. Assume the mean income of this population is $65,000 with a standard deviation of $8,000. The following histogram is a sketch of the density curve for the distribution of income.

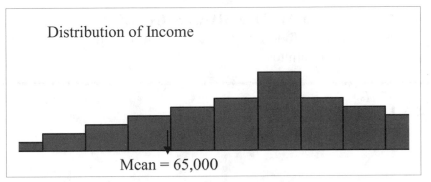

Fig. 7.49. Reading a histogram.

Circle one choice:
(I) What is the distribution of X?
(a) Symmetric (b) Skewed to the right (c) Skewed to the left
(II) What can you say about the median of the distribution?
(a) Less than $65,000 (b) Equal to $65,000 (c) More than $65,000
(III) What can you say about the mode of the distribution?
(a) Less than $65,000 (b) Equal to $65,000 (c) More that $65,000

LUDI 7.13. (THINK ABOUT POOR PEOPLE) Let X represent the income in $ of the employees at a university campus. Assume the median salary of this population is $6,500. The following histogram is a sketch of the density curve for the distribution of salaries.

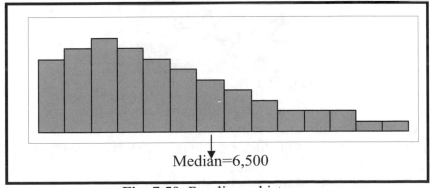

Fig. 7.50. Reading a histogram.

Circle one choice:
(I) What is the distribution of X?
(a) Symmetric (b) Skewed to the right (c) Skewed to the left
(II) What can you say about the mean of the distribution?
(a) Less than \$6,500 (b) Equal to \$6,500 (c) More than \$6,500

LUDI 7.14. (FOLLOW THE RULES IN AIRPORTS) Consider the following density function giving the waiting time of an airplane at a local airport in minutes:

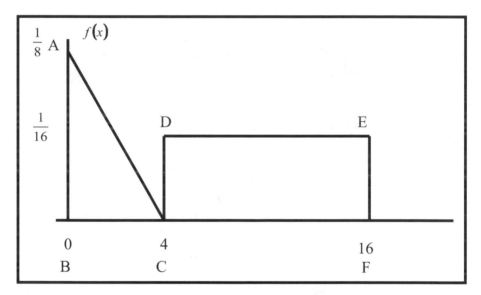

Fig. 7.51. Continuous distribution.

(a) Are you sure that this gives you a density function? Justify your answer.

Fig. 7.52. Boarding aircraft.

(b) What is the maximum waiting time after entering the airport?
(c) What is the minimum waiting time after entering the airport?
(d) What proportion of the time do you have to wait more than 4 minutes?
(e) The value of X = 4 minutes is called - - - - - - -
          (i) Q_1 (ii) Q_2 (iii) Q_3 (iv) None

LUDI 7.15. (PRACTICE WITH THE Z-TABLE) Use **Table II** to find:

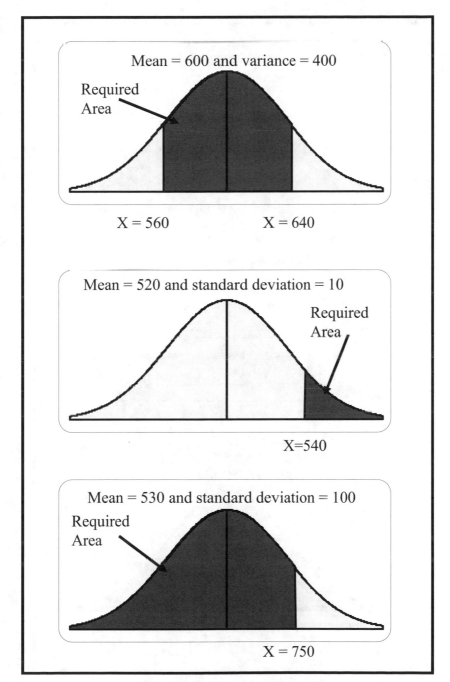

Fig. 7.53. Area under a normal curve.

LUDI 7.16. (THE STANDARD NORMAL TABLE IS USEFUL IN HIGHER STATISTICS COURSES) Find the Z-values from **Table II** for each of the following situations:

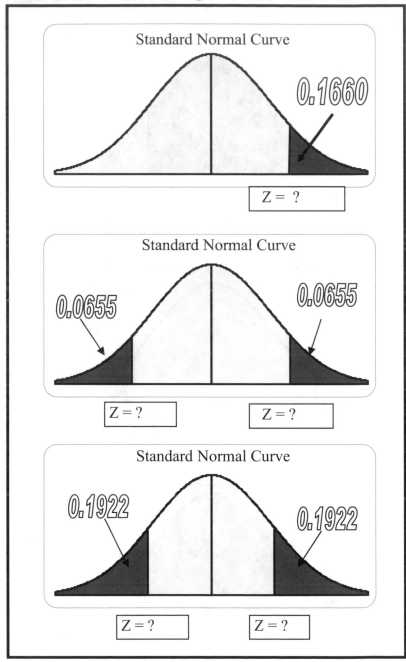

Fig. 7.54. Area under a normal curve.

LUDI 7.17. (FARMER IN A FIELD) An engineer developed a sprinkler that distributes seeds uniformly in a field and keeps the distance from plant to plant 25 cm to 55 cm. Experience shows that a minimum distance of 30 cm allows the plants to grow well.

Fig. 7.55. A farmer watching his plants.

(a) If 10,000 seeds are distributed in a field, then find the number of plants that may not grow well.
(b) Find the average distance between the plants.
(c) Find the standard deviation of the distance between the plants.

LUDI 7.18. (WHERE DO I STAND IN THE CLASS?) The TOEFL has been widely used to predict the performance of applicants. The range of possible scores on a TOEFL is 200 to 900. The statistics department at a university finds that the scores of its applicants on the TOEFL are approximately normal with mean $\mu = 545$ and variance $\sigma^2 = 10,000$. Find the relative frequency of applicants whose score X satisfies each of the following conditions.
(a) X > 710
(b) X < 400
(c) 400 < X < 710

LUDI 7.19. (GOOD MARKS ARE ESSENTIAL)
(i) The scores of students on an examination follow a **normal** distribution with a mean of 750 and variance of 576.
(a) What proportion of students have scores between 694.80 and 787.44? Sketch it.
(b) What proportion of students have scores higher than 787.44? Sketch it.
(ii) In another analytical ability test the scores are normally distributed with a mean of 600 and a variance of 400.
(a) Sketch the normal distribution plot.
(b) Jessica and Melissa have scores of 650 and 660.
What proportion of students has scores between Jessica and Melissa?

LUDI 7.20. (CONFIDENTIAL DATA) On a special examination, the scores of the students are normally distributed. Due to some unavoidable circumstance, it was decided that the true score of any student will not be disclosed, thus the standard Z-scores were displayed on the board. Answer the following:

(a) Amy has a Z-score of +2.86. Sketch a standard normal curve and find the proportion of students having scores higher than Amy's score.

Fig.7.56. Amy.

(b) Chris and Matt have the Z-scores −1.67 and +2.85. Sketch and find the proportion of students having scores between Chris and Matt's scores.

Chris   Matt

Fig.7.57.

(c) Bob has a Z-score of -2.35. What proportion of students has scores below Bob's score? Sketch.

Fig.7.58. Bob.

(d) Don and Eric have -2.45 and -2.54 standard Z-scores. Sketch and find the proportion of students having scores between Don and Eric's scores.

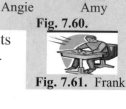

Don   Eric

Fig.7.59.

(e) Angie and Amy have scores of 1.85 and 2.86. Find the proportion of students having scores between Angie and Amy's scores.

Angie   Amy

Fig. 7.60.

(f) Frank was told that almost 25% of the students have scores less than his score. What is Frank's Z-score? Sketch it and name it if possible.

Fig. 7.61. Frank.

LUDI 7.21. (FISHING CAN BE FUN) A farmer believes that the weight of the fish on his fishery farm is normally distributed with an average weight of 700 gm and a variance of 400 gm^2.

(a) What proportion of fish weigh more than 650 gm? Sketch it.

(b) What proportion of fish weigh less than 750 gm? Sketch it.

(c) What proportion of fish weigh between 650 gm and 750 gm? Sketch it.

Fig. 7.62. A big fish.

8. SAMPLING DISTRIBUTIONS

8.1 INTRODUCTION

In this chapter we discuss sampling distribution of sample mean, sample proportion, standard error of sample mean and sample proportion, the central limit theorem, and point and confidence interval estimators.

8.2 SAMPLING DISTRIBUTION OF A STATISTIC

Sampling distribution of a statistic is the distribution of the values of the statistic in all possible samples of the same size n taken from the same population.

8.3 SAMPLING DISTRIBUTION OF SAMPLE MEAN

Consider a class consisting of 4 students with their names and scores as given in the following table:

 Teacher	 Amy	 Bob	 Chris	 Don
Score	80	50	60	70

Fig. 8.1. A class of four students.

Here $N = 4$, $X_1 = 80$, $X_2 = 50$, $X_3 = 60$, and $X_4 = 70$. The variable of interest, score, is a quantitative variable. The average score, population mean, of the whole class is given by:

$$\text{Population Mean} = \mu = \frac{\sum_{I=1}^{N} X_I}{N} = \frac{\sum_{I=1}^{4} X_I}{4} = \frac{X_1 + X_2 + X_3 + X_4}{4}$$

$$= \frac{80 + 50 + 60 + 70}{4} = \frac{260}{4} = 65 \text{ (Parameter)}.$$

Remember that a parameter is an unknown quantity and we try to estimate it by taking a random sample from the population. Consider a sample of $n = 2$ students. Thus, the total number of simple random and with replacement (SRSWR) samples will be:

$$N^n = 4^2 = 4 \times 4 = 16.$$

Let us now construct those 16 samples as follows:

Sample No.	Sampled Students		Score of selected students		Estimate $\bar{x} = \dfrac{\sum_{i=1}^{n} x_i}{n}$
1	Amy	Amy	80	80	$\dfrac{80+80}{2} = \dfrac{160}{2} = 80$
2	Amy	Bob	80	50	$\dfrac{80+50}{2} = \dfrac{130}{2} = 65$
3	Amy	Chris	80	60	$\dfrac{80+60}{2} = \dfrac{140}{2} = 70$
4	Amy	Don	80	70	$\dfrac{80+70}{2} = \dfrac{150}{2} = 75$
5	Bob	Amy	50	80	$\dfrac{50+80}{2} = \dfrac{130}{2} = 65$
6	Bob	Bob	50	50	$\dfrac{50+50}{2} = \dfrac{100}{2} = 50$
7	Bob	Chris	50	60	$\dfrac{50+60}{2} = \dfrac{110}{2} = 55$
8	Bob	Don	50	70	$\dfrac{50+70}{2} = \dfrac{120}{2} = 60$
9	Chris	Amy	60	80	$\dfrac{60+80}{2} = \dfrac{140}{2} = 70$
10	Chris	Bob	60	50	$\dfrac{60+50}{2} = \dfrac{110}{2} = 55$
11	Chris	Chris	60	60	$\dfrac{60+60}{2} = \dfrac{120}{2} = 60$
12	Chris	Don	60	70	$\dfrac{60+70}{2} = \dfrac{130}{2} = 65$
13	Don	Amy	70	80	$\dfrac{70+80}{2} = \dfrac{150}{2} = 75$
14	Don	Bob	70	50	$\dfrac{70+50}{2} = \dfrac{120}{2} = 60$
15	Don	Chris	70	60	$\dfrac{70+60}{2} = \dfrac{130}{2} = 65$
16	Don	Don	70	70	$\dfrac{70+70}{2} = \dfrac{140}{2} = 70$

Thus, we have the following frequency distribution table:

Mean Estimates $\bar{x}_i$	Tally Marking	Frequency f_i	Relative Frequency $RF_i = f_i / \Sigma f_i = p_i$
50	|	1	$\dfrac{1}{16} = 0.0625$
55	||	2	$\dfrac{2}{16} = 0.1250$
60	|||	3	$\dfrac{3}{16} = 0.1875$
65	||||	4	$\dfrac{4}{16} = 0.2500$
70	|||	3	$\dfrac{3}{16} = 0.1875$
75	||	2	$\dfrac{2}{16} = 0.1250$
80	|	1	$\dfrac{1}{16} = 0.0625$
	Sum	16	1.00

The above frequency distribution table can be displayed with a histogram as follows:

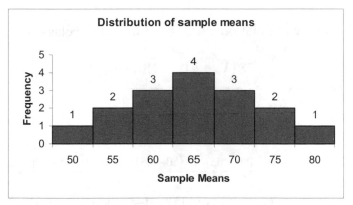

Fig. 8.2. Distribution of estimates of the population mean.

The above graph shows that the distribution of the estimates of the population mean is bell shaped and symmetric, or normal. Let the random variable X take values $\bar{x}_i$ with probabilities $p_i = RF_i$ for $i = 1,2,3,4,5,6,7$. The expected value of $X = \bar{x}$ is given by:

$$E(X) = \mu_{\bar{x}} = E(\bar{x}) = \sum_{i=1}^{7} p_i \bar{x}_i$$

$$= p_1\bar{x}_1 + p_2\bar{x}_2 + p_3\bar{x}_3 + p_4\bar{x}_4 + p_5\bar{x}_5 + p_6\bar{x}_6 + p_7\bar{x}_7$$

$$= \frac{1}{16} \times 50 + \frac{2}{16} \times 55 + \frac{3}{16} \times 60 + \frac{4}{16} \times 65 + \frac{3}{16} \times 70 + \frac{2}{16} \times 75 + \frac{1}{16} \times 80$$

$$= \frac{50}{16} + \frac{110}{16} + \frac{180}{16} + \frac{260}{16} + \frac{210}{16} + \frac{150}{16} + \frac{80}{16}$$

$$= \frac{1,040}{16} = 65 = \mu \quad \text{(Population mean)}.$$

Thus,

$$E(\bar{x}) = \mu_{\bar{x}} = \mu$$

which implies that the sample mean estimator $\bar{x}$ is unbiased for the population mean μ. By the computing formula, the variance of the sample mean $\bar{x}$ is given by:

$$\sigma_{\bar{x}}^2 = \left[\sum_{i=1}^{7} \left(p_i \bar{x}_i^2 \right) \right] - (\mu_{\bar{x}})^2$$

$$= \left[p_1\bar{x}_1^2 + p_2\bar{x}_2^2 + p_3\bar{x}_3^2 + p_4\bar{x}_4^2 + p_5\bar{x}_5^2 + p_6\bar{x}_6^2 + p_7\bar{x}_7^2 \right] - (\mu)^2$$

$$= \left[\frac{1}{16} \times 50^2 + \frac{2}{16} \times 55^2 + \frac{3}{16} \times 60^2 + \frac{4}{16} \times 65^2 + \frac{3}{16} \times 70^2 + \frac{2}{16} \times 75^2 + \frac{1}{16} \times 80^2 \right] - (65)^2$$

$$= 62.5.$$

Also note that the population variance of the whole class is given by:

$$\text{Population variance} = \sigma^2 = \frac{\sum_{I=1}^{N} (X_I - \mu)^2}{N} = \frac{\sum_{I=1}^{4} (X_I - \mu)^2}{4}$$

$$= \frac{(X_1 - \mu)^2 + (X_2 - \mu)^2 + (X_3 - \mu)^2 + (X_4 - \mu)^2}{N}$$

$$= \frac{(80 - 65)^2 + (50 - 65)^2 + (60 - 65)^2 + (70 - 65)^2}{4}$$

$$= \frac{(+15)^2 + (-15)^2 + (-5)^2 + (+5)^2}{4}$$

$$= \frac{225 + 225 + 25 + 25}{4} = \frac{500}{4} = 125 \,(\text{Parameter}).$$

Thus,

$$\frac{\sigma^2}{n} = \frac{125}{2} = 62.5 = \sigma_{\bar{x}}^2.$$

Thus, the variance of the sample mean, $\bar{x}$, while using simple random and with replacement (SRSWR) sampling is given by:

$$\sigma_{\bar{x}}^2 = \frac{\sigma^2}{n}$$

and the standard deviation of the sample mean, $\bar{x}$, is given by:

$$\sigma_{\bar{x}} = \sqrt{\sigma_{\bar{x}}^2} = \sqrt{\frac{\sigma^2}{n}} = \frac{\sigma}{\sqrt{n}}$$

In practice σ is unknown, thus the standard deviation of the sample mean, $\sigma_{\bar{x}}$, is estimated by the standard error of the sample mean given by:

$$\hat{\sigma}_{\bar{x}} = \frac{s}{\sqrt{n}}$$

where s is the sample standard deviation of the variable x.

Thus, we have the following results:

8.3.1 SOME IMPORTANT RESULTS

(a) The value of the sample mean $\bar{x}$ varies from sample to sample.

(b) The shape of the distribution of the sample mean $\bar{x}$ value is approximately normal, or bell shaped and symmetric.

(c) $\mu_{\bar{x}} = \mu$, the average of all possible sample means is the true population mean μ. We say that the sample mean $\bar{x}$ is an unbiased estimator of the population mean μ.

(d) $\sigma_{\bar{x}}^2 = \frac{\sigma^2}{n}$, the variance of the sample mean $\bar{x}$ decreases as sample size n increases.

(e) $\sigma_{\bar{x}} = \frac{\sigma}{\sqrt{n}}$, the standard deviation (SD) of the sample mean $\bar{x}$ decreases as the sample size n increases.

(f) $\hat{\sigma}_{\bar{x}} = \frac{s}{\sqrt{n}}$, the standard error (SE) of the sample mean $\bar{x}$ decreases as the sample size n increases.

(g) $\bar{x} \sim N\left(\mu, \; \dfrac{\sigma}{\sqrt{n}}\right)$, that is, the distribution of the sample mean $\bar{x}$ is approximately normal, and the value of standard Z-score is given by:

$$Z = \frac{(\bar{x} - \mu)}{\sigma_{\bar{x}}} = \frac{(\bar{x} - \mu)}{\sigma/\sqrt{n}}$$

which is called the Central Limit Theorem (CLT) for the sample mean. If the sample size is large $(n > 30)$, we can always replace the unknown parameter σ with s.

8.4 SAMPLING DISTRIBUTION OF SAMPLE PROPORTION

Consider a class consisting of 4 students with their names and majors given in the following table:

 Teacher	 Amy	 Bob	 Chris	 Don
Major	English	Math	English	Math
	(E)	(M)	(E)	(M)

Fig. 8.3. A class of four students.

Count the number of students in the population, $N = 4$. Note that here the variable of interest (major) is qualitative.

Count the number of students with an English major: COUNT = 2

Population proportion of English students:

$$P = \frac{\text{COUNT}}{N} = \frac{2}{4} = 0.5 \;\; \text{(Parameter)}.$$

We know that a parameter is an unknown quantity and we try to estimate it by taking a random sample from the population. Consider a sample of $n = 2$ students. Thus the total number of simple random and with replacement (SRSWR) samples will be:

$$N^n = 4^2 = 4 \times 4 = 16.$$

Let us now construct those 16 samples as follows:

Sample No.	Sampled Students		Major of selected students		# of English students	Estimate $\hat{p} = \dfrac{count}{n}$
1	Amy	Amy	E	E	2	$\dfrac{2}{2} = 1.0$
2	Amy	Bob	E	M	1	$\dfrac{1}{2} = 0.5$
3	Amy	Chris	E	E	2	$\dfrac{2}{2} = 1.0$
4	Amy	Don	E	M	1	$\dfrac{1}{2} = 0.5$
5	Bob	Amy	M	E	1	$\dfrac{1}{2} = 0.5$
6	Bob	Bob	M	M	0	$\dfrac{0}{2} = 0.0$
7	Bob	Chris	M	E	1	$\dfrac{1}{2} = 0.5$
8	Bob	Don	M	M	0	$\dfrac{0}{2} = 0.0$
9	Chris	Amy	E	E	2	$\dfrac{2}{2} = 1.0$
10	Chris	Bob	E	M	1	$\dfrac{1}{2} = 0.5$
11	Chris	Chris	E	E	2	$\dfrac{2}{2} = 1.0$
12	Chris	Don	E	M	1	$\dfrac{1}{2} = 0.5$
13	Don	Amy	M	E	1	$\dfrac{1}{2} = 0.5$
14	Don	Bob	M	M	0	$\dfrac{0}{2} = 0.0$
15	Don	Chris	M	E	1	$\dfrac{1}{2} = 0.5$
16	Don	Don	M	M	0	$\dfrac{0}{2} = 0.0$

Thus, the frequency distribution table of the estimates of the population proportion is given below:

Proportion Estimates $\hat{p}_i$	Tally Marking	Frequency f_i	Relative Frequency $RF_i = f_i / \Sigma f_i = p_i$
0.0	||||	4	$\dfrac{4}{16} = 0.25$
0.5	$\cancel{\|\|\|\|}\|\|\|$	8	$\dfrac{8}{16} = 0.50$
1.0	||||	4	$\dfrac{4}{16} = 0.25$
	Sum	16	1.00

The above frequency distribution table can be displayed in a histogram as follows:

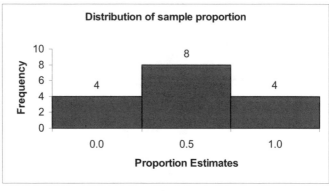

Fig. 8.4. Distribution of estimates of proportion.

The above graph shows that the distribution of the estimates of the population proportion is bell shaped and symmetric, or normal. This may not be true if the population proportion P is not equal to 0.5. In practice, the distribution of sample proportion is expected to be bell shaped only if the sample size n is large, more than 30. Let the random variable X take the values $\hat{p}_i$ with probabilities $p_i = RF_i$ for $i = 1, 2, 3$. Thus, the expected value of $X = \hat{p}$ is given by:

$$E(X) = \mu_{\hat{p}} = E(\hat{p}) = \sum_{i=1}^{3} p_i \hat{p}_i = p_1 \hat{p}_1 + p_2 \hat{p}_2 + p_3 \hat{p}_3$$

$$= \frac{4}{16} \times 0.0 + \frac{8}{16} \times 0.50 + \frac{4}{16} \times 1.0$$

$$= 0.0 + 0.25 + 0.25$$

$$= 0.5$$

$$= P \quad \text{(Population proportion).}$$

Thus,

$$E(\hat{p}) = \mu_{\hat{p}} = P$$

which implies that the sample proportion estimator $\hat{p}$ is unbiased for the population proportion P. By the computing formula, the variance of the sample proportion $\hat{p}$ is given by:

$$\sigma_{\hat{p}}^2 = \left[\sum_{i=1}^{3} \left(p_i \hat{p}_i^2 \right) \right] - \left(\mu_{\hat{p}} \right)^2 = \left[p_1 \hat{p}_1^2 + p_2 \hat{p}_2^2 + p_3 \hat{p}_3^2 \right] - (P)^2$$

$$= \left[\frac{4}{16} \times 0.0^2 + \frac{8}{16} \times 0.50^2 + \frac{4}{16} \times 1.0^2 \right] - (0.5)^2$$

$$= [0.000 + 0.125 + 0.250] - (0.250)$$

$$= 0.375 - 0.250$$

$$= 0.125.$$

Also, note that:

$$\frac{P(1-P)}{n} = \frac{0.5(1-0.5)}{2} = \frac{0.25}{2} = 0.125.$$

Thus, the variance of the estimator of population proportion using simple random and with replacement (SRSWR) sampling is given by:

$$\sigma_{\hat{p}}^2 = \frac{P(1-P)}{n}$$

and the standard deviation of the sample proportion, $\hat{p}$, is given by:

$$\sigma_{\hat{p}} = \sqrt{\sigma_{\hat{p}}^2} = \sqrt{\frac{P(1-P)}{n}}$$

In practice P is unknown, thus the standard deviation of $\hat{p}$ can be estimated by the standard error of $\hat{p}$ given by:

$$\hat{\sigma}_{\hat{p}} = \sqrt{\frac{\hat{p}(1-\hat{p})}{n}}$$

Thus, we have the following results:

8.4.1 SOME IMPORTANT RESULTS

(a) The sample proportion $\hat{p}$ varies from sample to sample.

(b) The shape of the distribution of the sample proportion $\hat{p}$ values is approximately bell shaped and symmetric or normal, if n is large.

(c) $\mu_{\hat{p}} = P$, the average of all the possible sample proportion $\hat{p}$ estimates is the true population proportion P. This means that the sample proportion $\hat{p}$ is an unbiased estimator of the population proportion P.

(d) $\sigma_{\hat{p}}^2 = \dfrac{P(1-P)}{n}$, the variance of the sample proportion $\hat{p}$ decreases as sample size n increases.

(e) $\sigma_{\hat{p}} = \sqrt{\dfrac{P(1-P)}{n}}$, the standard deviation (SD) of the sample proportion $\hat{p}$ decreases as sample size n increases.

(f) $\hat{\sigma}_{\hat{p}} = \sqrt{\dfrac{\hat{p}(1-\hat{p})}{n}}$, the standard error (SE) of the sample proportion $\hat{p}$ decreases as sample size n increases.

(g) $\hat{p} \sim N\left(P, \ \sqrt{\dfrac{P(1-P)}{n}}\right)$, the distribution of the sample proportion $\hat{p}$ is approximately normal, and the value of the standard Z-score is:

$$Z = \frac{\hat{p} - P}{\sqrt{\dfrac{P(1-P)}{n}}}$$

which is called the Central Limit Theorem (CLT) for the proportion.

8.5 UNBIASEDNESS

A statistic is unbiased if the center of its sampling distribution is equal to the corresponding population parameter value.

For example:

(i) the sample proportion $\hat{p}$ is an unbiased estimator of the population proportion P.

Thus $\mu_{\hat{p}} = P$

(ii) the sample mean $\bar{x}$ is an unbiased estimator of the population mean μ.

Thus $\mu_{\bar{x}} = \mu$

8.6 STANDARD DEVIATION AND STANDARD ERROR OF A STATISTIC

Note that the variability of a statistic corresponds to the spread of its sampling distribution. It is generally measured in terms of standard deviation or standard error of a statistic. Further, note that standard deviation is a parameter and standard error is a statistic.

The formulae used for standard deviations and standard errors:

(i) standard deviation and standard error of the sample proportion $\hat{p}$ are, respectively, given by:

$$\sigma_{\hat{p}} = \sqrt{\frac{P(1-P)}{n}} \quad \text{and} \quad \hat{\sigma}_{\hat{p}} = \sqrt{\frac{\hat{p}(1-\hat{p})}{n}}$$

(ii) standard deviation and standard error of the sample mean $\bar{x}$ are, respectively, given by:

$$\sigma_{\bar{x}} = \frac{\sigma}{\sqrt{n}} \quad \text{and} \quad \hat{\sigma}_{\bar{x}} = \frac{s}{\sqrt{n}}$$

Example 8.1. (WHY A LARGE SAMPLE SIZE?) A population has a standard deviation of 10. What is the standard deviation of the sample mean $\bar{x}$ for a random sample of size:

(a) $n = 25$? (b) $n = 400$? (c) Comment.

Solution. We are given a standard deviation of X as $\sigma = 10$, thus:
(a) If $n = 25$, then the standard deviation of the sample mean is:

$$\sigma_{\bar{x}} = \frac{\sigma}{\sqrt{n}} = \frac{10}{\sqrt{25}} = \frac{10}{5} = 2.$$

(b) If $n = 400$, then the standard deviation of the sample mean is:

$$\sigma_{\bar{x}} = \frac{\sigma}{\sqrt{n}} = \frac{10}{\sqrt{400}} = \frac{10}{20} = 0.5.$$

(c) If the sample size increases, the standard deviation of the sample mean decreases, which increases the precision of the sample mean.

Example 8.2. (STOP SMOKING) In a particular community 80% of the people hate smoking. What is the standard deviation of the sample proportion $\hat{p}$ for a random sample of size:

(a) $n = 40$? (b) $n = 400$? (c) Comment.

Solution. We are given the population proportion $P = 0.80$, thus:

(**a**) If $n = 40$, then the standard deviation of $\hat{p}$ is:

$$\sigma_{\hat{p}} = \sqrt{\frac{P(1-P)}{n}} = \sqrt{\frac{0.80(1-0.80)}{40}} = 0.063 \, .$$

(**b**) If $n = 400$, then the standard deviation of $\hat{p}$ is:

$$\sigma_{\hat{p}} = \sqrt{\frac{P(1-P)}{n}} = \sqrt{\frac{0.80(1-0.80)}{400}} = 0.02 \, .$$

(**c**) If the sample size increases, the standard deviation of the sample proportion decreases, which increases its precision.

Example 8.3. (SOCIAL SERVICE) Assume that a survey was conducted in a certain locality to ask the question of whether or not to celebrate a "Non-Smoking Day" every year on March 13 at an international level. The proportion of people who favor the observance of a "Non-Smoking Day" in the locality was 0.95.

Fig. 8.5. Smoking is a bad habit.

What is the standard error of the sample proportion $\hat{p}$ for a random sample of size: (a) $n = 25$? (b) $n = 400$? (c) Comment.

Solution. We are given the sample proportion $\hat{p} = 0.95$, thus:

(**a**) If $n = 25$, then the standard error of the sample proportion $\hat{p}$ is:

$$\hat{\sigma}_{\hat{p}} = \sqrt{\frac{\hat{p}(1-\hat{p})}{n}} = \sqrt{\frac{0.95(1-0.95)}{25}} = 0.04359.$$

(**b**) If $n = 400$, then the standard error of the sample proportion $\hat{p}$ is:

$$\hat{\sigma}_{\hat{p}} = \sqrt{\frac{\hat{p}(1-\hat{p})}{n}} = \sqrt{\frac{0.95(1-0.95)}{400}} = 0.01089.$$

(**c**) Note that as the sample size increases, the standard error of the sample proportion $\hat{p}$ decreases, which implies that the precision of the sample proportion $\hat{p}$ increases.

Example 8.4. (SOME JOBS ARE FLEXIBLE) Given that the sample standard deviation is 12, what is the standard error of a sample mean $\bar{x}$ for a random sample of size: (a) $n = 25$? (b) $n = 400$?
Solution. We are given $s = 12$, thus:

(a) If $n = 25$, the standard error of the sample mean $\bar{x}$ is given by:

$$\hat{\sigma}_{\bar{x}} = \frac{s}{\sqrt{n}} = \frac{12}{\sqrt{25}} = \frac{12}{5} = 2.4.$$

(b) If $n = 400$, the standard error of the sample mean $\bar{x}$ is given by:

$$\hat{\sigma}_{\bar{x}} = \frac{s}{\sqrt{n}} = \frac{12}{\sqrt{400}} = \frac{12}{20} = 0.6.$$

8.7 CENTRAL LIMIT THEOREM

The sampling distribution of the sample mean $\bar{x}$ can be approximated by a normal distribution when the sample size n is large, irrespective of the shape of the population distribution.

8.8 POINT ESTIMATE AND ESTIMATOR

A single numerical value obtained from the sample information is called a point estimate. For example, consider the problem of estimating the true average GPA of a entire class, say Stat 193, based on a random sample of a couple of students. Based on the information collected from the selected students, let the sample mean of the GPAs be 3.2, then $\bar{x} = 3.2$ is a point estimate of the population mean μ .

The formula or method used to obtain the point estimate is called a point estimator. For example, $\bar{x} = \frac{1}{n} \sum_{i=1}^{n} x_i$ is a point estimator of the population mean μ .

Note that only a couple of students will be happy with a GPA point estimate of 3.2, because some students will have a GPA more than 3.2 and others less than 3.2. Thus, a point estimate may not be a very convincing estimate.

8.9 INTERVAL ESTIMATE AND ESTIMATOR

Two numerical values obtained from the sample information, in which the true parameter value is expected to lie, are called an interval estimate. For example, if we say the true average GPA of the class lies between 2.5 and 3.8, then note that many students will fall in this interval, and will be happy to know about it. Then, it is called an interval estimate of the population mean (μ).

Any formula or method used to obtain an interval estimate is called an interval estimator. For example, the empirical rule $\bar{x} \pm ks$, $k = 1,2,3$ is a good example of an interval estimator.

8.10 LARGE SAMPLE CONFIDENCE INTERVAL ESTIMATE AND ESTIMATOR

If a certain confidence is associated with the interval estimate to contain a particular parameter, it is called a confidence interval estimate. For example, if we say that we are 95% sure that the true average GPA of the class lies between 2.5 and 3.8, then more students will trust it, because we are ensuring with 95% confidence!

Any formula or method used to obtain a confidence interval estimate is called a confidence interval estimator. A general formula for a confidence interval estimator based on a large sample is given by:

$$(\text{Estimator}) \pm Z_{\alpha/2}(\text{Standard Error of the Estimator})$$

or

$$(\text{Estimator}) \pm (\text{Margin of Error})$$

8.10.1 POPULATION MEAN

Assuming that the sample size is large (say, $n > 30$), then a $(1-\alpha)100\%$ confidence interval estimator for the population mean, μ, is given by:

$$\bar{x} \pm Z_{\frac{\alpha}{2}}(\sigma_{\bar{x}}) \qquad \text{or} \qquad \bar{x} \pm Z_{\frac{\alpha}{2}}\left(\frac{\sigma}{\sqrt{n}}\right)$$

If $\alpha = 0.05$, this implies a $(1-\alpha)100\% = (1-0.05)100\% = 95\%$ confidence interval estimator of the population mean, μ, is given by:

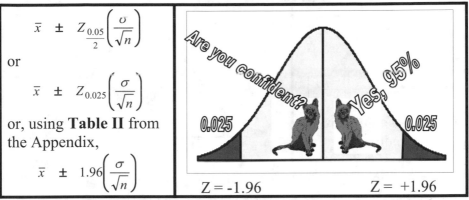

Fig. 8.6. Constructing a 95% confidence interval estimate.

If $\alpha = 0.01$, this implies a $(1-\alpha)100\% = (1-0.01)100\% = 99\%$ confidence interval estimator of the population mean, μ, is given by:

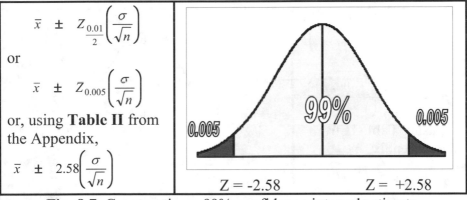

Fig. 8.7. Constructing a 99% confidence interval estimate.

8.10.2 POPULATION PROPORTION

Assuming that the sample size is large (say $n > 30$), then a $(1-\alpha)100\%$ confidence interval estimator for the population proportion P is:

$$\hat{p} \pm Z_{\frac{\alpha}{2}}(\hat{\sigma}_{\hat{p}}) \qquad \text{or} \qquad \hat{p} \pm Z_{\frac{\alpha}{2}}\left(\sqrt{\frac{\hat{p}(1-\hat{p})}{n}}\right)$$

If $\alpha = 0.05$, this implies a $(1-\alpha)100\% = (1-0.05)100\% = 95\%$ confidence interval estimator of the population proportion, P, is given by:

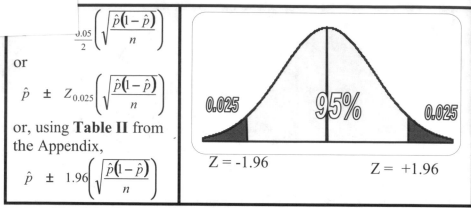

$$\frac{0.05}{2}\left(\sqrt{\frac{\hat{p}(1-\hat{p})}{n}}\right)$$

or

$$\hat{p} \pm Z_{0.025}\left(\sqrt{\frac{\hat{p}(1-\hat{p})}{n}}\right)$$

or, using **Table II** from the Appendix,

$$\hat{p} \pm 1.96\left(\sqrt{\frac{\hat{p}(1-\hat{p})}{n}}\right)$$

0.025 95% 0.025

$Z = -1.96$ $Z = +1.96$

Fig. 8.8. Constructing a 95% confidence interval estimate.

If $\alpha = 0.01$, this implies a $(1-\alpha)100\% = (1-0.01)100\% = 99\%$ confidence interval estimator of the population proportion, P, is given by:

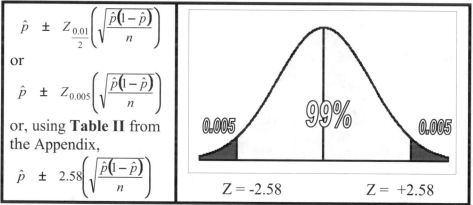

$$\hat{p} \pm Z_{\frac{0.01}{2}}\left(\sqrt{\frac{\hat{p}(1-\hat{p})}{n}}\right)$$

or

$$\hat{p} \pm Z_{0.005}\left(\sqrt{\frac{\hat{p}(1-\hat{p})}{n}}\right)$$

or, using **Table II** from the Appendix,

$$\hat{p} \pm 2.58\left(\sqrt{\frac{\hat{p}(1-\hat{p})}{n}}\right)$$

0.005 99% 0.005

$Z = -2.58$ $Z = +2.58$

Fig. 8.9. Constructing a 99% confidence interval estimate.

Example 8.5. (MAKE YOUR KIDS CONFIDENT) Consider the problem of estimating the population mean μ (may be average score, weight etc.) based on a random sample of size n from the population. Compute the 95% confidence interval (CI) estimate, and state the margin of error in each of the following cases:

(a) $n = 138$, $\sigma = 22$, $\bar{x} = 33.5$.

(b) $n = 50$, $s = 7.2$, $\bar{x} = 10.7$.

(c) $n = 60$, $\sum x_i = 852$, $\sum(x_i - \bar{x})^2 = 215$.

(d) $n = 40$, $\sum x_i = 203$, $\sum x_i^2 = 1{,}250$.

(e) $n = 9$, $\sigma = 6$, $\bar{x} = 15$.

Fig. 8.10. Good hit.

Solution. (a) We are given $n = 138$, $\sigma = 22$, and $\bar{x} = 33.5$. Note that the population standard deviation σ is known, so a 95% confidence interval estimate of the population mean, μ, is given by:

$$\bar{x} \pm 1.96 \frac{\sigma}{\sqrt{n}}$$

or $33.5 \pm 1.96 \dfrac{22}{\sqrt{138}}$ or 33.5 ± 3.67

or $(33.5 - 3.67, \quad 33.5 + 3.67)$ or $(29.83, \quad 37.17)$.

Interpretation: This means that we are 95% sure that the true population mean, μ, lies between 29.83 and 37.17.

Margin of error: The margin of error is 3.67.

(b) We are given $n = 50$, $s = 7.2$, and $\bar{x} = 10.7$. Note that σ is unknown, but the sample size n is more than 30. Thus, the 95% confidence interval estimate of the population mean, μ, is given by:

$$\bar{x} \pm 1.96 \frac{s}{\sqrt{n}}$$

or $10.7 \pm 1.96 \dfrac{7.2}{\sqrt{50}}$ or 10.7 ± 1.99

or $(10.7 - 1.99, \quad 10.7 + 1.99)$ or $(8.71, \quad 12.69)$.

Interpretation: This means that we are 95% sure that the true population mean, μ, lies between 8.71 and 12.69.

Margin of error: The margin of error is 1.99.

(c) We are given $n = 60$, $\Sigma x_i = 852$, $\Sigma(x_i - \bar{x})^2 = 215$. Note that σ is unknown, but the sample size n is more than 30. Thus the 95% confidence interval estimate of the population mean, μ, is given by:

$$\bar{x} \pm 1.96 \frac{s}{\sqrt{n}}.$$

Here

$$\bar{x} = \frac{\Sigma x_i}{n} = \frac{852}{60} = 14.2, \quad s^2 = \frac{\Sigma(x_i - \bar{x})^2}{n-1} = \frac{215}{60-1} = \frac{215}{59} = 3.64,$$

and $s = \sqrt{s^2} = \sqrt{3.64} = 1.91$.

Thus, the 95% CI estimate of the population mean becomes:

$$14.2 \pm 1.96 \frac{1.91}{\sqrt{60}} \qquad \text{or} \qquad 14.2 \pm 0.48$$

or $(14.2 - 0.48, \ 14.2 + 0.48)$ or $(13.72, \ 14.68)$.

Interpretation: This means that we are 95% sure that the true population mean, μ, lies between 13.72 and 14.68.

Margin of error: The margin of error is 0.48.

(d) We are given $n = 40$, $\sum x_i = 203$, and $\sum x_i^2 = 1,250$. Note that σ is unknown, but the sample size n is more than 30. Thus, the 95% confidence interval estimate of the population mean, μ, is given by:

$$\bar{x} \pm 1.96 \frac{s}{\sqrt{n}}.$$

Here $\bar{x} = \dfrac{\sum x_i}{n} = \dfrac{203}{40} = 5.075,$

$$s^2 = \frac{n\left(\sum x_i^2\right) - \left(\sum x_i\right)^2}{n(n-1)} = \frac{40(1,250) - (203)^2}{40(40-1)} = 5.635,$$

and, $s = \sqrt{s^2} = \sqrt{5.635} = 2.374.$

Thus the 95% CI estimate of the population mean becomes:

$$5.075 \pm 1.96 \frac{2.374}{\sqrt{40}} \qquad \text{or} \qquad 5.075 \pm 0.736$$

or $(5.075 - 0.736, \ 5.075 + 0.736)$ or $(4.339, \ 5.811)$.

Interpretation: This means that we are 95% sure that the true population mean, μ, lies between 4.339 and 5.811.

Margin of error: The margin of error is 0.736.

(e) We are given $n = 9$, $\sigma = 6$ and $\bar{x} = 15$. Although the sample size $n = 9$ is not large, but because the population standard deviation $\sigma = 6$ is known, thus the 95% confidence interval estimate is given by:

$$\bar{x} \pm 1.96 \frac{\sigma}{\sqrt{n}} \quad \text{or} \quad 15 \pm 1.96 \frac{6}{\sqrt{9}} \quad \text{or} \quad 15 \pm 3.92$$

or $(15 - 3.92, \ 15 + 3.92)$ or $(11.08, \ 18.92)$.

Interpretation: This means that we are 95% sure that the true population mean, μ, lies between 11.08 and 18.92.

Margin of error: The margin of error is 3.92.

Example 8.6. (MAKE YOUR KIDS CONFIDENT) The following table shows the examination scores of 40 students:

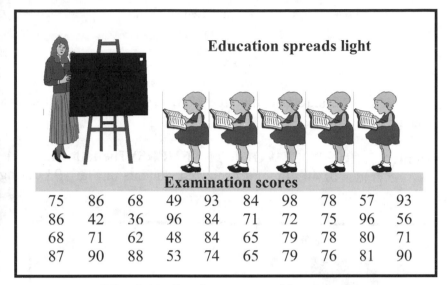

Education spreads light									
Examination scores									
75	86	68	49	93	84	98	78	57	93
86	42	36	96	84	71	72	75	96	56
68	71	62	48	84	65	79	78	80	71
87	90	88	53	74	65	79	76	81	90

Fig. 8.11. Good scores need hard work.

(a) Obtain the point estimate of the average examination score.
(b) Construct a 95% confidence interval estimate and interpret it.
Solution. We are given:

$$n = 40, \quad \sum_{i=1}^{n} x_i = 2{,}984, \text{ and } \sum_{i=1}^{n} x_i^2 = 231{,}728.$$

(a) **Point estimate**: The sample mean is given by:

$$\bar{x} = \frac{\sum_{i=1}^{n} x_i}{n} = \frac{2{,}984}{40} = 74.6.$$

(b) **Interval estimate**: The sample variance is given by:

$$s^2 = \frac{n\left(\sum_{i=1}^{n} x_i^2\right) - \left(\sum_{i=1}^{n} x_i\right)^2}{n(n-1)} = \frac{40(231{,}728) - (2{,}984)^2}{40(40-1)} = 233.89.$$

The sample standard deviation is given by:

$$s = \sqrt{s^2} = \sqrt{233.89} = 15.29.$$

Thus, the 95% confidence interval estimate of the population mean score is given by:

$$\bar{x} \pm 1.96 \frac{s}{\sqrt{n}} \quad \text{or} \quad 74.6 \pm 1.96 \times \frac{15.29}{\sqrt{40}}$$

or 74.6 ± 4.74 or $(69.86, \ 79.34)$.

Thus, we are 95% sure that the true average score of the class lies between 69.86 and 79.34.

Example 8.7. (RESPECT YOUR TEACHERS) The following table shows the grades of 40 randomly selected students in Stat 193 during a summer course:

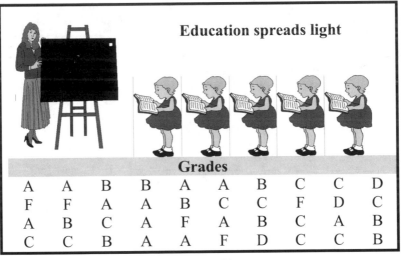

A	A	B	B	A	A	B	C	C	D
F	F	A	A	B	C	C	F	D	C
A	B	C	A	F	A	B	C	A	B
C	C	B	A	A	F	D	C	C	B

Fig. 8.12. Class.

(a) Obtain the point estimate of the proportion of "A" grades.
(b) Construct a 95% confidence interval estimate and interpret it.
(c) Construct a 99% confidence interval estimate and interpret it.

Solution. We are given the total number of students $n = 40$.

Count the number of students who have "A" grades $\text{count} = 12$.

(**a**) The point estimate of the proportion of students who have "A" grades is given by:

$$\hat{p} = \frac{\text{count}}{n} = \frac{12}{40} = 0.30.$$

(**b**) The 95% confidence interval estimate of the proportion of students having "A" grades is given by:

$$\hat{p} \pm 1.96\left(\sqrt{\frac{\hat{p}(1-\hat{p})}{n}}\right) \quad \text{or} \quad 0.30 \pm 1.96\left(\sqrt{\frac{0.30(1-0.30)}{40}}\right)$$

or $\quad 0.30 \pm 1.96\left(\sqrt{0.00525}\right) \quad$ or $\quad 0.30 \pm 1.96\left(0.0725\right)$

or $\quad 0.30 \pm 0.14 \quad$ or $\left(0.30-0.14, \; 0.30+0.14\right)$ or $\left(0.16, \; 0.44\right).$

Interpretation: This means that we are 95% sure that the proportion of students with "A" grades lies between 0.16 and 0.44.
(**c**) The 99% confidence interval estimate of the proportion of students with "A" grades is given by:

$$\hat{p} \pm 2.58\left(\sqrt{\frac{\hat{p}(1-\hat{p})}{n}}\right) \quad \text{or } 0.30 \pm 2.58\left(\sqrt{\frac{0.30(1-0.30)}{40}}\right)$$

or $\quad 0.30 + 2.58\left(\sqrt{0.00525}\right)$ or $0.30 \pm 2.58\left(0.0725\right)$

or $\quad 0.30 \pm 0.19$ or $\left(0.30-0.19, \; 0.30+0.19\right)$ or $\left(0.11, \; 0.49\right).$

Interpretation: This means that we are 99% sure that the proportion of students with "A" grades lies between 0.11 and 0.49.

Example 8.8. (LEARN TRICKS) Given that the 95% confidence interval estimate of the population proportion P is $(0.40, 0.70)$, find:
(a) the sample proportion $\hat{p}$.
(b) the sample size n.
(c) the standard error of $\hat{p}$.

Fig. 8.13. Puzzle pieces.

Solution. We are given:

$$\hat{p}-1.96\sqrt{\frac{\hat{p}(1-\hat{p})}{n}} = 0.40 \quad \text{and} \quad \hat{p}+1.96\sqrt{\frac{\hat{p}(1-\hat{p})}{n}} = 0.70$$

(**a**) Adding these two equations, we have:
$\qquad 2\hat{p} = 1.1.$

Thus, the sample proportion $\hat{p}$ is given by:

$$\hat{p} = \frac{1.1}{2} = 0.55 .$$

(**b**) Thus, the margin of error is given by:

$$1.96\sqrt{\frac{\hat{p}(1-\hat{p})}{n}} = 0.15 \quad \text{or} \quad 1.96\sqrt{\frac{0.55(1-0.55)}{n}} = 0.15$$

or $\quad 1.96\sqrt{\frac{0.2475}{n}} = 0.15 \quad$ or $\quad \sqrt{\frac{0.2475}{n}} = \frac{0.15}{1.96} .$

Squaring both sides, we have:

$$\frac{0.2475}{n} = \left(\frac{0.15}{1.96}\right)^2$$

Thus, the sample size is given by:

$$n = \frac{0.2475}{(0.15/1.96)^2} = 42.25 \approx 42.$$

(**c**) The standard error of the sample proportion $\hat{p}$ is given by:

$$\hat{\sigma}_{\hat{p}} = \sqrt{\frac{\hat{p}(1-\hat{p})}{n}} = \sqrt{\frac{0.55(1-0.55)}{42}} = 0.07676.$$

8.11   SMALL SAMPLE SIZE CONFIDENCE INTERVAL ESTIMATE AND ESTIMATOR

If the sample size is small (say $n \le 30$), then a general formula to construct a $(1-\alpha)100\%$ confidence interval estimate, called a confidence interval estimator of the population mean, μ, is given by:

$$(\text{Estimator}) \pm \left[t_{\alpha/2}(\text{df} = n-1)\right](\text{Standard error of the estimator})$$

where $\text{df} = (n-1)$ stands for the degree of freedom, *which is generally defined as the number of observations minus the number of parameters estimated (or constraints imposed).*

Here $t_{\alpha/2}(\text{df} = n-1)$ denotes the t-value from **Table III,** given in the Appendix, at α level of significance and with $\text{df} = (n-1)$.

For example, if $n = 10$ and $\alpha = 0.05$, then from **Table III,** we have:

$$t_{\alpha/2}(\text{df} = n-1) = t_{0.05/2}(\text{df} = 10-1) = t_{0.025}(\text{df} = 9) = 2.262 .$$

In other words, in **Table III** we have $c = 0.025$, thus we can say:

$$t_c = t_{0.025}(\text{df} = 9) = 2.262$$

Note that the t-distribution is also symmetric for a given sample size n around the mean as shown below:

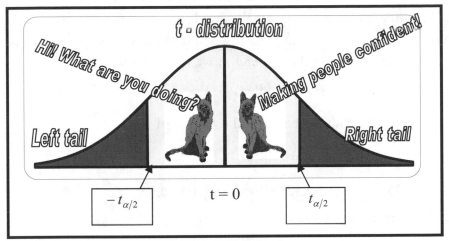

Fig. 8.14. t-distribution

8.11.1 SMALL SAMPLE SIZE CONFIDENCE INTERVAL ESTIMATOR FOR A SINGLE POPULATION MEAN

Assuming that the sample size is small (say $n \le 30$) and the population standard deviation σ is unknown, then a $(1-\alpha)100\%$ confidence interval estimator for the population mean, μ, is given by:

$$\bar{x} \pm \left[t_{\alpha/2}(\text{df} = n-1)\right](\hat{\sigma}_{\bar{x}}), \quad \text{or} \quad \bar{x} \pm \left[t_{\alpha/2}(\text{df} = n-1)\right]\left(\frac{s}{\sqrt{n}}\right)$$

Note carefully that if σ is known and the sample size is small, then we use the previous formula, given by:

$$\bar{x} \pm Z_{\alpha/2}\left(\frac{\sigma}{\sqrt{n}}\right)$$

For more details refer to the **Example 8.5 (e)**.

Example 8.9. (A LITTLE INFORMATION MAY BE WORTH IT) Compute the following confidence interval (CI) estimates of the population mean, μ, based on the information provided in each case:
(a) 95% CI estimate for $n = 10$, $\hat{\sigma}_{\bar{x}} = 2.5$, and $\bar{x} = 33.5$.
(b) 99% CI estimate for $n = 15$, $s = 7.2$, and $\bar{x} = 10.7$.

Fig. 8.15. A confident boy.

(c) 90% CI estimate for $n = 20$, $\Sigma x_i = 852$, and $\Sigma (x_i - \bar{x})^2 = 215$.

(d) 95% CI estimate for $n = 25$, $\Sigma x_i = 203$, and $\Sigma x_i^2 = 2,450$.

Solution. (a) Here $n = 10$ (small sample), $\hat{\sigma}_{\bar{x}} = 2.5$, and $\bar{x} = 33.5$.

Now $(1 - \alpha)100\% = 95\%$ implies $(1 - \alpha) = 0.95$, thus $\alpha = 1 - 0.95 = 0.05$.

Using **Table III** from the Appendix, we have:

$$t_{\alpha/2}(\text{df} = n - 1) = t_{0.05/2}(\text{df} = 10 - 1) = t_{0.025}(\text{df} = 9) = 2.262.$$

Thus, the 95% confidence interval estimate of μ is given by:

$$\bar{x} \; \pm \; \left[t_{\alpha/2}(\text{df} = n - 1) \right] (\hat{\sigma}_{\bar{x}})$$

or $33.5 \; \pm \; (2.262)(2.5)$ or $33.5 \; \pm \; 5.655$

or $(33.5 - 5.655, \; 33.5 + 5.655)$ or $(27.845, \; 39.155)$.

Thus, we are 95% sure that the true population mean lies between 27.845 and 39.155.

(b) Here $n = 15$ (small sample), $s = 7.2$, and $\bar{x} = 10.7$.

Now $(1 - \alpha)100\% = 99\%$ implies $(1 - \alpha) = 0.99$, thus $\alpha = 1 - 0.99 = 0.01$.

Using **Table III** from the Appendix, we have:

$$t_{\alpha/2}(\text{df} = n - 1) = t_{0.01/2}(\text{df} = 15 - 1) = t_{0.005}(\text{df} = 14) = 2.977.$$

Thus, the 99% confidence interval estimate of μ is given by:

$$\bar{x} \; \pm \; \left[t_{\alpha/2}(\text{df} = n - 1) \right] \left(\frac{s}{\sqrt{n}} \right) \quad \text{or} \; \bar{x} \; \pm \; \left[t_{0.005}(\text{df} = 14) \right] \left(\frac{s}{\sqrt{n}} \right)$$

or $10.7 \; \pm \; (2.977) \left(\dfrac{7.2}{\sqrt{15}} \right)$ or $10.7 \; \pm \; 5.53$

or $(10.70 - 5.53, \; 10.70 + 5.53)$ or $(5.17, \; 16.23)$.

Thus, we are 99% sure that the true population mean lies between 5.17 and 16.23.

(c) Here $n = 20$ (small sample), $\Sigma x_i = 852$, and $\Sigma (x_i - \bar{x})^2 = 215$. Thus,

$$\bar{x} = \frac{\sum\limits_{i=1}^{n} x_i}{n} = \frac{852}{20} = 42.6 \, , \; s^2 = \frac{\sum\limits_{i=1}^{n} (x_i - \bar{x})^2}{n - 1} = \frac{215}{20 - 1} = \frac{215}{19} = 11.32 \, ,$$

and $s = \sqrt{s^2} = \sqrt{11.32} = 3.36$.

Now $(1-\alpha)100\% = 90\%$ implies $(1-\alpha) = 0.90$, thus $\alpha = 1 - 0.90 = 0.10$, Using **Table III** from the Appendix, we have:

$$t_{\alpha/2}(\mathrm{df} = n-1) = t_{0.10/2}(\mathrm{df} = 20-1) = t_{0.05}(\mathrm{df} = 19) = 1.729.$$

Thus, the 90% confidence interval estimate of μ is given by:

$$\bar{x} \pm \left[t_{\alpha/2}(\mathrm{df} = n-1)\right]\left(\frac{s}{\sqrt{n}}\right) \quad \text{or} \quad \bar{x} \pm \left[t_{0.05}(\mathrm{df} = 19)\right]\left(\frac{s}{\sqrt{n}}\right)$$

or $42.60 \pm (1.729)\left(\dfrac{3.36}{\sqrt{20}}\right)$ or 42.60 ± 1.30

or $(42.60 - 1.30, \quad 42.60 + 1.30)$ or $(41.30, \quad 43.90)$.

Thus, we are 90% sure that the true population mean lies between 41.30 and 43.90.

(**d**) Here $n = 25$ (small sample), $\Sigma x_i = 203$, and $\Sigma x_i^2 = 2{,}450$. Thus,

$$\bar{x} = \frac{\sum_{i=1}^{n} x_i}{n} = \frac{203}{25} = 8.12,$$

$$s^2 = \frac{n\left(\sum_{i=1}^{n} x_i^2\right) - \left(\sum_{i=1}^{n} x_i\right)^2}{n(n-1)} = \frac{25(2{,}450) - (203)^2}{25(25-1)} = 33.40,$$

and $s = \sqrt{s^2} = \sqrt{33.40} = 5.78$.

Now $(1-\alpha)100\% = 95\%$ implies $(1-\alpha) = 0.95$, thus $\alpha = 1 - 0.95 = 0.05$. Using **Table III** from the Appendix, we have:

$$t_{\alpha/2}(\mathrm{df} = n-1) = t_{0.05/2}(\mathrm{df} = 25-1) = t_{0.025}(\mathrm{df} = 24) = 2.064$$

Thus, the 95% confidence interval estimate of μ is given by:

$$\bar{x} \pm \left[t_{\alpha/2}(\mathrm{df} = n-1)\right]\left(\frac{s}{\sqrt{n}}\right)$$

or $8.12 \pm (2.064)\left(\dfrac{5.78}{\sqrt{25}}\right)$ or 8.12 ± 2.39

or $(8.12 - 2.39, \quad 8.12 + 2.39)$ or $(5.73, \quad 10.51)$.

Thus, we are 95% sure that the true population mean lies between 5.73 and 10.51.

8.12 CONFIDENCE INTERVAL ESTIMATOR FOR THE DIFFERENCE BETWEEN TWO MEANS

Consider the two populations shown below:

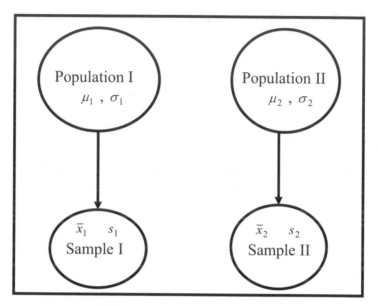

Fig. 8.16. Two populations.

Here we would like to discuss two cases:

8.12.1 LARGE SAMPLE SIZE

Let μ_1 and μ_2 be the two population means. Let σ_1^2 and σ_2^2 be the known variances of populations I and II, respectively. Let $\bar{x}_1$ and $\bar{x}_2$ be the two sample means based on the two independent random samples of large sizes n_1 and n_2 taken from the two populations. Then a $(1-\alpha)100\%$ confidence interval estimator of the difference between two population means $(\mu_1 - \mu_2)$ is given by:

$$(\bar{x}_1 - \bar{x}_2) \pm Z_{\alpha/2}\sqrt{\frac{\sigma_1^2}{n_1} + \frac{\sigma_2^2}{n_2}}$$

Note that if σ_1^2 and σ_2^2 are unknown they can be estimated by s_1^2 and s_2^2 respectively.

Example 8.10. (CONFIDENCE EARNS GOOD GRADES) An experienced instructor wants to know the difference between the average scores of two large sections of Stat 229. He collected the following information from two independent random samples as well as from his past five years experience of teaching the same class:

Class	Sample size	Sample mean	Population variance
A	32	85	25
B	35	82	16

Fig. 8.17. An old instructor's story.

Construct the 95% confidence interval estimate for the difference between the average scores of the two classes.

Solution. Given $n_1 = 32$, $n_2 = 35$, $\bar{x}_1 = 85$, $\bar{x}_2 = 82$, $\sigma_1^2 = 25$, and $\sigma_2^2 = 16$. Now a $(1-\alpha)100\% = 95\%$ implies $\alpha = 0.05$, thus using **Table II** from the Appendix, we have $Z_{\alpha/2} = Z_{0.05/2} = Z_{0.025} = 1.96$. Hence the 95% confidence interval estimate of the difference between two population means $(\mu_1 - \mu_2)$ is given by:

$$(\bar{x}_1 - \bar{x}_2) \pm 1.96\sqrt{\frac{\sigma_1^2}{n_1} + \frac{\sigma_2^2}{n_2}}$$

or $\qquad (85-82) \pm 1.96\sqrt{\dfrac{25}{32} + \dfrac{16}{35}}$

or $\qquad (0.819, 5.181)$.

Thus, the instructor is 95% sure that the difference between the average scores of the two sections is 0.819 to 5.181.

8.12.2 SMALL SAMPLE SIZE

Let μ_1 and μ_2 be the two population means. Let $\bar{x}_1$ and $\bar{x}_2$ be the two sample means based on the two independent random samples of small sizes n_1 and n_2 taken from the two populations having common variance σ^2. In other words, assume $\sigma_1^2 = \sigma_2^2 = \sigma^2$. Let s_1^2 and s_2^2 be the first and second sample variances, respectively.

Then a $(1-\alpha)100\%$ confidence interval estimator of the difference between the two population means $(\mu_1 - \mu_2)$ is given by:

$$(\bar{x}_1 - \bar{x}_2) \pm \left[t_{\alpha/2}(\text{df} = n_1 + n_2 - 2)\right] \sqrt{s_p^2\left(\frac{1}{n_1} + \frac{1}{n_2}\right)}$$

where $s_p^2 = \dfrac{(n_1 - 1)s_1^2 + (n_2 - 1)s_2^2}{n_1 + n_2 - 2}$ is called pooled sample variance.

Note that the pooled sample variance s_p^2 is an estimator of the common population variance σ^2 of the two populations. Further note that we are estimating two means, thus the *degree of freedom* will be total number of observations minus two, that is, $\text{df} = n_1 + n_2 - 2$.

Example 8.11. (HARD WORK BRINGS CONFIDENCE) A new instructor wants to know the difference between the average scores of two large sections of Stat 229. She collected the following information from two independent random samples from the two classes:

	Class	Sample size	Sample mean	Sample variance
	A	16	86	16
	B	10	90	9

Fig. 8.18. A new instructor's story.

Construct the 95% confidence interval estimate for the difference between the average scores of the two classes.

Solution. Given $n_1 = 16$, $n_2 = 10$, $\bar{x}_1 = 86$, $\bar{x}_2 = 90$, $s_1^2 = 16$, and $s_2^2 = 9$. Note that the instructor is new and she does not have any past information. Thus, let us find the pooled sample variance as:

$$s_p^2 = \frac{(n_1 - 1)s_1^2 + (n_2 - 1)s_2^2}{n_1 + n_2 - 2} = \frac{(16-1)\times 16 + (10-1)\times 9}{16 + 10 - 2} = 13.375.$$

Now a $(1-\alpha)100\% = 95\%$ implies $\alpha = 0.05$, thus from **Table III** in the Appendix, we have:

$$t_{\frac{\alpha}{2}}(\text{df} = n_1 + n_2 - 2) = t_{0.05}{\frac{}{2}}(\text{df} = 16 + 10 - 2) = t_{0.025}(\text{df} = 24) = 2.064.$$

Hence the 95% confidence interval estimate of the difference between two population means $(\mu_1 - \mu_2)$ is given by:

$$(\bar{x}_1 - \bar{x}_2) \pm 2.064 \sqrt{s_p^2 \left(\frac{1}{n_1} + \frac{1}{n_2} \right)}$$

or $(86 - 90) \pm 2.064 \sqrt{13.375 \left(\frac{1}{16} + \frac{1}{10} \right)}$

or $(-4) \pm 3.04$ or $(-7.04, -0.96)$.

Thus, the instructor is 95% sure that the difference between the true average scores of the two sections is −7.04 to −0.96. Note that the difference between two means can be negative.

8.13 CONFIDENCE INTERVAL ESTIMATOR FOR THE DIFFERENCE BETWEEN TWO PROPORTIONS

Consider P_1 and P_2 to be the two population proportions.

Let $\hat{p}_1 = \frac{x_1}{n_1}$ and $\hat{p}_2 = \frac{x_2}{n_2}$ be the two sample proportions based on the two independent random samples of large sizes n_1 and n_2 taken from the two populations. Then a $(1 - \alpha)100\%$ confidence interval estimator of the difference between two population proportions $(P_1 - P_2)$ is given by:

$$(\hat{p}_1 - \hat{p}_2) \pm Z_{\alpha/2} \sqrt{\frac{\hat{p}_1(1 - \hat{p}_1)}{n_1} + \frac{\hat{p}_2(1 - \hat{p}_2)}{n_2}}$$

Example 8.12. (GOOD GRADES BRING CONFIDENCE) An instructor wants to know the difference between the proportions of the number of "B" grades in two large sections of Stat 229. He collected the following information from two independent random samples as:

	Class	Sample size	No. of "B" grades
	I	32	12
	II	40	9

Fig. 8.19. Good instructors care about results.

Construct the 95% confidence interval estimate for the difference between the proportions of the number of "B" grades in the two sections.

Solution. We are given that $n_1 = 32$, $n_2 = 40$, $x_1 = 12$, and $x_2 = 9$. Thus,

$$\hat{p}_1 = \frac{x_1}{n_1} = \frac{12}{32} = 0.375 \quad \text{and} \quad \hat{p}_2 = \frac{x_2}{n_2} = \frac{9}{40} = 0.225$$

Now

$$(1-\alpha)100\% = 95\% \text{ implies } \alpha = 0.05$$

thus using **Table II** from the Appendix, we have:

$$Z_{\alpha/2} = Z_{0.05/2} = Z_{0.025} = 1.96.$$

Hence the 95% confidence interval estimate of the difference between two population proportions $(P_1 - P_2)$ is given by:

$$(\hat{p}_1 - \hat{p}_2) \pm Z_{\alpha/2}\sqrt{\frac{\hat{p}_1(1-\hat{p}_1)}{n_1} + \frac{\hat{p}_2(1-\hat{p}_2)}{n_2}}$$

or $$(0.375 - 0.225) \pm 1.96\sqrt{\frac{0.375(1-0.375)}{32} + \frac{0.225(1-0.225)}{40}}$$

or $0.15 \pm 1.96\sqrt{0.0011684}$ or $(-0.062, 0.362)$.

Thus, the instructor is 95% sure that the difference between the true proportions of students with B grades in the two sections is -0.062 to 0.362. Note that proportion itself can never be negative, but the difference between two proportions can be negative.

Example 8.13. (SOMETIMES OVERCONFIDENCE KILLS) Three students Amy, Bob, and Chris were asked to do independent surveys to estimate the true average score in a summer class of Stat 193. All the three students came with the following three different confidence interval estimates: Amy claims that she is 95% sure that the average score of the class lies between 70% to 85%; Bob claims that he is 100% sure that the average score of the class lies between 0% to 100% inclusive; and Chris claims that he is 95% sure that the average score the class lies between 67% to 92%. Whose confidence interval estimate will be preferred and why?

Solution. Note that Amy's confidence interval estimate is narrower than Chris' estimate at the same 95% level of confidence. Thus Amy's estimate will be preferred over Chris' estimate. Although Bob is 100% sure, his estimate cannot be preferred because everyone knows that the true score always lies between 0% and 100%.

8.14 SOME IMPORTANT RESULTS

(a) The standard deviation of the difference between two sample means is given by:

$$\sigma_{(\bar{x}_1 - \bar{x}_2)} = \sqrt{\frac{\sigma_1^2}{n_1} + \frac{\sigma_2^2}{n_2}}$$

(b) The **pooled standard error** of the difference between two sample means is given by:

$$\hat{\sigma}_{(\bar{x}_1 - \bar{x}_2)} = \sqrt{s_p^2 \left(\frac{1}{n_1} + \frac{1}{n_2} \right)} \quad \text{where } s_p^2 = \frac{(n_1 - 1)s_1^2 + (n_2 - 1)s_2^2}{n_1 + n_2 - 2}.$$

(c) The **standard error** of the difference between two sample proportions $\hat{p}_1$ and $\hat{p}_2$ is given by:

$$\hat{\sigma}_{(\hat{p}_1 - \hat{p}_2)} = \sqrt{\frac{\hat{p}_1(1 - \hat{p}_1)}{n_1} + \frac{\hat{p}_2(1 - \hat{p}_2)}{n_2}}$$

(d) The **pooled standard error** of the difference between two sample proportions $\hat{p}_1$ and $\hat{p}_2$ is given by:

$$\hat{\sigma}_{(\hat{p}_1 - \hat{p}_2)(\text{pooled})} = \sqrt{\hat{p}\hat{q} \left(\frac{1}{n_1} + \frac{1}{n_2} \right)}$$

where

$$\hat{p} = \frac{n_1\hat{p}_1 + n_2\hat{p}_2}{n_1 + n_2} = \frac{x_1 + x_2}{n_1 + n_2} \quad \text{is called the } \textbf{pooled sample}$$

proportion and

$$\hat{q} = (1 - \hat{p}).$$

Note that the pooled standard error of the difference between two sample proportions cannot be used to construct a confidence interval estimate for the difference between two population proportions.

LUDI 8.1. (HOW HEAVY ARE PUMPKINS?)

Consider the problem of estimating the average weight (μ) of pumpkins on a farm based on a random sample of size n. Compute the standard deviation of the sample mean (or standard error of the sample mean), margin of error, and construct the 99% confidence interval estimate in each one of the following cases, and interpret your findings assuming weight in lbs:

(a) $n = 138$, $\sigma = 22$, $\bar{x} = 15.5$.

(b) $n = 50$, $s = 7.2$, $\bar{x} = 10.7$.

(c) $n = 60$, $\sum x_i = 852$, $\sum (x_i - \bar{x})^2 = 215$.

(d) $n = 40$, $\sum x_i = 600$, $\sum x_i^2 = 9,000$.

Fig. 8.20. Pumpkin.

(e) $n = 30$, $\bar{x} = 16.8$, $\sum x_i^2 = 8,500$.

LUDI 8.2. (ORCHARDS NEED ATTENTION)

Consider the problem of estimating the population proportion, P, of rotten apples in an orchard based on a random sample of size n.

(I) Compute the 95% (and 99%) confidence interval estimates in each of the following cases:

(a) $n = 100$, $\hat{p} = 0.50$.

(b) $n = 100$, $x = 20$.

(c) $n = 100$, $\hat{p} = 0.30$.

(d) $n = 100$, $\hat{p} = 0.70$.

Fig. 8.21. Apple and worm.

(II) Compute the standard error and the margin of error of sample proportion $\hat{p}$ in each of the above cases.

(III) Repeat (I) and (II) with $n = 200$ and $n = 50$. Comment on the results.

LUDI 8.3. (ALWAYS KEEP YOUR UMBRELLA AT HOME)

Assume that the variance of rainy day temperatures in the last 20 years is known to be 400 $^{\circ}F^2$ in a country. What is the standard deviation of the sample mean $\bar{x}$ for a random sample of: (a) $n = 25$? (b) $n = 100$? (c) $n = 400$?

Fig. 8.22. Raining.

LUDI 8.4. (BE OBEDIENT TO YOUR TEACHERS) The following table shows the examination scores of 30 students:

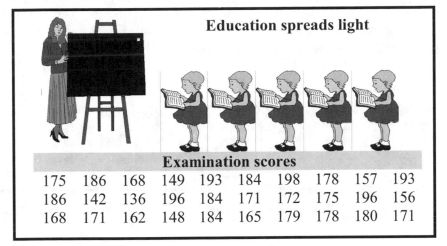

Education spreads light									
Examination scores									
175	186	168	149	193	184	198	178	157	193
186	142	136	196	184	171	172	175	196	156
168	171	162	148	184	165	179	178	180	171

Fig. 8.23. Class outlook.

(a) Obtain the point estimate of the mean of the examination score.
(b) Construct a 95% confidence interval estimate and interpret it.

LUDI 8.5. (PARTICIPATE IN CLASS DISCUSSIONS) A class consists of four students Amy, Bob, Chris, and Don with their scores as: 8, 2, 4, and 6.

Teacher	Amy	Bob	Chris	Don
Score	8	2	4	6

Fig. 8.24. Class.

Consider drawing a random sample of size 2 with replacement.
(a) List all the possible samples and evaluate the sample means $\bar{x}$ in each case.
(b) Write down the sampling distribution of sample mean $\bar{x}$.
(c) Write down the population distribution and calculate its mean μ and standard deviation σ.
(d) Calculate the mean and standard deviation of the sampling distribution $\bar{x}$ obtained in part (b) and verify that these agree with μ and $\sigma/\sqrt{2}$, respectively.

LUDI 8.6. (TOBACCO SMOKE FREE DAY-MARCH 13)
Assume that a survey was conducted in a certain location asking the question of whether or not to observe a "Non-Smoking Day" every year on March 13 at an international level. The proportion of people who favor the observance of a "Non-Smoking Day" in that location is found to be 0.99.

Fig. 8.25. The habit of smoking may put you down among your friends.

What is the standard error of sample proportion $\hat{p}$ for a random sample of: (a) $n = 25$? (b) $n = 100$? (c) $n = 400$?

LUDI 8.7. (LOST AND FOUND) Either fill in the blank or state whether or not the statement is true or false.

(1) The sampling distribution of a statistic is the distribution of the values of the statistic in all possible samples, as long as ....
(a) the samples are all of the same size n.
(b) the samples are all taken from the same population.
(c) (a) and (b).
(d) (a) but not (b).

(2) A statistic is said to be....if the mean of the sampling distribution of a statistic is equal to its population parameter value.
(a) efficient
(b) sufficient
(c) unbiased
(d) biased

(3) An unbiased estimator for the population proportion , P , is ....
(a) sample mean, $\bar{x}$.
(b) sample standard deviation, s.
(c) sample proportion, $\hat{p}$.
(d) population standard deviation, σ.

(4) The central limit theorem states that if the original population is normally distributed and the sample size n is large, then the distribution of the sample mean $\bar{x}$ is also normal with the original population mean.
(a) True
(b) False

(5) An unbiased estimator for the population mean , μ , is …
(a) sample mean, $\bar{x}$.
(b) sample standard deviation, s .
(c) sample proportion, $\hat{p}$.
(d) population standard deviation, σ .

(6) If we increase the sample size, the standard deviation of the sample proportion $\hat{p}$ will increase as well.
(a) True
(b) False

(7) At St. Cloud State University, a random sample of 20 students is taken. The sample proportion of US citizens is 0.92. If the process were repeated (take another sample of $n = 20$ size), the sample proportion will always be 0.92.
(a) True
(b) False

(8) At St. Cloud State University, 65% of girls are taller than 170 cms. Suppose 200 different random samples, each of 30 girls from St. Cloud State University, are taken and for each of them the sample proportion of girls taller than 170 cm is calculated. The average value of all those sample proportions will be approximately 65%.
(a) True
(b) False
(c) Can't tell because I miss my classes too often.

(9) Assume that on average every St. Cloud State University student buys 15 notebooks per year. If we took 135 samples each of 60 students, we would expect the average value of all the sample means to be …..
(a) 15
(b) 135
(c) 120
(d) Can't tell because I attended only a few classes.

LUDI 8.8. (PROTECT YOUR TEETH) Eighty percent of the kids of all the faculty members at St. Cloud State University prefer to buy chocolate. In a random sample of 250 kids of St. Cloud State University faculty members, 75% were found to be buyers of chocolate in a particular week.

(a) What is the value of the parameter?

(b) How many of the sampled kids bought chocolate?

(c) What is the distribution of $\hat{p}$?

(d) What is the standard error of $\hat{p}$?

(e) What is the average value of $\hat{p}$?

Fig. 8.26. Brush your teeth everyday.

(f) Construct a 95% confidence interval estimate of the proportion and interpret it.

(g) Construct a 99% confidence interval estimate of the proportion and interpret it.

LUDI 8.9. (GOOD GRADES ARE IMPORTANT) The following table shows the grades of 50 randomly selected students in Stat 193 during a summer course:

Education spreads light

Grades									
A	A	B	B	A	A	B	C	C	D
F	F	A	A	B	C	C	F	D	C
A	B	C	A	F	A	B	C	A	B
C	C	B	A	A	F	D	C	C	B
A	B	A	B	C	C	C	F	F	F

Fig. 8.27. Class.

(a) Obtain the point estimate of the proportion of "A" grades.

(b) Construct a 95% confidence interval estimate and interpret it.

(c) Construct a 99% confidence interval estimate and interpret it.

LUDI 8.10. (SOMETIMES A LITTLE INFORMATION HELPS)
Consider the problem of the estimation of the population mean, μ, based on a random sample of size n from a population. Find the standard deviation/error of the sample mean and margin of error. Construct and interpret the following confidence interval (CI) estimates:

(a) 95% CI estimate for $n = 8$, $\hat{\sigma}_{\bar{x}} = 22$, and $\bar{x} = 33.5$.

(b) 90% CI estimate for $n = 15$, $s = 7.2$, and $\bar{x} = 10.7$.

(c) 99% CI estimate for $n = 12$, $\Sigma x_i = 852$, and $\Sigma(x_i - \bar{x})^2 = 215$.

(d) 99% CI estimate for $n = 5$, $\Sigma x_i = 23$, and $\Sigma x_i^2 = 1{,}250$.

(e) 90% CI estimate for $n = 16$, $\bar{x} = 6.8$, and $\Sigma x_i^2 = 1{,}400$.

(f) 99% CI estimate for $n = 5$, $\bar{x} = 12.46$, and $\sigma = 1.2$.

(g) 98.4% CI estimate for $n = 8$, $\bar{x} = 14.76$ and $\sigma^2 = 0.81$.

(h) 95% CI estimate for $n = 15$, $\bar{x} = 5{,}000$ and $\sigma_{\bar{x}} = 500$.

LUDI 8.11. (OLD INSTRUCTORS ARE STRICT) An experienced instructor wants to know the difference between the average scores of two large sections of Stat 229. He collected the following information from two independent random samples as well as from his past ten years of experience of teaching the same class as follows:

	Class	Sample size	Sample mean	Population variance
	A	35	88	16
	B	39	82	25

Fig. 8.28. An old instructor's story.

Construct the 99% confidence interval estimate for the difference between the average scores of the two classes and interpret it.

LUDI 8.12. (GENERALLY TOWNHOUSES ARE HANDY)
A random sample of 9 townhouses from your local newspaper has the monthly rents (dollars):

500, 600, 505, 450, 550, 515, 495, 650, 395

(a) Construct a 95% confidence interval estimate for the average rent and interpret it.

Fig. 8.29. Townhouse.

(b) Construct a 99% confidence interval estimate for the average rent and interpret it.

LUDI 8.13. (LADIES ARE TEACHERS BY BIRTH) A new instructor wants to know the difference between the average scores of two large sections of Stat 229. She collected the following information from two independent random samples from the two classes.

	Class	Sample size	Sample mean	Sample variance
	A	12	90	25
	B	8	85	16

Fig. 8.30. A new instructor's story.

Construct the 99% confidence interval estimate for the difference between the average scores of the two classes and interpret it.

LUDI 8.14. (YOUNG INSTRUCTORS MAKE MISTAKES ON THE BOARD) An instructor wants to know the difference between the proportion of the number of "A" grades in two large sections of Stat 229. He collected the following information from two random and independent samples as follows:

	Class	Sample size	No. of "A" grades
	I	36	14
	II	45	18

Fig. 8.31. Good instructors care about results.

Construct the 99% confidence interval estimate of the difference between the proportions of the number of "A" grades in the two classes.

LUDI 8.15. (INDIAN FOOD IS SPICY) Based on a random sample of n chilies taken from a farm, the 95% confidence interval estimate of the proportion of red-hot chilies is (0.25, 0.65).
(a) Find the sample proportion of red-hot chilies $\hat{p}$.
(b) Find the margin of error and interpret it.
(c) Find the value of the standard error of the sample proportion of red-hot chilies.

Fig. 8.32. Chili plant.

9. THE IDEA OF HYPOTHESES TESTING

9.1 INTRODUCTION

In this chapter, we discuss the need of hypotheses testing, two types of errors, one-tailed and two-tailed tests, tests for single mean, single proportion, the difference between two means, the difference between two proportions, chi-square tests, and a one way ANOVA.

9.2 THE NEED OF HYPOTHESES TESTING

As we discussed in **Chapter 1**, we now provide information about testing the hypotheses.

Elephants need ropes   **Ships need ropes**

Fig. 9.1. Demand of ropes.

Consider a rope seller and a rope buyer in a market. Assume that the rope buyer is a circus owner and wants to buy good ropes to control his elephants in the circus or to control his ship.

Consider the situation where the rope seller says, "The proportion of defective ropes is 0.25."

Seller says   proportion of defective ropes is 0.25   **Buyer listens**

Fig. 9.2. Null hypothesis.

If the buyer believes the seller, then there is nothing to test. However, a good buyer will test before making any decision, because the buyer may not trust the seller.

Assume that the buyer says, "The proportion of defective ropes is **not equal** to 0.25"

Fig. 9.3. Alternative hypothesis.

Thus, there are two statements:
(a) One statement is given by the seller about all the ropes in his shop (population).
(b) The second statement is given by the buyer about all the ropes in the shop (again population).

Note that, "Any statement about the value of a parameter is called a **null hypothesis**." It is denoted by H_0.

Thus, the null hypothesis is the statement given by the seller as follows:

H_0 : The proportion of defective ropes is equal to 0.25

or equivalently

$H_0 : P = 0.25$

Further, note that, "Any statement which contradicts the null hypothesis is called an **alternative hypothesis**." It is denoted by H_a or H_1.

Thus, the alternative hypothesis is a statement given by the buyer:

H_1 : The proportion of defective ropes is not equal to 0.25

or equivalently

$H_1 : P \neq 0.25$ (two-tailed test)

Note that according to the buyer, the proportion of defective ropes may be less than 0.25 or may be more than 0.25. When there are two such possibilities, the test is called a **two-tailed test**.

Further note that a good buyer will be worried about the ropes if the proportion of defective ropes is more than 0.25. Thus, the buyer will be interested in testing the alternative hypothesis.

H_1 : The proportion of defective ropes is more than 0.25

or equivalently

$H_1 : P > 0.25$ (right-tailed test)

Now the buyer is focusing only on the right tail. When there is only one such possibility, the test is called a **one-tailed test**.

Now the buyer wants to know the true proportion of defective ropes. Consider there are 1,000 ropes in the shop with the population size $N = 1,000$. If the buyer breaks all 1,000 ropes to find the proportion of defective ropes, then the buyer will have nothing to buy and the seller will have nothing to sell! Thus, a good buyer will buy a few ropes, say a sample of $n = 30$, to test before making any decision.

Can a buyer make an error in making his decision?

Assume the buyer randomly selected $n = 30$ ropes and found that the proportion of defective ropes is 0.29. From this information, the buyer concludes that the proportion of defective ropes is more than 0.25. Now the following question arises:

"Is the buyer 100% correct?"

The buyer may or may not be right. The reason is that the buyer made his decision only on the basis of information obtained from 30 ropes.

Now there are two possibilities:

First Possibility:

Assume the true proportion of defective ropes is 0.25, but due to some error, the buyer concludes that it is **not** equal to 0.25. Note that the null hypothesis $H_0 : P = 0.25$ was true, but due to some error the buyer rejected it. Such an error is called a **Type I error** and it is denoted by α (alpha).

Thus:

$\text{Type I error} = \alpha = \text{Prob.}[\text{Rejecting } H_0 \text{ when } H_0 \text{ is true}]$

The probability of rejecting the null hypothesis when it is true is called a **Type I error** or the **level of significance**.

Second Possibility:

Assume that the true proportion of the defective ropes is **not** equal to 0.25, but due to some error, the buyer concludes that it is equal to 0.25. This means the null hypothesis $H_0 : P = 0.25$ was **not** true, but the buyer accepted it due to some error. Such an error is called a **Type II error**, and is denoted by β.

Thus:

$\text{Type II error} = \beta = \text{Prob.}[\text{Accepting } H_0 \text{ when } H_0 \text{ is not true}]$

The probability of accepting the null hypothesis when it is not true is called a **Type II error**.

Note that we can summarize the above results as follows:

Decision	Hypothesis (H_0)	
	True	**False (or not true)**
Reject H_0	Type I error (α)	Correct decision
Accept H_0	Correct decision	Type II error (β)

9.3 SIGNIFICANCE LEVEL

The significance level is a number α and is the chance of **Type I error**, that is, the probability of rejecting the null hypothesis when it is true.

9.4 DECISION RULE

A decision rule helps to reject or accept the null hypothesis (H_0).

9.5 REJECTION REGION

A rejection region is the set of values for which we would reject the null hypothesis H_0 at a certain level of significance.

9.6 ACCEPTANCE REGION

The acceptance region is the set of values for which we would accept the null hypothesis H_0 at a certain level of significance.

9.7 CRITICAL VALUE

A critical value is the starting point of a set of values that form the rejection region. It is generally denoted by Z_c or t_c etc.

9.8 LARGE SAMPLE TEST FOR SINGLE PROPORTION

Consider P_0 to be the true proportion of an attribute to be tested in a population. There are generally six steps to apply any test.

Step I. Set the null hypothesis: The true proportion is equal to P_0, thus:

$H_0 : P = P_0$, (where P_0 is any fixed value)

Step II. Set the alternative hypothesis: Under the alternative hypothesis, there are three possibilities. The true proportion, P, may be: (i) not equal to; (ii) less than; or (iii) more than the guessed value P_0 as shown below:

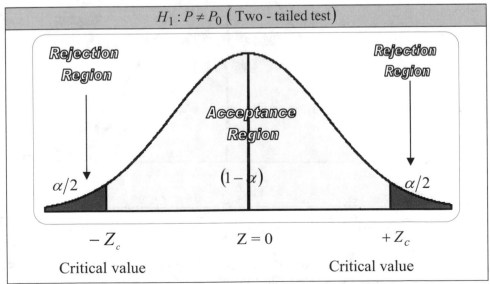

Fig. 9.4. Two-tailed test has two critical values.

or

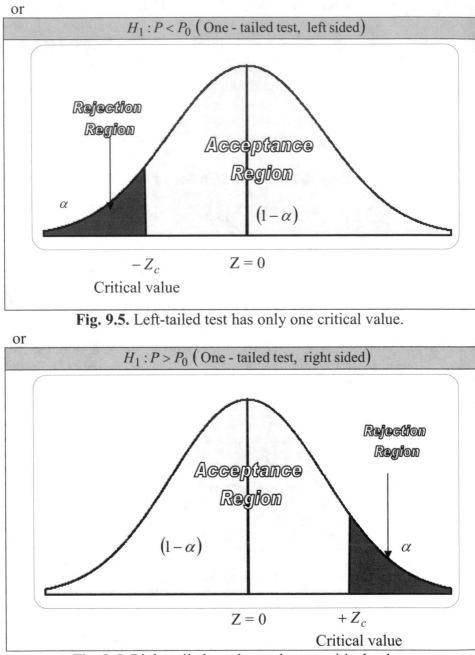

Fig. 9.5. Left-tailed test has only one critical value.

or

Fig. 9.6. Right-tailed test has only one critical value.

Note that the choice of alternative hypothesis depends upon the statement under the alternative hypothesis.

Step III. Set the level of significance: On the basis of the desired level of accuracy of the results, decide:

$\alpha = 0.05$, or $\alpha = 0.01$, ... etc.

Step IV. Compute the Z-Statistic: Let $n \geq 30$ (large sample), then calculate the Z-statistic:

$$Z_{cal} = \frac{\hat{p} - P_0}{\sqrt{\dfrac{P_0(1 - P_0)}{n}}}$$

where n is the sample size and $\hat{p} = x/n$ is an observed proportion of sampled units possessing the attribute of interest.

Step V. Find critical value(s): For a large sample test, the critical value(s) are based on two things: (i) the level of significance α, and (ii) the type of the alternative hypothesis.

For example, (a) if $\alpha = 0.05$ and we are using a two-tailed (or non-directional) test, using **Table II** from the Appendix, there are two critical values given by: $Z_c - \pm 1.96$.

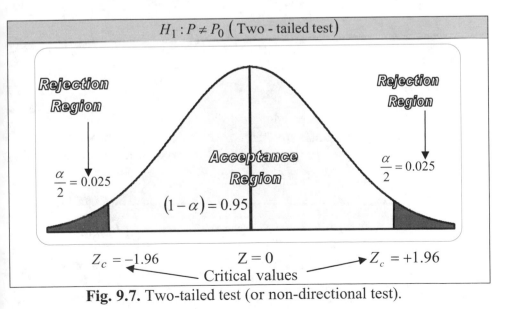

Fig. 9.7. Two-tailed test (or non-directional test).

(b) if $\alpha = 0.05$ and we are using a one-tailed (or left sided) test, using **Table II** from the Appendix, there is only one critical value given by: $Z_c = -1.645$.

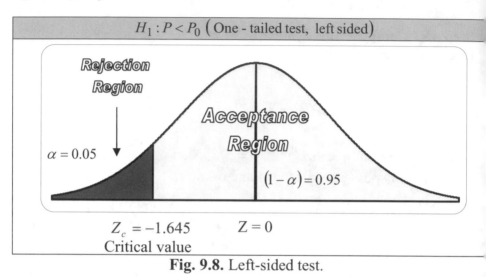

$H_1 : P < P_0$ (One - tailed test, left sided)

Fig. 9.8. Left-sided test.

(c) If $\alpha = 0.05$ and we are using a one-tailed (or right sided) test, using **Table II** from the Appendix, there is only one critical value given by: $Z_c = +1.645$.

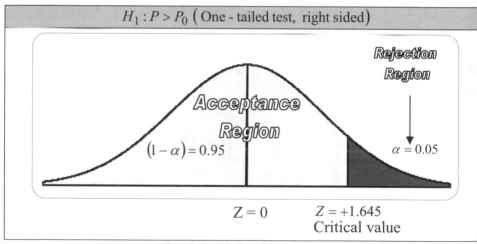

$H_1 : P > P_0$ (One - tailed test, right sided)

Fig. 9.9. Right sided test.

Step VI. Decision: If the value of Z_{cal} lies in the rejection region, we are $(1 - \alpha)100\%$ sure that we should reject the null hypothesis H_0 at $\alpha\%$ level of significance.

Example 9.1. (DOG LOVERS) A rope seller makes a statement that 20% of the leashes he sells are defective. A woman wants to buy a leash to control her dog and claims that the proportion of defective leashes is **more than** 20%. She takes a random sample of $n = 500$ leashes and finds that $x = 125$ are defective. Test her claim at a 5% level of significance.

Buyer says: I doubt it.   **Seller says**: 20% leashes are defective.

Buyer   Ropes   Seller

Fig. 9.10. A buyer and a seller.

Solution. Step I. Set the null hypothesis: Remember that the statement given by the seller corresponds to the null hypothesis. Thus, set the null hypothesis as:

$$H_0 : \text{Proportion of defective leashes, } P = P_0 = 0.20$$

Step II. Set the alternative hypothesis: Remember that an alternative hypothesis is a statement given by the buyer that contradicts the null hypothesis. Thus, set the alternative hypothesis as:

$$H_1 : \text{Proportion of defective leashes, } P > 0.20 \text{ (Right-tailed test)}$$

Step III. Level of significance: Given $\alpha = 0.05\,(\,5\%\,)$.

Step IV. Calculate the Z-statistic: Given $x = 125$ and $n = 500$, the observed sample proportion of defective leashes is:

$$\hat{p} = \frac{x}{n} = \frac{125}{500} = 0.25 .$$

Thus, the calculated Z-statistic is given by:

$$Z_{cal} = \frac{\hat{p} - P_0}{\sqrt{\dfrac{P_0(1 - P_0)}{n}}} = \frac{0.25 - 0.20}{\sqrt{\dfrac{0.20(1 - 0.20)}{500}}}$$

$$= \frac{+0.05}{\sqrt{0.00032}} = \frac{+0.05}{0.01789} = +2.79.$$

Step V. Find the critical value(s): Remember that the critical value depends upon two things: (i) level of significance, and (ii) type of hypothesis. Here we are given $\alpha = 0.05$ and we are using a one-tailed (right sided) test, thus there is only one critical value given by:

$$Z_c = +1.645.$$

Step VI. Decision: Note that $Z_{cal} = +2.79$ lies in the rejection region. Thus, we reject the null hypothesis H_0 at a 5% level of significance.

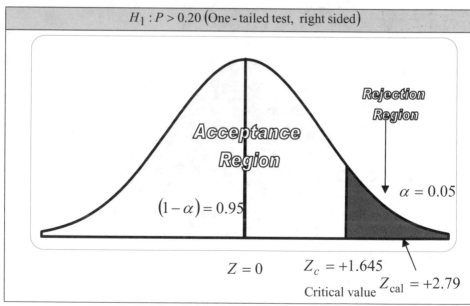

Fig. 9.11. Right-sided test.

Conclusion: The conclusion is that the proportion of defective leashes is **more than** 20%. The woman is happy that she is correct, and she may decide not to buy the leash, because she is 95% sure that there is more than a 20% chance that the dog may run away from her.

9.9 LARGE SAMPLE TEST FOR SINGLE MEAN

Consider μ_0 be the true and unknown mean, say the average breaking strength (BS), of ropes of the variable of interest in a population.

Step I. Set the null hypothesis: Let μ_0 be the true population mean, then set the null hypothesis as:

$H_0 : \mu = \mu_0$, (where μ_0 is a fixed value)

Assume a situation where the rope seller says, "The true average breaking strength of the ropes is equal to 25 kg, or equivalently $H_0 : \mu = 25 \text{ kg}$"

Fig. 9.12. Null hypothesis.

Step II. Set the alternative hypothesis: If the buyer believes the seller, then there is nothing to test. However, a good buyer will test before making any decision, because the buyer may not trust the seller. The buyer can make any one of the following three statements:

"The true average breaking strength of the ropes is **not equal** to 25 kg or equivalently:

$H_1 : \mu \neq 25 \text{ kg}$"

Fig. 9.13. Alternative hypothesis.

or

"The true average breaking strength of the ropes is **less than** 25 kg, or equivalently:

$$H_1 : \mu < 25 \text{ kg}"$$

or

"The true average breaking strength of the ropes is **more than** 25 kg, or equivalently:

$$H_1 : \mu > 25 \text{ kg}"$$

Again there are three possibilities, that is, the true mean may be: (i) not equal to; (ii) less than; or (iii) more than, the true value μ_0 as discussed below:

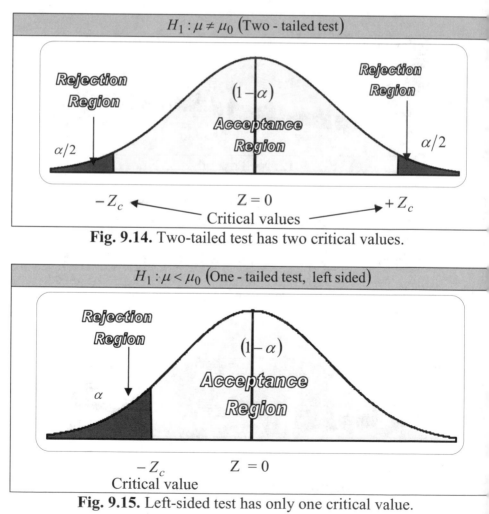

$$H_1 : \mu \neq \mu_0 \text{ (Two - tailed test)}$$

Fig. 9.14. Two-tailed test has two critical values.

$$H_1 : \mu < \mu_0 \text{ (One - tailed test, left sided)}$$

Fig. 9.15. Left-sided test has only one critical value.

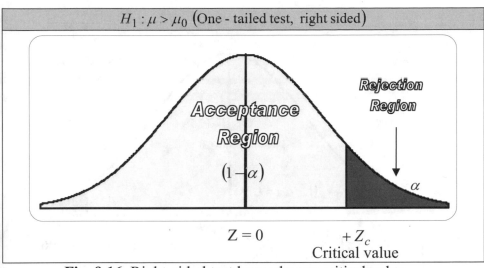

Fig. 9.16. Right-sided test has only one critical value.

Note that the choice of alternative hypothesis depends upon the statement under the alternative hypothesis.

Step III. Set the level of significance: On the basis of the level of the required accuracy of the results, decide:

$\alpha = 0.05$, or $\alpha = 0.01$, ..., etc.

Step IV. Compute the Z-statistic: Let n ≥ 30 (large sample), then calculate the Z - statistic:

$$Z_{cal} = \frac{(\bar{x} - \mu_0)}{\sigma/\sqrt{n}}$$

where n is the sample size, $\bar{x}$ is the sample mean, and σ is the population standard deviation. Note that if σ is unknown, then it can be estimated with the sample standard deviation s.

Step V. Find the critical value(s): The critical value(s) depend upon two things: (i) the level of significance and (ii) the type of the alternative hypothesis.

For example: (a) If $\alpha = 0.05$ and we are using a two-tailed test, using **Table II** from the Appendix, there are two critical values given by: $Z_c = \pm 1.96$.

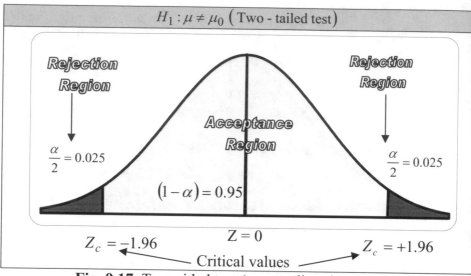

$$H_1 : \mu \neq \mu_0 \ (\text{Two - tailed test})$$

Rejection Region

Acceptance Region

Rejection Region

$\frac{\alpha}{2} = 0.025$

$\frac{\alpha}{2} = 0.025$

$(1 - \alpha) = 0.95$

$Z_c = -1.96$

$Z = 0$

$Z_c = +1.96$

Critical values

Fig. 9.17. Two-sided test (or non-directional test).

(b) If $\alpha = 0.05$ and we are using a one-tailed test (left sided), using **Table II** from the Appendix, there is only one critical value given by: $Z_c = -1.645$.

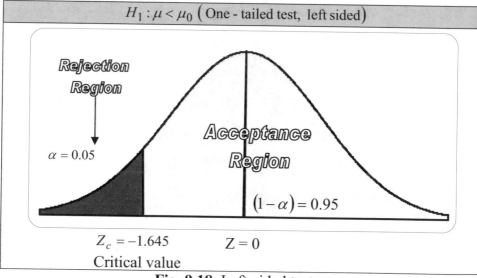

$$H_1 : \mu < \mu_0 \ (\text{One - tailed test, left sided})$$

Rejection Region

Acceptance Region

$\alpha = 0.05$

$(1 - \alpha) = 0.95$

$Z_c = -1.645$

$Z = 0$

Critical value

Fig. 9.18. Left-sided test.

(c) If $\alpha = 0.05$ and we are using a one-tailed test (right sided), using **Table II** from the Appendix, there is only one critical value given by: $Z_c = +1.645$.

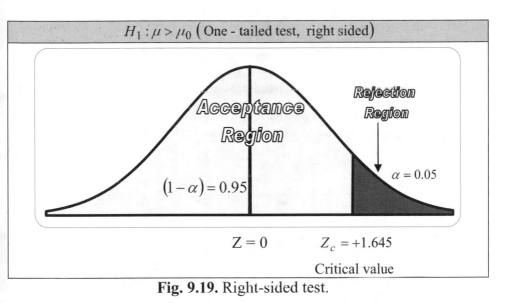

$H_1 : \mu > \mu_0$ (One - tailed test, right sided)

Acceptance Region

Rejection Region

$(1 - \alpha) = 0.95$

$\alpha = 0.05$

$Z = 0$ $Z_c = +1.645$

Critical value

Fig. 9.19. Right-sided test.

Step VI. Decision: If Z_{cal} falls in the rejection region, then reject the null hypothesis at $\alpha\%$ level of significance.

Example 9.2. (LISTEN TO YOUR PARENTS) Amy went to the market with her mom and asked her to buy a swing.

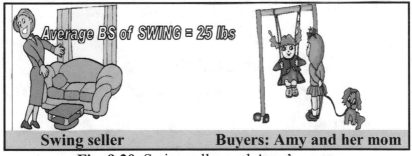

Average BS of SWING = 25 lbs

Swing seller   Buyers: Amy and her mom

Fig. 9.20. Swing seller and Amy's mom.

Amy's mom talks to the seller and the seller, having 10 years of experience, says that these swings' ropes have an average breaking strength (BS) of 25 lbs with a standard deviation of 5 lbs and are great for kids. Amy's mom talked to 35 of her friends who bought these swings for their kids and found the average breaking strength to be 22 lbs. Now Amy's mom claims that the average breaking strength of these swings' ropes is less than 25 lbs. Test Amy's mom's claim at a 5% level of significance.

Solution. Step I. Set the null hypothesis:
$$H_0 : \mu = 25$$

Step II. Set the alternative hypothesis:
$$H_1 : \mu < 25 \text{ (left-tailed test)}$$

Step III. Level of significance: Given $\alpha = 0.05$.

Step IV. Calculate the Z-statistic: Given $n = 35$, (large sample), $\bar{x} = 22$, and $\sigma = 5$. Thus the calculated Z-statistic is given by:

$$Z_{cal} = \frac{(\bar{x} - \mu_0)}{\left(\dfrac{\sigma}{\sqrt{n}}\right)} = \frac{22 - 25}{\dfrac{5}{\sqrt{35}}} = -3.55.$$

Step V. Critical value(s): We are given $\alpha = 0.05$ and we are using a one-tailed test (left sided), so the critical value is given by:
$$Z_c = -1.645.$$

Step VI. Decision: The calculated Z-value falls in the rejection (or critical) region as shown below:

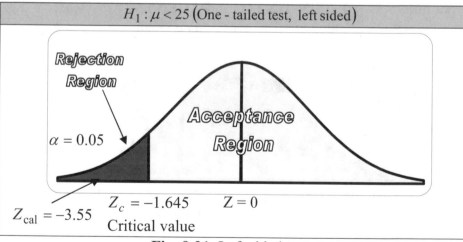

Fig. 9.21. Left sided test.

Thus, we are 95% sure that we should reject the null hypothesis H_0. In other words, the average breaking strength of the swings is less than 25 lbs at a 5% level of significance and Amy's mom is 95% correct. Just for fun we can say that we are 95% sure that kids should listen to their parents' advise while making any decision. Thus, Amy should look for a better swing instead of arguing with her mom.

Example 9.3. (WEIGHING PUMPKINS IS FUN) It is a cumbersome and expensive job to weigh all the pumpkins one by one on a big farm. Last year, each pumpkin was weighed one by one and the average weight of all the pumpkins was $\mu = 12$ pounds with a standard deviation of $\sigma = 6$ pounds. This year, due to a shortage of funds, the farmer (or seller) could not weigh all the pumpkins one by one. A buyer took a random sample of $n = 36$ pumpkins and found that the average weight was $\bar{x} = 10$ pounds, and claimed that the average weight is less than last year. Test the buyer's claim at a 5% level of significance.

Solution. Let us do six steps.

Step I. Set the null hypothesis:

$$H_0 : \mu = 12$$

Step II. Set the alternative hypothesis:

$$H_1 : \mu < 12 \text{ (One-tailed test)}$$

Step III. Level of significance:

Given $\alpha = 0.05$.

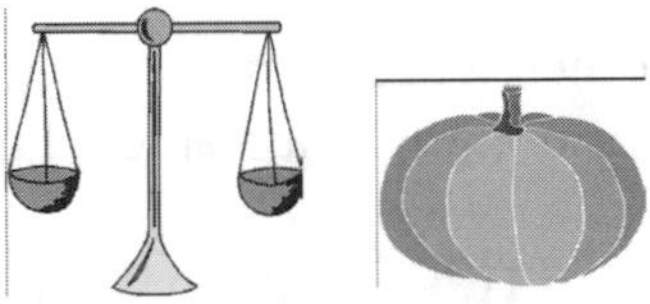

Balance   Pumpkin

Fig. 9.22. Weighing pumpkins.

Step IV. Calculate the Z-statistic: Given $n = 36$, (large sample), $\bar{x} = 10$, and $\sigma = 6$. Thus, the calculated Z-statistic is given by:

$$Z_{cal} = \frac{(\bar{x} - \mu_0)}{\left(\dfrac{\sigma}{\sqrt{n}}\right)} = \frac{10 - 12}{\dfrac{6}{\sqrt{36}}} = \frac{-2}{1} = -2.$$

Step V. Critical value(s): We are given $\alpha = 0.05$ and we are using a one-tailed test (or left sided test), so the critical value is given by:

$$Z_c = -1.645.$$

Step VI. Decision: The calculated Z-value, $Z_{cal} = -2$, lies in the rejection region, as shown in **Figure 9.23** and hence we reject the null hypothesis at a 5% level of significance.

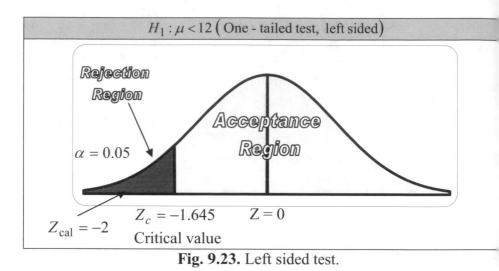

Fig. 9.23. Left sided test.

Thus, we are 95% sure that the statement made by the buyer is true. In other words, the average weight of this year's pumpkins is less than 12 pounds at a 5% level of significance.

9.10 IDEA OF p-VALUE USING Z-SCORE

The p-value is the minimum level of significance at which the null hypothesis can be rejected. Also, note that the smaller the p-value, the stronger the evidence is to reject the null hypothesis. If the p-value is less than the α value, then the result is statistically significant. Note that if one result is significant at a 1% level, then it is automatically significant at a 5% level. In other words, if we are 99% sure we should reject a null hypothesis, then we are automatically 95% sure we should reject that null hypothesis.

9.10.1 p-VALUE FOR LEFT-TAILED TEST USING Z-SCORE

Consider a left-tailed test being applied to test any statement under the null hypothesis based on the Z-score such as:

H_0 : Donkeys and horses are the same height.
H_1 : Donkeys are shorter than horses (left-tailed test).

Let

$$Z_{cal} = -2.35$$

Fig. 9.24. Donkey and Horse.

Then, the p-value is the area to the left of the calculated Z-score as shown:

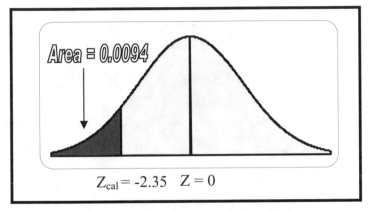

Fig. 9.25. p-value using a left-tailed test.

Thus, using **Table II** from the Appendix, the p-value is given by:

p-value = 0.0094

Here the p-value 0.0094 means we are (1-0.0094)100% = 99.06% sure that we should reject the null hypothesis. In other words, we are 99.06% sure we should accept the statement that donkeys are shorter than horses.

9.10.2 p-VALUE FOR RIGHT-TAILED TEST USING Z-SCORE

Assume that we are applying a right-tailed test to test any statement under the null hypothesis based on the Z-score such as:

H_0 : Dogs and cats are of the same height.
H_1 : Dogs are taller than cats (right-tailed test).

Let

$$Z_{cal} = +1.98$$

Fig. 9.26. Cat and Dog.

Then, the p-value is the area to the right of the calculated Z-score as shown:

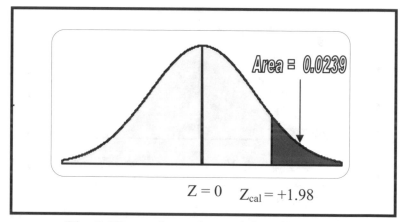

Area = 0.0239

$Z = 0$ $Z_{cal} = +1.98$

Fig. 9.27. p-value using a right-tailed test.

Thus, using **Table II** from the Appendix, the p-value is given by:

p-value = 0.0239

Here, the p-value 0.0239 means we are $(1-0.0239)100\% = 97.61\%$ sure that we should reject the null hypothesis. In other words, we are 97.61% sure that we should accept the statement that dogs are taller than cats.

9.10.3 p-VALUE FOR TWO-TAILED TEST USING Z-SCORE

Consider a two-tailed test being applied to test any statement under the null hypothesis based on the Z-score such as:

H_0 : Elephants and camels are the same height.
H_1 : Elephants and camels are not the same height (two-tailed test).

Let

$$Z_{cal} = +1.32$$

Fig. 9.28. Camel and Elephant.

Then, the p-value is the total area to the right of the calculated Z-score and its image on the left side.

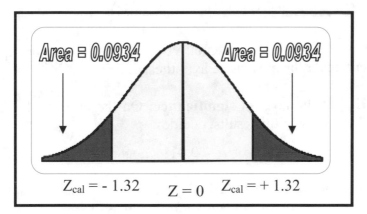

Area = 0.0934   Area = 0.0934

$Z_{cal} = -1.32$ $Z = 0$ $Z_{cal} = +1.32$

Fig. 9.29. p-value using a two-tailed test.

Thus, using **Table II** from the Appendix, the p-value is given by:

p-value = 0.0934 + 0.0934 = 0.1868

Here, the p-value 0.1868 means, we are $(1-0.1868)100\% = 81.32\%$ sure that we should reject the null hypothesis. In other words, we are 81.32% sure that elephants and camels are not the same height.

9.11 SMALL SAMPLE TEST FOR SINGLE MEAN

Consider μ_0 be the true and unknown mean, μ, of the variable of interest in a population under study.

Step I. Set the null hypothesis:

$H_0 : \mu = \mu_0$, (where μ_0 is the true or a guessed value of μ)

Step II. Set the alternative hypothesis: Again there are three possibilities, that is, the true mean may be: (i) not equal to, (ii) less than, or (iii) more than the guessed value μ_0 as discussed below.

$$H_1 : \mu \neq \mu_0 \text{ (Two-tailed test)}$$

or

$$H_1 : \mu < \mu_0 \text{ (One-tailed test, left sided)}$$

or

$$H_1 : \mu > \mu_0 \text{ (One-tailed test, right sided)}$$

Note that the choice of alternative hypothesis depends upon the statement under the alternative hypothesis.

Step III. Set the level of significance: On the basis of the required level of accuracy of the results, decide:

$$\alpha = 0.05, \text{ or } \alpha = 0.01, ..., \text{ etc.}$$

Step IV. Compute the t-statistic: Let n < 30 (small sample), then calculate the t-score (*or Student's t-statistic*) as:

$$t_{cal} = \frac{(\bar{x} - \mu_0)}{s/\sqrt{n}}$$

where n is the sample size, $\bar{x}$ is the sample mean, and s is the sample standard deviation.

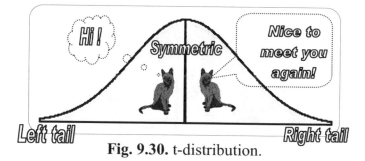

Fig. 9.30. t-distribution.

Step V. Find the critical value(s): Note that for a small sample test the critical value(s) depend upon three things: (i) the value of the level of significance, (ii) type of alternative hypothesis, and (iii) degree of freedom.

For cxamplc:
(a) if $\alpha = 0.05$ and we are using a two-tailed test, using **Table III** from the Appendix, there are two critical values given by:

$$t_c = \pm t_{\frac{\alpha}{2}}(df = n-1)$$

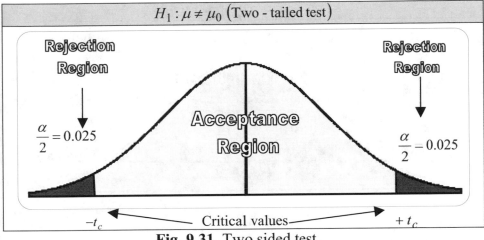

Fig. 9.31. Two sided test.

(b) if $\alpha = 0.05$ and we are using a one-tailed test (left sided), using **Table III** from the Appendix, there is only one critical value given by:

$$t_c = -t_\alpha(df = n-1)$$

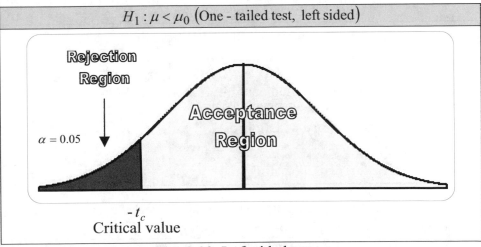

Fig. 9.32. Left sided test.

(c) if $\alpha = 0.05$ and we are using a one-tailed test (right sided), using **Table III** from the Appendix, there is only one critical value given by:

$$t_c = +t_\alpha\left(\mathrm{df} = n - 1\right)$$

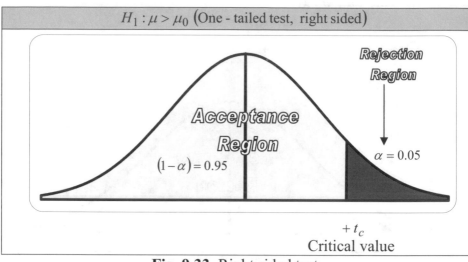

$$H_1 : \mu > \mu_0 \ \left(\text{One - tailed test, right sided}\right)$$

Rejection Region

Acceptance Region

$\left(1 - \alpha\right) = 0.95$

$\alpha = 0.05$

$+ t_c$

Critical value

Fig. 9.33. Right sided test.

Step VI. Decision: If the value of t_{cal} lies in the rejection region, then we should reject the null hypothesis H_0 with $\left(1 - \alpha\right)100\%$ confidence at $\alpha\%$ level of significance.

Example 9.4. (TRAFFIC RULES ARE FOR YOUR SAFETY) Due to a sharp turn in a road, the traffic administration decided to reduce the speed limit to 15 miles/hour. A random sample of 10 vehicles shows the following speeds at the time of turning:

15	18
14	16
22	28
16	15
18	20

Fig. 9.34. Obey the traffic rules.

(a) Test at a 5% level of significance that the vehicles are following the traffic rules. Interpret your results and give any suggestions.

(b) Construct a 95% confidence interval estimate of the average speed. Does the recommended speed of 15 miles/hour fall in it?

Solution. (a) Apply the six steps:

(i) Set the null hypothesis:
$$H_0 : \mu = 15$$

(ii) Set the alternative hypothesis:
$$H_a : \mu \neq 15 \text{ (Two-tailed test)}$$

Assume that the cars traveling under and over speeds are violating the traffic rules.

(iii) Level of significance: Given $\alpha = 0.05$.

(iv) Calculate the t-statistic: From the data set, we have:

	x_i	$(x_i - \bar{x})$	$(x_i - \bar{x})^2$
	15	-3.2	10.24
	14	-4.2	17.64
	22	3.8	14.44
	16	-2.2	4.84
	18	-0.2	0.04
	18	-0.2	0.04
	16	-2.2	4.84
	28	9.8	96.04
	15	-3.2	10.24
	20	1.8	3.24
Sum	182	0.00	161.6

From the above table, we have:
$$n = 10, \ \sum_{i=1}^{n} x_i = 182, \ \sum_{i=1}^{n}(x_i - \bar{x}) = 0.00, \text{ and } \sum_{i=1}^{n}(x_i - \bar{x})^2 = 161.6.$$

Thus, we have:
$$\bar{x} = \frac{\sum_{i=1}^{n} x_i}{n} = \frac{182}{10} = 18.2, \ s^2 = \frac{\sum_{i=1}^{n}(x_i - \bar{x})^2}{n-1} = \frac{161.6}{10-1} = 17.96,$$

and
$$s = \sqrt{s^2} = \sqrt{17.96} = 4.24.$$

The calculated t-score (*or Student's t-statistic*) is given by:

$$t_{cal} = \frac{(\bar{x} - \mu_0)}{s/\sqrt{n}} = \frac{(18.2 - 15)}{4.24/\sqrt{10}} = +2.39.$$

(v) **Critical values(s):** Remember that for a small sample test, the critical value(s) depend upon three things: (i) level of significance, (ii) degree of freedom, and (iii) type of hypothesis.

Here $\alpha = 0.05$, df $= n - 1 = 10 - 1 = 9$ and we are applying a two-tailed test, using **Table III** from the Appendix, there are two critical values given by:

$$t_c = \pm t_{\alpha/2}(df = n - 1) = \pm t_{0.05/2}(df = 10 - 1) = \pm t_{0.025}(df = 9) = \pm 2.262.$$

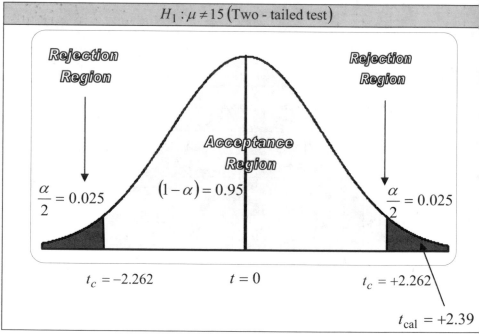

Fig. 9.35. Two sided test.

(vi) **Decision:** The $t_{cal} = +2.39$ falls in the rejection region, thus we should reject the null hypothesis at a 5% level of significance. This means we are 95% sure that the vehicles are not following the traffic rules at this sharp turn. Thus, installing a speed camera may be recommended based on this conclusion.

(b) Confidence interval estimate: We know that a $(1-\alpha)100\%$ confidence interval estimate of the population mean μ is given by:

$$\bar{x} \pm \left[t_{\alpha/2}(\text{df} = n-1)\right]\left(\frac{s}{\sqrt{n}}\right)$$

Now for $\alpha = 0.05$ and $n = 10$, using **Table III** from the Appendix, we have:

$$t_{\alpha/2}(\text{df} = n-1) = t_{0.05/2}(\text{df} = 10-1) = t_{0.025}(\text{df} = 9) = 2.262.$$

Thus, the 95% confidence interval estimate of the true average speed is given by:

$$18.2 \ \pm \ (2.262)\left(\frac{4.24}{\sqrt{10}}\right)$$

or $18.2 \ \pm \ 3.03$ or $(18.2 - 3.03, \ 18.2 + 3.03)$ or $(15.17, \ 21.23)$.

The recommended speed of 15 miles/hour does not fall in the 95% confidence interval estimate. One of the interpretations is that 95% of the vehicles passing through the sharp turn are going between 15.17 miles/hour and 21.23 miles/hour.

Example 9.5. (POMEGRANATE LOVERS) A gardener claims that if the weight of a pomegranate becomes more than 0.50 lbs, then it is ready to eat. The gardener took a random sample of $n = 6$ pomegranates and recorded their weights in lbs as:

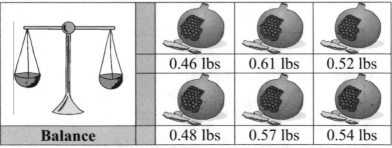

Fig. 9.36. Weighing pomegranates.

Do the six measurements present sufficient evidence to indicate that the average weight of the pomegranates in the orchard is more than 0.5 lbs?

Solution. (i) **Set the null hypothesis:**
$$H_0 : \mu = 0.5$$

(ii) **Set the alternative hypothesis:**
$$H_a : \mu > 0.5 \quad \text{(Right-tailed test)}$$

(iii) **Level of significance:** Let us decide $\alpha = 0.05$.

(iv) **Calculate the t-statistic:.** We are given:

	x_i	x_i^2
	0.46	0.2116
	0.61	0.3721
	0.52	0.2704
	0.48	0.2304
	0.57	0.3249
	0.54	0.2916
Sum	**3.18**	**1.7010**

From the above table, we have:
$$n = 6, \ \sum_{i=1}^{n} x_i = 3.18 \text{ and } \sum_{i=1}^{n} x_i^2 = 1.7010.$$

Now the sample mean is given by:
$$\bar{x} = \frac{1}{n} \sum_{i=1}^{n} x_i = \frac{3.18}{6} = 0.53.$$

and the sample variance is given by:

$$s^2 = \frac{n\left(\sum_{i=1}^{n} x_i^2\right) - \left(\sum_{i=1}^{n} x_i\right)^2}{n(n-1)} = \frac{6(1.7010) - (3.18)^2}{6(6-1)} = 0.00312.$$

Thus, the sample standard deviation is given by:
$$s = \sqrt{s^2} = \sqrt{0.00312} = 0.0559.$$

Then the calculated t-statistic is given by:
$$t_{cal} = \frac{\bar{x} - \mu_0}{s/\sqrt{n}} = \frac{0.53 - 0.50}{0.0559/\sqrt{6}} \approx +1.31.$$

(v) **Critical value(s):** Note that we are using a one-tailed test and $\alpha = 0.05$. For a small sample test, the critical value is given by:

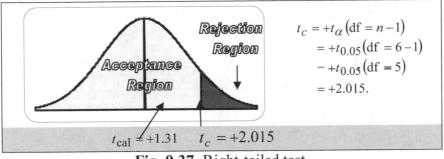

$$t_c = +t_\alpha (\text{df} = n-1)$$
$$= +t_{0.05}(\text{df} = 6-1)$$
$$- +t_{0.05}(\text{df} = 5)$$
$$= +2.015.$$

$t_{\text{cal}} = +1.31 \qquad t_c = +2.015$

Fig. 9.37. Right-tailed test.

(vi) **Decision:** Note that $t_{\text{cal}} = +1.31$ lies in the acceptance region, so we do not reject the null hypothesis H_0 at a 5% level of significance. Thus, we are 95% sure that the gardener should wait to harvest the crop.

An Important Remark: Note carefully that if the sample size is small and the population variance is known, then we prefer to use the Z-score test instead of the t-score test.

9.12 LARGE SAMPLE TEST FOR THE DIFFERENCE BETWEEN TWO MEANS

Let us now consider a situation of two kinds of ropes: red ropes and green ropes.

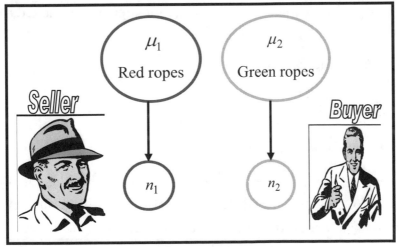

Fig. 9.38. Two types of ropes.

The seller says, "The average breaking strength of the red ropes is the same as that of the green ropes."

Let μ_1 be the true average breaking strength of the red ropes and μ_2 be the true average breaking strength of the green ropes.

Step I. Set the null hypothesis:
$$H_0 : \mu_1 - \mu_2 = 0$$

Step II. Set the alternative hypothesis: The buyer can claim any one of the following three possibilities, given by:

$$H_a : \mu_1 - \mu_2 \neq 0 \text{ (Two-tailed test)}$$

or

$$H_a : \mu_1 - \mu_2 > 0 \text{ (Right-tailed test)}$$

or

$$H_a : \mu_1 - \mu_2 < 0 \text{ (Left-tailed test)}.$$

Step III. Level of significance: Decide $\alpha = 0.05$ or $\alpha = 0.01$ etc.

Step IV. Calculate the Z-statistic: Let the buyer select two independent and random samples of n_1 red ropes and n_2 green ropes, respectively. Let $\bar{x}_1$ and $\bar{x}_2$ be the estimates of the average breaking strengths of the red ropes and green ropes. Let σ_1^2 and σ_2^2 be the known population variances of the breaking strengths of the red ropes and green ropes, respectively.

Then calculate the Z_{cal} score as:

$$Z_{cal} = \frac{(\bar{x}_1 - \bar{x}_2) - 0}{\sqrt{\dfrac{\sigma_1^2}{n_1} + \dfrac{\sigma_2^2}{n_2}}}$$

Note that $Z_{cal} \sim N(0,1)$. This is also called the Central Limit Theorem (CLT) for the difference between two means. If the population variances σ_1^2 and σ_2^2 are unknown, then these can be estimated by sample variances s_1^2 and s_2^2 respectively, provided that both sample sizes are large.

Step V. Critical value: Based on the given value of α and the type of hypothesis, find the Z_c value(s) as:

$$Z_c - \begin{cases} \pm Z_{\alpha/2}, & \text{for a two - tailed test} \\ -Z_{\alpha}, & \text{for a left - tailed test} \\ +Z_{\alpha}, & \text{for a right - tailed test} \end{cases}$$

Step VI. Decision: Reject the null hypothesis H_0 at $\alpha\%$ level of significance if the calculated Z_{cal} falls in the rejection region.

Step VII. It is always possible to find the p-value based on the computed statistic and type of hypothesis.

Example 9.6. (QUALITY CONTROL) Assume that we select a random sample of $n_1 = 40$ red ropes and another independent random sample of $n_2 = 35$ green ropes to study the breaking strengths (kg) of the ropes. From the sample information, we found $\bar{x}_1 = 13.57$ and $\bar{x}_2 = 12.30$.

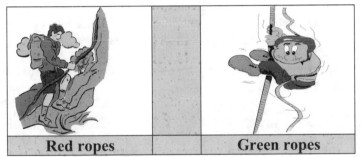

| Red ropes | | Green ropes |

Fig. 9.39. Rock climbers.

Test the hypothesis that the average breaking strength of the red ropes is the same as that of the green ropes at a 5% level of significance. Given: $\sigma_1^2 = 6.6$ and $\sigma_2^2 = 2.5$.

Solution. Let μ_1 be the true average breaking strength of the red ropes and μ_2 be the true average breaking strength of the green ropes.

(i) **Set the null hypothesis:**

$$H_0 : \mu_1 - \mu_2 = 0$$

(ii) **Set the alternative hypothesis:**

$$H_a : \mu_1 - \mu_2 \neq 0 \text{ (Two-tailed test)}$$

(iii) **Level of significance:** Given $\alpha = 0.05$.

(iv) **Calculate the Z-statistic:** Here $n_1 = 40$, $\bar{x}_1 = 13.57$, $n_2 = 35$, and $\bar{x}_2 = 12.30$. Note that it is given: $\sigma_1^2 = 6.6$ and $\sigma_2^2 = 2.5$. Thus, the calculated Z_{cal} score is given by:

$$Z_{cal} = \frac{(\bar{x}_1 - \bar{x}_2) - 0}{\sqrt{\dfrac{\sigma_1^2}{n_1} + \dfrac{\sigma_2^2}{n_2}}} = \frac{(13.57 - 12.30) - 0}{\sqrt{\dfrac{6.6}{40} + \dfrac{2.5}{35}}}$$

$$= \frac{+1.27}{0.486} = +2.61.$$

(v) **Critical value(s):** Here $\alpha = 0.05$, and we are using a two-tailed test, using **Table II** from the Appendix, there are two critical values given by:

$$Z_c = \pm Z_{\alpha/2} = \pm Z_{0.05/2} = \pm Z_{0.025} = \pm 1.96.$$

(vi) **Decision:** Note that $Z_{cal} = +2.61$ falls in the rejection region, thus we should reject the null hypothesis H_0 at a 5% level of significance. Thus, we are 95% sure that the red and green ropes have different average breaking strengths.

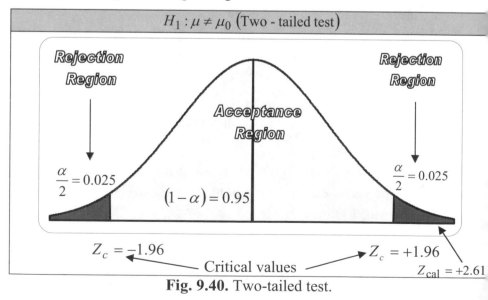

Fig. 9.40. Two-tailed test.

(vii) **p-value:** The p-value is called the exact level of significance.

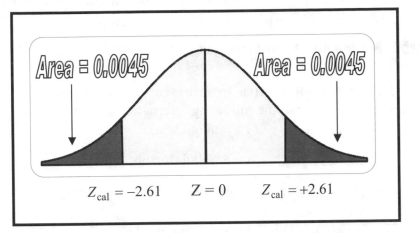

Fig. 9.41. The p-value with a two-tailed test.

Thus, the p-value is given by:

$$p - value = 0.0045 + 0.0045 = 0.0090.$$

Thus, we are $(1 - 0.0090)100\% = 99.1\%$ sure that we should reject the null hypothesis. Thus, we are 99.1% sure that the average breaking strength of the red ropes is different from that of the green ropes.

9.13 SMALL SAMPLE TEST FOR THE DIFFERENCE BETWEEN TWO POPULATION MEANS FOR INDEPENDENT RANDOM SAMPLES

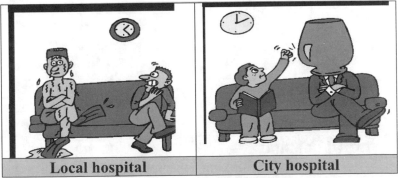

Fig. 9.42. Patients waiting areas.

A medical team claims that the average waiting time for the patients in a local hospital (say, μ_1) and in a city hospital (say, μ_2) differs by d. For example, if $d = 0$, then there is no difference.

Step I. Set the null hypothesis:
$$H_0 : \mu_1 - \mu_2 = d$$

Step II. Set the alternative hypothesis: A local consular may be interested in any one of the following alternative hypotheses:

$$H_a : \mu_1 - \mu_2 \neq d \quad \text{(Two-tailed test)}$$

or $\quad H_a : \mu_1 - \mu_2 > d \quad$ (One-tailed or right-tailed test)

or $\quad H_a : \mu_1 - \mu_2 < d \quad$ (One-tailed or left-tailed test)

Step III. Decide the level of significance: Based on the desired level of accuracy, decide if $\alpha = 0.05$ or $\alpha = 0.01$, or whatever you prefer if it is not given in the question.

Step IV. Calculate the t-score: Here we have the following situation:

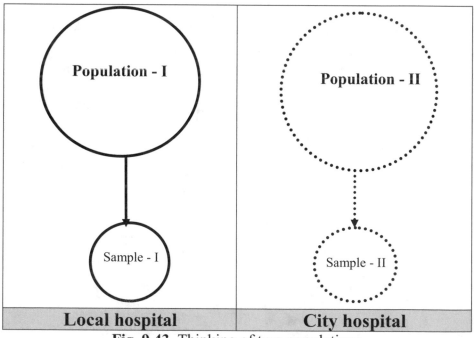

Fig. 9.43. Thinking of two populations.

Let $x_{11}, x_{12}, \ldots, x_{1m_1}$ be a random sample of n_1 data values taken from the population-I (the local hospital). Find the sample-I mean as:

$$\bar{x}_1 = \frac{1}{n_1} \sum_{i=1}^{m_1} x_{1i}$$

and the sample-I variance as:

$$s_1^2 = \frac{n_1 \left(\sum_{i=1}^{m_1} x_{1i}^2 \right) - \left(\sum_{i=1}^{m_1} x_{1i} \right)^2}{n_1 (n_1 - 1)}$$

Let $x_{21}, x_{22}, \ldots, x_{2m_2}$ be a random sample of n_2 data values taken from the population-II (the city hospital). Find the sample-II mean as:

$$\bar{x}_2 = \frac{1}{n_2} \sum_{i=1}^{n_2} x_{2i}$$

and the sample-II variance as:

$$s_2^2 = \frac{n_2 \left(\sum_{i=1}^{n_2} x_{2i}^2 \right) - \left(\sum_{i=1}^{n_2} x_{2i} \right)^2}{n_2 (n_2 - 1)}$$

Now if the ratio $\dfrac{Max\left(s_1^2, s_2^2\right)}{Min\left(s_1^2, s_2^2\right)} \leq 3$, then calculate the t-statistic:

$$t_{cal} = \frac{\left(\bar{x}_1 - \bar{x}_2\right) - d}{\sqrt{s_p^2 \left(\dfrac{1}{n_1} + \dfrac{1}{n_2} \right)}}$$

where

$$s_p^2 = \frac{(n_1 - 1)s_1^2 + (n_2 - 1)s_2^2}{n_1 + n_2 - 2}$$ is called the pooled sample variance.

Step V. Critical value(s): According to the value of α, the type of the alternative hypothesis, and degree of freedom, find the critical value(s) of the t-distribution from **Table III** given in the Appendix as:

$$t_c = \begin{cases} \pm t_{\frac{\alpha}{2}}\left(df = n_1 + n_2 - 2\right) = \text{-------} & \left(\text{for two - tailed test}\right) \\[2ex] -t_\alpha\left(df = n_1 + n_2 - 2\right) = \text{-----} & \left(\text{for left - tailed test}\right) \\[2ex] +t_\alpha\left(df = n_1 + n_2 - 2\right) = \text{-----} & \left(\text{for right - tailed test}\right) \end{cases}$$

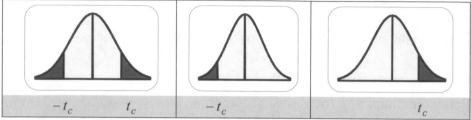

Fig. 9.44. Differentiating two-tailed, left-tailed, and right-tailed tests.

Step VI. Decision: If the calculated statistic t_{cal} falls in the rejection region (or critical region), then reject the null hypothesis H_0 at α % level of significance.

Remark: If the ratio $\dfrac{\text{Max}\left(s_1^2,\ s_2^2\right)}{\text{Min}\left(s_1^2,\ s_2^2\right)} > 3$, then calculate the t-statistic as:

$$t_{cal} = \frac{\left(\bar{x}_1 - \bar{x}_2\right) - d}{\sqrt{\dfrac{s_1^2}{n_1} + \dfrac{s_2^2}{n_2}}}$$

According to the value of α and the type of hypothesis, find the *magnitude* of the critical value from **Table III** given in the Appendix. We have:

$$t_c = \begin{cases} t_{\frac{\alpha}{2}}\left(df \approx \dfrac{\left(s_1^2/n_1 + s_2^2/n_2\right)^2}{\dfrac{\left(s_1^2/n_1\right)^2}{n_1-1} + \dfrac{\left(s_2^2/n_2\right)^2}{n_2-1}}\right) = ----- \ (\text{for two-tailed test}) \\[6em] t_{\alpha}\left(df \approx \dfrac{\left(s_1^2/n_1 + s_2^2/n_2\right)^2}{\dfrac{\left(s_1^2/n_1\right)^2}{n_1-1} + \dfrac{\left(s_2^2/n_2\right)^2}{n_2-1}}\right) = ----- (\text{for one-tailed test}) \end{cases}$$

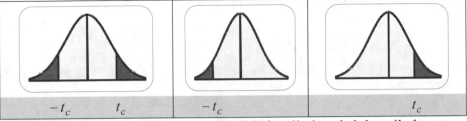

Fig. 9.45. Differentiating two-tailed, left-tailed and right-tailed tests.

If the calculated statistic t_{cal} falls in the rejection (or critical region), then we should reject the null hypothesis H_0 at a α % level of significance.

Example 9.7. (TIME IS IMPORTANT) Two independent random samples, each of nine patients, were selected from two walk-in-clinics. One was located in the city and another in the local area. The waiting times of these patients were noted in minutes as:

| Local clinic | 32 | 37 | 35 | 28 | 41 | 44 | 35 | 31 | 34 |
| City clinic | 35 | 31 | 29 | 25 | 34 | 40 | 27 | 32 | 31 |

Fig. 9.46. Time never comes back.

Does the data set present sufficient evidence to indicate that the mean waiting time at the city clinic is less than at the local clinic? Test at a 5% level of significance.

Solution. (i) **Set the null hypothesis:**

$$H_0 : \mu_1 - \mu_2 = 0$$

(ii) **Set the alternative hypothesis:**

$$H_a : \mu_1 - \mu_2 > 0 \quad \text{(one-tailed or right-tailed test)}$$

(iii) **Level of significance:** Given $\alpha = 0.05$.

(iv) **Calculate the t-score:** We have:

	Sample I Local Clinic			Sample II City Clinic	
	x_{1i}	x_{1i}^2		x_{2i}	x_{2i}^2
	32	1,024		35	1,225
	37	1,369		31	961
	35	1,225		29	841
	28	784		25	625
	41	1,681		34	1,156
	44	1,936		40	1,600
	35	1,225		27	729
	31	961		32	1,024
	34	1,156		31	961
Sum	317	11,361		284	9,122

From the above table, we have the following:

$$\bar{x}_1 = \frac{1}{n_1} \sum_{i=1}^{n_1} x_{1i} = \frac{317}{9} = 35.22, \quad \bar{x}_2 = \frac{1}{n_2} \sum_{i=1}^{n_2} x_{2i} = \frac{284}{9} = 31.56,$$

$$s_1^2 = \frac{n_1 \left(\sum_{i=1}^{n_1} x_{1i}^2 \right) - \left(\sum_{i=1}^{n_1} x_{1i} \right)^2}{n_1(n_1 - 1)} = \frac{9(11,361) - (317)^2}{9(9-1)} = 24.44,$$

$$s_2^2 = \frac{n_2 \left(\sum_{i=1}^{n_2} x_{2i}^2 \right) - \left(\sum_{i=1}^{n_2} x_{2i} \right)^2}{n_2(n_2 - 1)} = \frac{9(9,122) - (284)^2}{9(9-1)} = 20.03.$$

Now

$$\text{Ratio} = \frac{\text{Max}\left(s_1^2, s_2^2\right)}{\text{Min}\left(s_1^2, s_2^2\right)} = \frac{\text{Max}(24.44, 20.03)}{\text{Min}(24.44, 20.03)} = \frac{24.44}{20.03} = 1.22 \le 3.$$

Thus, the pooled sample variance is given by:

$$s_p^2 = \frac{(n_1 - 1)s_1^2 + (n_2 - 1)s_2^2}{n_1 + n_2 - 2} = \frac{(9-1)(24.44) + (9-1)(20.03)}{9 + 9 - 2}$$

$$= 22.235.$$

Now the calculated t-statistic is given by:

$$t_{cal} = \frac{(\bar{x}_1 - \bar{x}_2) - d}{\sqrt{s_p^2 \left(\frac{1}{n_1} + \frac{1}{n_2}\right)}} = \frac{(35.22 - 31.56) - 0}{\sqrt{22.235 \left(\frac{1}{9} + \frac{1}{9}\right)}} = +1.646$$

(v) **Critical value(s):** Here $\alpha = 0.05$ and we are using a right-tailed test, using **Table III** from the Appendix, the critical value is given by:

$$t_c = +t_\alpha (df = n_1 + n_2 - 2) = +t_{0.05}(df = 9 + 9 - 2)$$
$$= +t_{0.05}(df = 16) = +1.746 \text{ (for a one - tailed test)}$$

(vi) **Decision:**

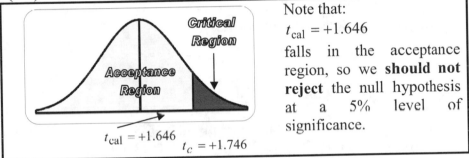

Note that:

$t_{cal} = +1.646$

falls in the acceptance region, so we **should not reject** the null hypothesis at a 5% level of significance.

$t_{cal} = +1.646$

$t_c = +1.746$

Fig. 9.47. Performing a right-tailed test.

Conclusion: We are 95% sure that the average waiting time in a local and city clinic is the same.

Example 9.8. (FRUITS ARE GOOD FOR YOUR HEALTH)
There are two types of oranges, sweet and sour. A vendor claims that the average weight of the sweet and sour oranges is the same.

Fig. 9.48. Vendor and customer.

A customer took two independent samples of sweet and sour oranges, and noted the information in the following table.

Information	Number of oranges selected	Average weight (gm)	Sample standard deviation
Sweet oranges	$n_1 = 10$	$\bar{x}_1 = 108$	$s_1 = 1.5$
Sour oranges	$n_2 = 6$	$\bar{x}_2 = 106$	$s_2 = 1.3$

Test the vendor's claim at a 95% level of confidence.

Solution. Let us do the six steps.
Step I. Set the null hypothesis:

$$H_0 : \mu_1 - \mu_2 = 0$$

Step II. Set the alternative hypothesis:

$$H_a : \mu_1 - \mu_2 \neq 0 \text{ (Two-tailed test)}$$

Step III. Level of significance: Given $\alpha = 0.05$.

Step IV. Calculate the t-score: Assuming that both kinds of oranges came from the same orchard (or the same population), the pooled sample variance is given by:

$$s_p^2 = \frac{(n_1 - 1)s_1^2 + (n_2 - 1)s_2^2}{n_1 + n_2 - 2} = \frac{(10 - 1)(1.5^2) + (6 - 1)(1.3^2)}{10 + 6 - 2} = 2.05.$$

The calculated t-score is given by:

$$t_{cal} = \frac{(\bar{x}_1 - \bar{x}_2) - 0}{\sqrt{s_p^2 \left(\frac{1}{n_1} + \frac{1}{n_2} \right)}} = \frac{(108 - 106) - 0}{\sqrt{2.05 \left(\frac{1}{10} + \frac{1}{6} \right)}}$$

$$= \frac{+2}{\sqrt{0.5467}} = \frac{+2}{0.739} = +2.71.$$

Step V. Critical region: Here $\alpha = 0.05$ and we are using a two-tailed test, thus using **Table III** from the Appendix, the critical region is formed by the two critical values given by:

$$t_c = \pm t_{\alpha/2}\left(df = n_1 + n_2 - 2\right) = \pm t_{0.025}\left(df = 10 + 6 - 2\right) = \pm 2.145$$

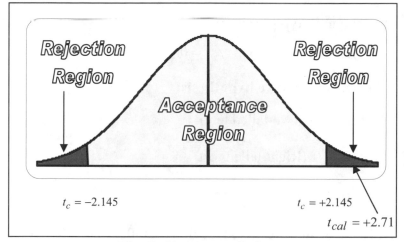

$$t_c = -2.145$$

$$t_c = +2.145$$

$$t_{cal} = +2.71$$

Fig. 9.49. Two-tailed test.

Step VI. Decision: Note that the calculated t-score $t_{cal} = 2.71$ falls in the rejection region. Thus, we must reject the null hypothesis at a 5% level of significance. In other words, the sweet and sour oranges have different weights at a 5% level of significance.

9.14 SMALL SAMPLE TEST FOR THE DIFFERENCE BETWEEN TWO MEANS FOR TWO DEPENDENT SAMPLES (PAIRED t-TEST)

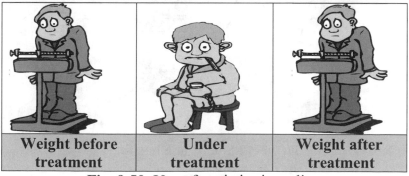

| Weight before treatment | Under treatment | Weight after treatment |

Fig. 9.50. Use of statistics in reality.

Let μ_D be the average weight of the change in two values of the ith unit in the population given by:

$$D_i = X_{2i} - X_{1i}, \quad i = 1, 2, ..., N.$$

Step I. Set the null hypothesis:

$$H_0 : \mu_1 - \mu_2 = 0 \quad \text{or} \quad H_0 : \mu_D = 0$$

Step II. Set the alternative hypothesis:

$$H_a : \mu_D \neq 0 \text{ (Two-tailed test)}$$

$$H_a : \mu_D > 0 \text{ (Right-tailed test)}$$

$$H_a : \mu_D < 0 \text{ (Left-tailed test)}$$

Step III. Level of significance: Decide $\alpha = 0.05$, or $\alpha = 0.01$ etc.

Step IV. Calculate the t-score: Consider the situation of admitting patients into a hospital with a particular disease.

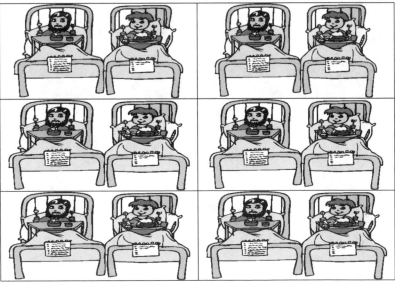

Fig. 9.51. Patients at a hospital.

The patients are not gaining weight due to some kind of disease. Thus, the weight of each patient was recorded before being admitted into the hospital. Then the patients were treated with a special kind of medicine and diet for a period of one week. Assume that no patient died or left the hospital. At the end of the week the weight of each patient was recorded again.

Let x_{1i}, $i = 1,2,..,n$ be the weight of the ith patient before being admitted to the hospital and x_{2i}, $i = 1,2,..,n$ be the weight of the ith patient after the period of a one-week treatment.

Obviously,

$$d_i = (x_{2i} - x_{1i})$$

is the change in weight of the ith patient. The structure of the data looks as follows:

Unit no.	Paired data values		Change $d_i = (x_{2i} - x_{1i})$	d_i^2
	First value x_{1i}	Second value x_{2i}		
1	x_{11}	x_{21}	$d_1 = (x_{21} - x_{11})$	d_1^2
2	x_{12}	x_{22}	$d_2 = (x_{22} - x_{12})$	d_2^2
n	x_{1n}	x_{2n}	$d_n = (x_{2n} - x_{1n})$	d_n^2
		Sum	?	?

From the above table, find:

$$n = ---, \quad \sum_{i=1}^{n} d_i = ---- \text{ and } \sum_{i=1}^{n} d_i^2 = ----$$

The average of the differences between paired values is given by:

$$\bar{d} = \frac{1}{n} \sum_{i=1}^{n} d_i$$

and the sample variance of the differences between paired values is given by:

$$s_d^2 = \frac{n\left(\sum_{i=1}^{n} d_i^2\right) - \left(\sum_{i=1}^{n} d_i\right)^2}{n(n-1)}$$

Calculate the t-statistic as:

$$t_{cal} = \frac{\bar{d} - 0}{s_d / \sqrt{n}}$$

Step V. Critical value(s): According to the value of α, the type of hypothesis and degree of freedom, find the standard (or critical) value(s) of the t-distribution from **Table III** given in the Appendix as:

$$t_c = \begin{cases} \pm t_{\frac{\alpha}{2}}(df = n-1) = - - - - & \text{(for a two - tailed test)} \\ -t_{\alpha}(df = n-1) = - - - - & \text{(for a left - tailed test)} \\ +t_{\alpha}(df = n-1) = - - - - & \text{(for a right - tailed test)} \end{cases}$$

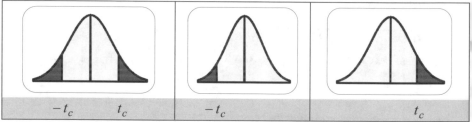

Fig. 9.52. Differentiating two-tailed, left-tailed and right-tailed tests.

Step VI. Decision: If the calculated statistic t_{cal} (from Step IV) falls in the rejection region (or critical region), then reject the null hypothesis H_0 at α % level of significance.

Example 9.9. (EXERCISE IS IMPORTANT FOR YOUR HEALTH) An experiment was conducted to reduce body size (measured in $100(\text{weight/height}^2)$). Ten patients were included in an experiment to give some special treatment, and their responses were recorded before and after the treatment as follows:

	A patient weighing before and after the treatment									
Patient No.	1	2	3	4	5	6	7	8	9	10
Before x_{1i}	824	866	841	770	829	764	857	831	846	759
After x_{2i}	702	725	744	663	792	708	747	685	742	610

Fig. 9.53. Two observations from each patient.

Does the data set present sufficient evidence to indicate a difference in mean reduction in body size of the patients?

Solution. (i) **Set the null hypothesis:**

$$H_0 : \mu_D = 0$$

(ii) **Set the alternative hypothesis:**

$$H_a : \mu_D \neq 0 \text{ (Two-tailed test)}$$

(iii) **Level of significance:** Let $\alpha = 0.05$.

(iv) **Calculate the paired t-score:** We have:

Patient No.	Before Treatment x_{1i}	After Treatment x_{2i}	Change $d_i = (x_{2i} - x_{1i})$	d_i^2
1	824	702	702-824 = -122	14,884
2	866	725	725-866 = -141	19,881
3	841	744	744-841 = -097	9,409
4	770	663	663-770 = -107	11,449
5	829	792	792-829 = -037	1,369
6	764	708	708-764 = -056	3,136
7	857	747	747-857 = -110	12,100
8	831	685	685-831 = -146	21,316
9	846	742	742-846 = -104	10,816
10	759	610	610-759 = -149	22,201
			Sum **-1,069**	**126,561**

From the above table, we have:

$$n = 10, \quad \sum_{i=1}^{n} d_i = -1,069, \quad \text{and} \quad \sum_{i=1}^{n} d_i^2 = 126,561.$$

Thus, we have:

$$\bar{d} = \frac{1}{n} \sum_{i=1}^{n} d_i = \frac{-1,069}{10} = -106.9$$

and

$$s_d^2 = \frac{n \left(\sum_{i=1}^{n} d_i^2 \right) - \left(\sum_{i=1}^{n} d_i \right)^2}{n(n-1)} = \frac{10(126,561) - (-1,069)^2}{10(10-1)} = 1,364.99$$

therefore,

$$s_d = \sqrt{s_d^2} = \sqrt{1,364.99} = 36.946.$$

Thus, the test statistic t_{cal} is given by:

$$t_{cal} = \frac{\overline{d}-0}{s_d/\sqrt{n}} = \frac{-106.9-0}{36.946/\sqrt{10}} = -9.150 \,.$$

(v) **Critical value(s):** Given $\alpha = 0.05$, $df = n-1 = 10-1 = 9$, and we are using a two-tailed test, using **Table III** from the Appendix, the two critical values are given by:

$$t_c = \pm t_{\frac{\alpha}{2}}\left(df = n-1\right) = \pm t_{0.025}\left(df = 10-1\right) = \pm 2.262 \,.$$

(vi) **Decision:**

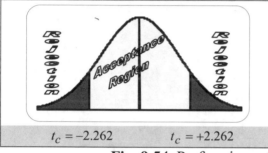

Clearly

$t_{cal} = -9.150$

falls in the left sided critical (or rejection) region. Thus, we should **reject** the null hypothesis H_0 at a 5% level of significance.

$t_c = -2.262$ $t_c = +2.262$

Fig. 9.54. Performing a two-tailed test.

9.15 IDEA OF p-VALUE USING t -SCORE

Again note that the p-value is the minimum level of significance at which the null hypothesis can be rejected. Also, note that the smaller the p-value, the stronger the evidence there is to reject the null hypothesis. If the p-value is less than the α value, then the result is statistically significant. Note that if one result is significant at a 1% level, then it is automatically significant at a 5% level. In other words, if we are 99% sure we should reject a null hypothesis, then we are automatically 95% sure we should reject that null hypothesis.

9.15.1 p-VALUE FOR LEFT-TAILED TEST USING t-SCORE

Consider applying a left-tailed test to test any statement under the null hypothesis based on a t-score such as:

H_0 : Donkeys and horses have the same height.
H_1 : Donkeys are shorter than horses (left-tailed test).

Let
$$t_{cal} = -2.35$$
and
$$df = 8$$

Fig. 9.55. Donkey and Horse.

From **Table III** given in the Appendix, the

$$|t_{cal}| = 2.35 \text{ for } df = 8$$

lies between $t_{c_1} = 2.306$ and $t_{c_2} = 2.752$.   Note that $|\,|$ is called the absolute value.

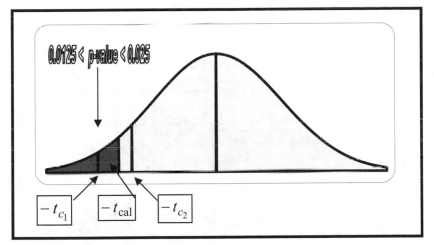

Fig. 9.56. p-value using the left-tailed test.

Thus, using **Table III** from the Appendix, the p-value is given by:

$$0.0125 < p - value < 0.025$$

Note that from **Table III**, we cannot find the exact area to the left of t_{cal}, so we say that the p-value lies between 0.025 and 0.05, which means that our confidence that we should reject the null hypothesis H_0 lies between (1-0.025)100% = 97.50% and (1-0.0125)100% = 98.75% depending how far is $-t_{cal}$ from $-t_{c_2}$ towards $-t_{c_1}$.

9.15.2 p-VALUE FOR RIGHT-TAILED TEST USING t-SCORE

Consider applying a right-tailed test to test any statement under the null hypothesis based on the t-score such as:

H_0 : Dogs and cats are the same height.
H_1 : Dogs are bigger than cats (right-tailed test).

Let

$$t_{cal} = +1.98$$

and

$$df = 24$$

Fig. 9.57. Cat and Dog.

From **Table III** given in the Appendix, the

$$t_{cal} = +1.98 \text{ for } df = 24$$

lies between $t_{c_1} = 1.711$ and $t_{c_2} = 2.064$.

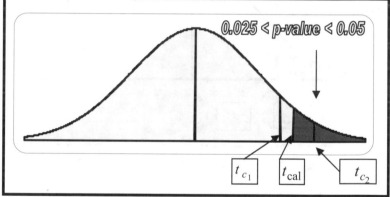

Fig. 9.58. p-value using a right-tailed test.

Thus, using **Table III** from the Appendix, the p-value is given by:

$$0.025 < p - value < 0.05$$

Note that from **Table III**, we can not find the exact area to the right of t_{cal}, so we say that the p-value lies between 0.025 and 0.05, which means that our confidence that we should reject the null hypothesis H_0 lies between $(1-0.05)100\% = 95\%$ and $(1-0.025)100\% = 97.50\%$ depending how far t_{cal} is from t_{c_1} towards t_{c_2}.

9.15.3 p-VALUE FOR TWO-TAILED TEST USING t-SCORE

Consider applying a two-tailed test to test any statement under the null hypothesis based on the t-score such as:

H_0 : Elephants and camels are the same height.
H_1 : Elephants and camels are not the same height (two-tailed test).

Let

$$t_{cal} = +1.32$$

and

$$df = 27$$

Fig. 9.59. Camel and Elephant.

From **Table III** given in the Appendix, the

$$t_{cal} = +1.32 \text{ for } df = 27$$

lies between $t_{c_1} = 1.314$ and $t_{c_2} = 1.703$.

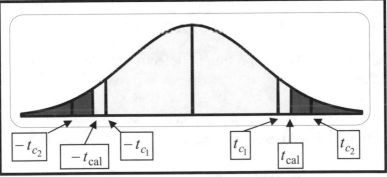

Fig. 9.60. p-value using a two-tailed test.

Thus, using **Table III** from the Appendix, the p-value is given by:

$$0.05 < \frac{(\text{p} - \text{value})}{2} < 0.10, \quad \text{or} \quad 0.10 < (\text{p} - \text{value}) < 0.20 .$$

Note that from **Table III**, we cannot find the exact area to the right of t_{cal} (or to the left of $-t_{cal}$), so we say that the p-value lies between 0.10 and 0.20. This means that our confidence that we should reject the null hypothesis H_0 lies between $(1-0.20)100\% = 80\%$ and $(1-0.10)100\% = 90\%$ depending how far $|t_{cal}|$ is from $|t_{c_1}|$ towards $|t_{c_2}|$.

9.16 LARGE SAMPLE TEST FOR THE DIFFERENCE BETWEEN TWO PROPORTIONS

Let us now consider the situation of two kinds of ropes: red ropes and green ropes.

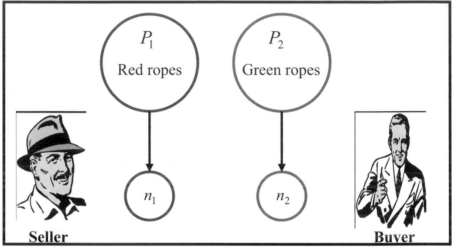

Fig. 9.61. Two types of ropes.

The seller says, "The proportion of defective red ropes is the same as that of green ropes."

Let P_1 be the true proportion of defective red ropes and P_2 be the true proportion of defective green ropes.

Step I. Set the null hypothesis:

$$H_0 : P_1 - P_2 = 0$$

Step II. Set the alternative hypothesis: A buyer can claim any one of the following three possibilities, given by:

$$H_a : P_1 - P_2 \neq 0 \text{ (Two-tailed test)}$$

or

$$H_a : P_1 - P_2 > 0 \text{ (Right-tailed test)}$$

or

$$H_a : P_1 - P_2 < 0 \text{ (Left-tailed test)}$$

Step III. Level of significance: Decide $\alpha = 0.05$ or $\alpha = 0.01$,... etc.

Step IV. Calculate the Z-score: Suppose the buyer selected a random sample of n_1 red ropes and found that x_1 are defective. Thus, the observed proportion of defective red ropes is given by:

$$\hat{p}_1 = \frac{x_1}{n_1}$$

Let x_2 be the number of defective green ropes in another independent random sample of n_2 green ropes. Thus, the observed proportion of defective green ropes is given by:

$$\hat{p}_2 = \frac{x_2}{n_2}$$

The observed pooled proportion of defective ropes is given by:

$$\hat{p} = \frac{x_1 + x_2}{n_1 + n_2} = \frac{n_1 \hat{p}_1 + n_2 \hat{p}_2}{n_1 + n_2}$$

The observed pooled proportion of non-defective ropes is given by:

$$\hat{q} = (1 - \hat{p})$$

Then, calculate the Z_{cal} score as:

$$Z_{cal} = \frac{(\hat{p}_1 - \hat{p}_2) - 0}{\sqrt{\hat{p}\hat{q}\left(\frac{1}{n_1} + \frac{1}{n_2}\right)}}$$

Step V. Critical value(s): Based on the given value of α and the type of alternative hypothesis, find the Z_c values as:

$$Z_c = \begin{cases} \pm Z_{\alpha/2}, & \text{for a two - tailed test} \\ -Z_\alpha, & \text{for a left - tailed test} \\ +Z_\alpha, & \text{for a right - tailed test} \end{cases}$$

Step VI. Decision: Reject the null hypothesis H_0 at $\alpha \%$ level of significance if the calculated Z_{cal} falls in the rejection region.

Step VII. p-value: It is always possible to find the p-value based on the computed Z_{cal} statistic and the type of alternative hypothesis.

Example 9.10. (QUALITY OF PRODUCTS) Suppose we selected 50 red ropes and 50 green ropes to study the quality of the ropes. The collected information has been classified into different categories into a 2x2 contingency table as follows:

Ropes	Quality Defective	Quality Non-defective	Total
Red	40	10	50
Green	30	20	50

Apply the Z-score to test the hypothesis that the proportion of defective red ropes is the same as that of green ropes at a 5% level of significance.

Solution. Let P_1 be the true proportion of defective red ropes and P_2 be the true proportion of defective green ropes.

(i) **Set the null hypothesis:**

$$H_0 : P_1 - P_2 = 0$$

(ii) **Set the alternative hypothesis:**

$$H_a : P_1 - P_2 \neq 0 \text{ (Two-tailed test)}$$

(iii) **Level of significance:** Given $\alpha = 0.05$.

(iv) **Calculate the Z score:** Here $n_1 = 50$, $x_1 = 40$, $n_2 = 50$, and $x_2 = 30$. Thus,

$$\hat{p}_1 = \frac{x_1}{n_1} = \frac{40}{50} = 0.80 \text{ and } \hat{p}_2 = \frac{x_2}{n_2} = \frac{30}{50} = 0.60.$$

Note that:

$$\hat{p} = \frac{x_1 + x_2}{n_1 + n_2} = \frac{40 + 30}{50 + 50} = 0.70 \text{ and } \hat{q} = 1 - \hat{p} = 1 - 0.70 = 0.30.$$

Thus, the computed Z_{cal} score is given by:

$$Z_{cal} = \frac{(\hat{p}_1 - \hat{p}_2) - 0}{\sqrt{\hat{p}\hat{q}\left(\frac{1}{n_1} + \frac{1}{n_2}\right)}} = \frac{(0.8 - 0.6) - 0}{\sqrt{0.70 \times 0.30\left(\frac{1}{50} + \frac{1}{50}\right)}} = +2.182.$$

(v) **Critical value(s):** Here $\alpha = 0.05$, and we are using a two-tailed test, using **Table II** from the Appendix, the two critical values are given by:

$$Z_c = \pm Z_{\alpha/2} = \pm Z_{0.05/2} = \pm Z_{0.025} = \pm 1.96.$$

(vi) **Decision:** Note that $Z_{cal} = +2.182$ falls in the rejection region, so we reject the null hypothesis H_0 at a 5% level of significance.

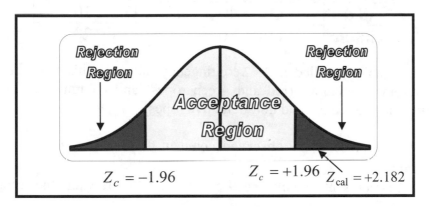

Fig. 9.62. Two-tailed test.

(vii) The **p-value** is the exact level of significance.

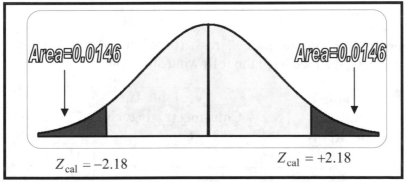

Fig. 9.63. The p-value with a two-tailed test.

Thus, the p-value is given by:

$$p - value = 0.0146 + 0.0146 = 0.0292.$$

Thus, we are $(1 - 0.0292)100\% = 97.08\%$ sure that we should reject the null hypothesis.

9.17 CONTINGENCY TABLE AND THE CHI-SQUARE TEST

A data set consisting of counts associated with categories is called categorical data. For example, say we selected 50 red ropes and 50 green ropes and classified them into different categories as follows:

Ropes	Quality	
	Defective	Non-defective
Red	$O_{11} = 40$	$O_{12} = 10$
Green	$O_{21} = 30$	$O_{22} = 20$

Such a table is called a 2×2 contingency table. In this table, rows represent populations (red and green ropes) and columns represent categories (defective and non-defective ropes).
Here

$O_{11} = 40$ represents the **observed** count in the first row and the first column,

$O_{12} = 10$ represents the **observed** count in the first row and the second column,

$O_{21} = 30$ represents the **observed** count in the second row and the first column, and

$O_{22} = 20$ represents the **observed** count in the second row and the second column.

In general, suppose there are R rows (populations) and C columns (categories), then we have the following situation:

Variable I Rows Populations	Variable II Columns (Categories)				Row Total
	1	2		C	
1	O_{11}	O_{12}		O_{1C}	R_1
2	O_{21}	O_{22}		O_{2C}	R_2
.					
R	O_{R1}	O_{R2}		O_{RC}	R_R
Column Total	C_1	C_2		C_C	N

where O_{ij}, $i = 1, 2, ..., R$ and $j = 1, 2, ..., C$ is the observed frequency in the (i, j)th cell. Such a table is called a $R \times C$ contingency table.

Let

$$P_{ij} = \frac{O_{ij}}{N}, \quad i = 1, 2, ..., R \text{ and } j = 1, 2, ..., C$$

be the proportion of counts in the (i, j) cell. Thus, we have the following situation:

Variable I Rows	Variable II Columns (Categories)			
Populations	1	2		C
1	P_{11}	P_{12}		P_{1C}
2	P_{21}	P_{22}		P_{2C}
.				
R	P_{R1}	P_{R2}		P_{RC}

Step I. Set the null hypothesis:

$$H_o : P_{11} = P_{21} = ... = P_{R1}$$
$$P_{12} = P_{22} = ... = P_{R2}$$
$$........................$$
$$P_{1C} = P_{2C} = ... = P_{RC}$$

This is called
a test of homogeneity

This is equivalent to setting the null hypothesis as:

H_0 : The rows (or populations) are homogeneous.

Step II. Set the alternative hypothesis:

H_a : The rows (or populations) are not homogeneous.

Note that a chi-square test can also be used for testing the independence of two variables:

We can set the null hypothesis:

H_0 : Variables I and II are independent.

This is called
a test of
independence

against the alternative hypothesis:

H_a : Variables I and II are not independent.

Step III. Level of significance: Decide $\alpha = 0.05$ or 0.01 etc.

Step IV. Calculate the chi-square statistic: Let $E_{ij} > 5 \ \forall \ i \, \& \, j$, that is, all the expected frequencies are greater than 5. If not, then merge a few rows or columns to make each expected frequency more than 5.

Let R_i denote the ith row total, C_j denote the jth column total, and N be the grand total. Then the expected frequency E_{ij} in the ith row and jth column is given by:

$$E_{ij} = \frac{R_i \times C_j}{N}, \ i = 1,2,...,R \ \text{and} \ j = 1,2,...,C.$$

Thus, we have the following table:

Variable I Rows Populations	Variable II Columns (Categories)			
	1	2		C
1	E_{11}	E_{12}		E_{1C}
2	E_{21}	E_{22}		E_{2C}
.				
R	E_{R1}	E_{R2}		E_{RC}

Then compute the chi-square test statistic given by:

$$X_{cal}^2 = \sum_{i=1}^{R} \sum_{j=1}^{C} \frac{\left(O_{ij} - E_{ij}\right)^2}{E_{ij}}$$

Note that the chi-square value can never be negative.

Step V. Critical value: From the chi-square **Table IV** given in the Appendix, find:

$$X_c^2 = X_\alpha^2 \left(\text{df} = (R-1)(C-1)\right)$$

Step VI. Decision: If $X_{cal}^2 > X_c^2$, then reject the null hypothesis at a $\alpha\%$ level of significance.

Note that the chi-square distribution is always skewed to the right. In other words, the right tail is longer than the left tail.

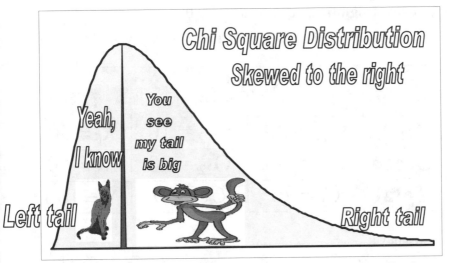

Fig. 9.64. Example of the shape of a chi-square distribution.

Example 9.11. (QUALITY MATTERS) Suppose we selected 50 red ropes and 50 green ropes to study the quality of the ropes. The collected information has been classified into different categories into a 2x2 contingency table as follows:

Ropes	Quality Defective	Quality Non-defective	Total
Red	40	10	50
Green	30	20	50

Apply the chi-square test to test the hypothesis that the proportion of red defective ropes is the same as that of green ropes at a 5% level of significance.

Solution. Let P_{11} be the proportion of defective red ropes in the population and P_{21} be the proportion of defective green ropes in the population.

(i) **Set the null hypothesis:**

$$H_0 : P_{11} = P_{21} \quad \textbf{(Test of homogeneity)}$$

(ii) **Set the alternative hypothesis:**

$$H_a : P_{11} \neq P_{21}.$$

(iii) **Level of significance:** Given $\alpha = 0.05$.

(iv) **Calculate the chi-square statistic:** We are given:

Ropes	Quality		Total
	Defective	Non-defective	
Red	$O_{11} = 40$	$O_{12} = 10$	$R_1 = 50$
Green	$O_{21} = 30$	$O_{22} = 20$	$R_2 = 50$
Total	$C_1 = 70$	$C_2 = 30$	$N = 100$

Thus, we have:

$$E_{11} = \frac{R_1 \times C_1}{N} = \frac{50 \times 70}{100} = 35, \ E_{12} = \frac{R_1 \times C_2}{N} = \frac{50 \times 30}{100} = 15$$

$$E_{21} = \frac{R_2 \times C_1}{N} = \frac{50 \times 70}{100} = 35, \ E_{22} = \frac{R_2 \times C_2}{N} = \frac{50 \times 30}{100} = 15$$

Thus we have:

Ropes	Quality		Total
	Defective	Non-defective	
Red	$E_{11} = 35$	$E_{12} = 15$	$R_1 = 50$
Green	$E_{21} = 35$	$E_{22} = 15$	$R_2 = 50$
Total	$C_1 = 70$	$C_2 = 30$	$N = 100$

Then, the calculated chi-square test statistic is given by:

$$X^2_{cal} = \sum_{i=1}^{R} \sum_{j=1}^{C} \frac{\left(O_{ij} - E_{ij}\right)^2}{E_{ij}} = \sum_{i=1}^{2} \sum_{j=1}^{2} \frac{\left(O_{ij} - E_{ij}\right)^2}{E_{ij}}$$

$$= \sum_{i=1}^{2} \left[\frac{(O_{i1} - E_{i1})^2}{E_{i1}} + \frac{(O_{i2} - E_{i2})^2}{E_{i2}} \right]$$

$$= \frac{(O_{11} - E_{11})^2}{E_{11}} + \frac{(O_{12} - E_{12})^2}{E_{12}} + \frac{(O_{21} - E_{21})^2}{E_{21}} + \frac{(O_{22} - E_{22})^2}{E_{22}}$$

$$= \frac{(40 - 35)^2}{35} + \frac{(10 - 15)^2}{15} + \frac{(30 - 35)^2}{35} + \frac{(20 - 15)^2}{15}$$

$$= \frac{25}{35} + \frac{25}{15} + \frac{25}{35} + \frac{25}{15} = 4.761 .$$

(v) **Critical value:** Using **Table IV** from the Appendix, we have:

$$X_c^2 = X_\alpha^2 \left(df = (R - 1)(C - 1) \right) = X_{0.05}^2 \left(df = (2 - 1)(2 - 1) \right)$$

$$= X_{0.05}^2 (df = 1) = 3.84 .$$

(vi) **Decision:** Note that $X_{cal}^2 = 4.761$ is greater than $X_c^2 = 3.84$. Thus, we should reject the null hypothesis H_0 at a 5% level of significance.

9.17.1 EQUIVALENCY OF CHI-SQUARE TEST AND Z-TEST

In **Example 9.10**, we had:

$$Z_{cal} = 2.182 .$$

Note that:

$$(Z_{cal})^2 = (2.182)^2 = 4.761 = X_{cal}^2$$

Also, for a two-tailed test and $\alpha = 0.05$, using **Table II** from the Appendix, the two critical values are given by:

$$Z_c = \pm Z_{\alpha/2} = \pm Z_{0.025} = \pm 1.96$$

Again, note that:

$$(Z_c)^2 = (\pm 1.96)^2 = 3.84 = X_c^2 = X_{0.05}^2 (df = 1)$$

Thus, the chi-square test is the same as the square of the Z-score test.

Example 9.12.(PETS AND BIRDS LIKE TO LIVE TOGETHER)
Assume that a survey of pet and bird owners was conducted for 650 families in a particular city. The study was conducted to find the opinions of animal and bird keepers in the city.

	Cat 	Dog 	Rabbit 	Total
Parrot 	36	55	109	200
Pigeon 	45	56	49	150
 Duck	54	78	168	300
Total	135	189	326	650

Fig. 9.65. Animal and bird lovers.

Apply the chi-square test to show that the choice of owners to keep birds and pets is not independent. Test at 5% level of significance.

Solution. (i) **Set the null hypothesis:**

H_0 : Bird owners choose pets regardless of the type of bird owned.

(ii) **Set the alternative hypothesis:**

H_a : Owners of certain birds choose pets different from owners of other birds.

(iii) **Level of significance:** Given $\alpha = 0.05$.

(iv) **Calculate the chi-square statistic:** We are given:

	Cat 	Dog 	Rabbit 	Row Total
Parrot 	$O_{11} = 36$	$O_{12} = 55$	$O_{13} = 109$	$R_1 = 200$
Pigeon 	$O_{21} = 45$	$O_{22} = 56$	$O_{23} = 49$	$R_2 = 150$
 Duck	$O_{31} = 54$	$O_{32} = 78$	$O_{33} = 168$	$R_3 = 300$
Column Total	$C_1 = 135$	$C_2 = 189$	$C_3 = 326$	Grand Total $N = 650$

Fig. 9.66. Assigning observed frequencies.

Now the expected frequencies are given by:

$$E_{11} = \frac{R_1 \times C_1}{N} = \frac{200 \times 135}{650} = 41.54, \quad E_{12} = \frac{R_1 \times C_2}{N} = \frac{200 \times 189}{650} = 58.15,$$

$$E_{13} = \frac{R_1 \times C_3}{N} = \frac{200 \times 326}{650} = 100.31, \quad E_{21} = \frac{R_2 \times C_1}{N} = \frac{150 \times 135}{650} = 31.15,$$

$$E_{22} = \frac{R_2 \times C_2}{N} = \frac{150 \times 189}{650} = 43.62, \quad E_{23} = \frac{R_2 \times C_3}{N} = \frac{150 \times 326}{650} = 75.23,$$

$$E_{31} = \frac{R_3 \times C_1}{N} = \frac{300 \times 135}{650} = 62.31, \quad E_{32} = \frac{R_3 \times C_2}{N} = \frac{300 \times 189}{650} = 87.23,$$

and

$$E_{33} = \frac{R_3 \times C_3}{N} = \frac{300 \times 326}{650} = 150.46.$$

Thus, we found:

	Cat 	Dog 	Rabbit 	Row Total
Parrot 	$E_{11} = 41.54$	$E_{12} = 58.15$	$E_{13} = 100.31$	$R_1 = 200$
Pigeon 	$E_{21} = 31.15$	$E_{22} = 43.62$	$E_{23} = 75.23$	$R_2 = 150$
 **Duck**	$E_{31} = 62.31$	$E_{32} = 87.23$	$E_{33} = 150.46$	$R_3 = 300$
Column Total	$C_1 = 135$	$C_2 = 189$	$C_3 = 326$	**Grand Total** $N = 650$

Fig. 9.67. Assigning expected frequencies.

Now the calculated chi-square test statistic is given by:

$$X_{cal}^2 = \sum_{i=1}^{R} \sum_{j=1}^{C} \frac{\left(O_{ij} - E_{ij}\right)^2}{E_{ij}} = \sum_{i=1}^{3} \sum_{j=1}^{3} \frac{\left(O_{ij} - E_{ij}\right)^2}{E_{ij}}$$

$$= \frac{(O_{11} - E_{11})^2}{E_{11}} + \frac{(O_{12} - E_{12})^2}{E_{12}} + \frac{(O_{13} - E_{13})^2}{E_{13}}$$

$$+ \frac{(O_{21} - E_{21})^2}{E_{21}} + \frac{(O_{22} - E_{22})^2}{E_{22}} + \frac{(O_{23} - E_{23})^2}{E_{23}}$$

$$+ \frac{(O_{31} - E_{31})^2}{E_{31}} + \frac{(O_{32} - E_{32})^2}{E_{32}} + \frac{(O_{33} - E_{33})^2}{E_{33}}$$

$$= \frac{(36 - 41.54)^2}{41.54} + \frac{(55 - 58.15)^2}{58.15} + \frac{(109 - 100.31)^2}{100.31}$$

$$+ \frac{(45 - 31.15)^2}{31.15} + \frac{(56 - 43.62)^2}{43.62} + \frac{(49 - 75.23)^2}{75.23}$$

$$+ \frac{(54 - 62.31)^2}{62.31} + \frac{(78 - 87.23)^2}{87.23} + \frac{(168 - 150.46)^2}{150.46}$$

$$= 0.739 + 0.171 + 0.753 + 6.158 + 3.514 + 9.145$$
$$+ 1.108 + 0.977 + 2.045$$
$$= 24.61.$$

(v) **Critical value(s):** Using **Table IV** from the Appendix, we have:

$$X_c^2 = X_\alpha^2 \left(df = (R-1)(C-1)\right)$$

$$= X_{0.05}^2 \left(df = (3-1)(3-1)\right)$$

$$= X_{0.05}^2 \left(df = 4\right) = 9.49 .$$

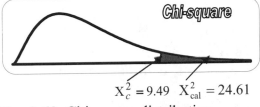

$$X_c^2 = 9.49 \quad X_{cal}^2 = 24.61$$

Fig. 9.68. Chi-square distribution.

(vi) **Decision:** Because $X_{cal}^2 = 24.61 > X_c^2 = 9.49$, we should reject the null hypothesis at a 5% level of significance. In other words, we are 95% sure that owners of certain birds choose pets different from the owners of other birds.

9.18 CHI-SQUARE TEST FOR SINGLE VARIANCE

The following steps are to be used while performing a chi-square test for testing the variance.

Step I. Set the null hypothesis:

$$H_0 : \sigma^2 = \sigma_0^2 \quad \text{(a fixed value)}$$

Step II. Set the alternative hypothesis:

$$H_a : \sigma^2 \neq \sigma_0^2 \quad \text{(Two-tailed test)}$$

or

$$H_a : \sigma^2 < \sigma_0^2 \quad \text{(Left-tailed test)}$$

or

$$H_a : \sigma^2 > \sigma_0^2 \quad \text{(Right-tailed test)}$$

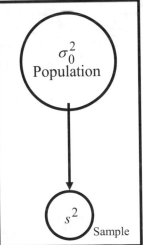

Fig. 9.69. Testing precision.

Step III. Level of significance: Decide $\alpha = 0.05$ or $0.01, \ldots,$ etc.

Step IV. Calculate the chi-square statistic: Let x_1, x_2, ..., x_n be a random sample of n units and let the sample variance be given by:

$$s^2 = (n-1)^{-1} \sum_{i=1}^{n} (x_i - \bar{x})^2$$

Then calculate the chi-square statistic given by:

$$X^2_{cal} = \frac{(n-1)s^2}{\sigma_0^2}$$

Step V. Critical value(s): (a) For a two-tailed test, the critical values are on both tails, but neither are negative, as shown in the following figure:

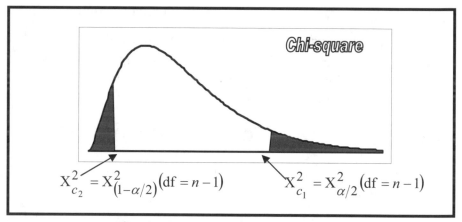

$$X^2_{c_2} = X^2_{(1-\alpha/2)}(df = n-1) \qquad X^2_{c_1} = X^2_{\alpha/2}(df = n-1)$$

Fig. 9.70. Two-tailed chi-square test.

(b) For a right-tailed test, the critical value is on the right tail as shown in the following figure:

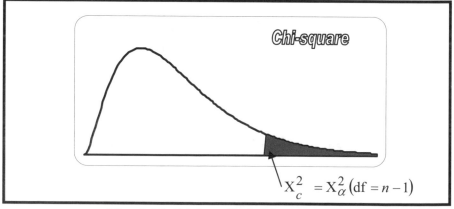

$$X^2_c = X^2_{\alpha}(df = n-1)$$

Fig. 9.71. Right-tailed chi-square test.

(c) For a left-tailed test, the critical value is on the left tail, with a positive sign, as shown in the following figure:

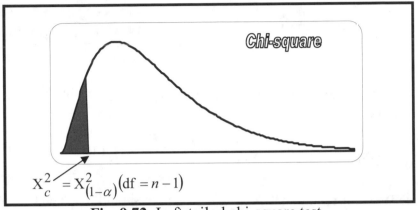

Chi-square

$$X_c^2 = X_{(1-\alpha)}^2 (df = n-1)$$

Fig. 9.72. Left-tailed chi-square test.

Step VI. Decision: If X_c^2 falls in the rejection region, then we should reject the null hypothesis at a α % level of significance.

Example 9.13. (QUALITY CONTROL)   It is believed that the precision (measured by the variance) of a drill press is no more than 10. Test at a 5% level of significance, given that the 7 observations taken on the same subject from the drill press have the values: 25, 24, 23, 27, 30, 32, and 28.

Fig. 9.73. Drill.

Solution. (i) **Set the null hypothesis:**

$$H_0 : \sigma^2 = 10$$

(ii) **Set the alternative hypothesis:**

$$H_a : \sigma^2 > 10 \text{ (one-tailed test)}$$

(iii) **Level of significance:**

Given $\alpha = 0.05$.

(iv) **Calculate the chi-square statistic:**

Now we have:

$$s^2 = \frac{n\left(\sum x_i^2\right) - \left(\sum x_i\right)^2}{n(n-1)}$$

$$= \frac{7(5,167) - (189)^2}{7(7-1)} = 10.667.$$

x_i	x_i^2
25	625
24	576
23	529
27	729
30	900
32	1,024
28	784
Sum 189	5,167

Thus, the calculated chi-square statistic value is given by:

From the above table:
$n = 7,$

$\sum x_i = 189,$

$\sum x_i^2 = 5{,}167.$

$$X_{\text{cal}}^2 = \frac{(n-1)s^2}{\sigma_0^2} = \frac{(7-1)(10.667)}{10} = 6.40 \, .$$

(v) **Critical value:** For the right-tailed test, using **Table IV** from the Appendix, we have:

$$X_c^2 = X_\alpha^2 \left(df = n-1 \right) = X_{0.05}^2 \left(df = 7-1 \right) = X_{0.05}^2 \left(df = 6 \right) = 12.59.$$

(vi) **Decision:** Note that $X_{\text{cal}}^2 < X_c^2$, so we should not reject the null hypothesis H_0 at a 5% level of significance.

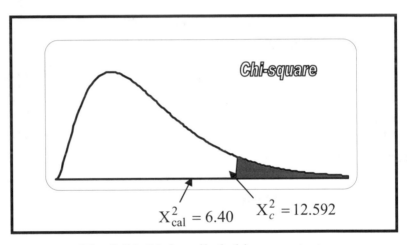

Fig. 9.74. Right-tailed chi-square test.

In other words, the precision of the drill press is not more than 10.

Example 9.14. (PRACTICE THE CHI-SQUARE TABLE) Use **Table IV** from the Appendix to find the values of $X_\alpha^2 \left(df = 7 \right)$ for $\alpha = 0.05$ and for $\alpha = 0.025$.
Solution. From **Table IV** given in the Appendix, we have:

$$X_{0.05}^2 \left(df = 7 \right) = 14.07 \quad \text{and} \quad X_{0.025}^2 \left(df = 7 \right) = 16.01.$$

9.19 F-TEST FOR TWO POPULATION VARIANCES

Step I. Set the null hypothesis:

$H_0 : \sigma_1^2 = \sigma_2^2$

Step II. Set the alternative hypothesis:

$H_a : \sigma_1^2 \neq \sigma_2^2$ (Two-tailed test)

or

$H_a : \sigma_1^2 > \sigma_2^2$ (Right-tailed test)

or

$H_a : \sigma_1^2 < \sigma_2^2$ (Left-tailed test)

Step III. Level of significance: Decide $\alpha = 0.05, 0.01$ etc.

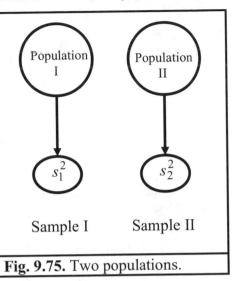

Fig. 9.75. Two populations.

Step IV. Calculated the F-ratio: Let s_1^2 and s_2^2 be the estimators of σ_1^2 and σ_2^2. Let $s_2^2 < s_1^2$. Then calculate the F-ratio as:

$$F_{cal} = \frac{s_1^2}{s_2^2}$$

Note that the F-ratio can never be negative.

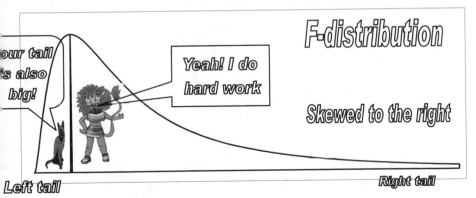

Fig. 9.76. F-distribution.

Step V. Critical value(s): Let

$v_1 = n_1 - 1 =$ degree of freedom for the numerator of the F-ratio

and

$v_2 = n_2 - 1 =$ degree of freedom for the denominator of the F-ratio.

(a) For a two-tailed test, the critical region is given by two critical values as:

$$F_{c_1} = F_{\alpha/2}(v_1, v_2) \text{ and } F_{c_2} = F_{(1-\alpha/2)}(v_1, v_2)$$

Note that:

$$F_{c_2} = 1/F_{c_1}$$

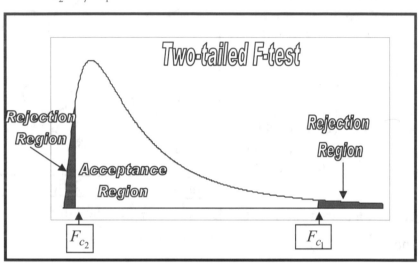

Fig 9.77. Two-tailed test using F-ratio.

(b) For a right-tailed test, the rejection region is given by one critical value as shown below:

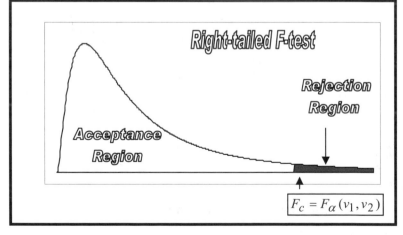

Fig. 9.78. Right-tailed test using F-ratio.

(c) For a left-tailed test, the rejection region is given by one critical value as shown below:

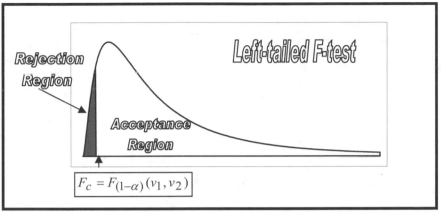

Fig. 9.79. Left-tailed test using F-ratio.

Step VI. Decision: If F_{cal} falls in the rejection region, then reject the null hypothesis at a α % level of significance.

Example 9.15. (COMPARING TWO INSTRUMENTS) It is believed that the precision of a drill is measured by the variance of measurements of certain kinds of subjects. For Drill A, in a sample of 8 data values of such subjects, the sum of the squares of the deviations from the sample mean was 84.4, and for Drill B it was 102.6 based on another random sample of 10 such subjects. Test whether both drills have the same precision at a 10% level of significance.

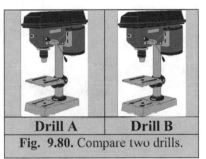

| Drill A | Drill B |

Fig. 9.80. Compare two drills.

Solution. (i) **Set the null hypothesis:**

$$H_0 : \sigma_1^2 = \sigma_2^2$$

(ii) **Set the alternative hypothesis:**

$$H_a : \sigma_1^2 \neq \sigma_2^2 \text{ (Two-tailed test)}$$

(iii) **Level of significance:** Given $\alpha = 0.10$.

(iv) **Calculate the F-ratio:** We are also given:

$$\sum_{i=1}^{n_1}\left(x_{1i} - \bar{x}_1\right)^2 = 84.4 \text{ and } \sum_{i=1}^{n_2}\left(x_{2i} - \bar{x}_2\right)^2 = 102.6$$

which implies:

$$s_1^2 = \frac{\sum\limits_{i=1}^{n_1}(x_{1i} - \bar{x}_1)^2}{n_1 - 1} = \frac{84.4}{8-1} = 12.057$$

and $$s_2^2 = \frac{\sum\limits_{i=1}^{n_2}(x_{2i} - \bar{x}_2)^2}{n_2 - 1} = \frac{102.6.}{10-1} = 11.4.$$

Note that $s_2^2 < s_1^2$, thus the calculated F-ratio is given by:

$$F_{cal} = \frac{s_1^2}{s_2^2} = \frac{12.057}{11.400} = 1.058.$$

(v) **Rejection region:** Here $\alpha = 0.10, v_1 = n_1 - 1 = 8 - 1 = 7$, $v_2 = n_2 - 1 = 10 - 1 = 9$ and for a two-tailed test, we have:

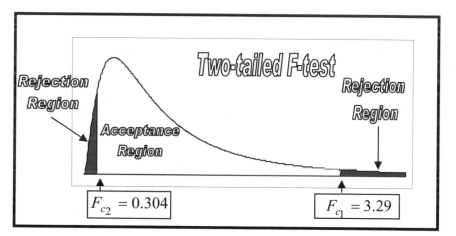

Fig. 9.81. Two-tailed test using F-ratio.

Using **Table V** from the Appendix, we have two critical values as:

$$F_{c_1} = F_{0.10/2}(v_1 = 8-1; v_2 = 10-1) = F_{0.05}(v_1 = 7, v_2 = 9) = 3.29$$

and

$$F_{c_2} = \frac{1}{F_{c_1}} = \frac{1}{3.29} = 0.304.$$

(vi) **Decision:** Note that the calculated value, $F_{cal} = 1.058$ falls in the acceptance region. Thus, we should accept the null hypothesis at a 10% level of significance. In other words, we are 90% sure we should accept the statement that both drills are equally efficient.

9.20 ANALYSIS OF VARIANCE (ANOVA)

In a field experiment, k treatments are applied to n plots in an experiment as shown below:

	Treatments			
	1	**2**		k
	x_{12}	x_{21}		x_{k1}
	x_{12}	x_{22}		x_{k2}
	•	•		•
	•	•		•
	x_{1n_1}	x_{2n_2}		x_{kn_k}
Total	T_1	T_2		T_k

Fig. 9.82. Farmers applying treatments.

Let μ_1, μ_2, ..., μ_k be the true average yield obtained after applying the first, second, ... and k th treatments, respectively. Our aim is to find out if there is any significant difference between the average yields obtained by applying any treatment. This can be analyzed with one way classification or completely randomized design (CRD) by applying the following **ten** steps:

Step I. Set the null hypothesis:

$$H_0 : \mu_1 = \mu_2 = ... = \mu_k$$

or

H_0 : All the population means are equal.

Step II. Set the alternative hypothesis:

H_1 : At least one of the population means is different.

Step III. Decide the level of significance:

$$\alpha = 0.05 \text{ or } \alpha = 0.01... \text{ etc.}$$

Step IV. Find the correction factor (CF):
Calculate the grand total (GT):

$$GT = \sum_{i=1}^{k} \sum_{j=1}^{n_i} x_{ij} = T_1 + T_2 + ... + T_k$$

and

$$n = \text{total number of data values} = \sum_{i=1}^{k} n_i = n_1 + n_2 + \dots + n_k.$$

Then, find the correction factor (CF) given by:

$$CF = \frac{(GT)^2}{n}$$

Step V. Calculate the total sum of squares (TSS):

$$TSS = \sum_{i=1}^{k} \sum_{j=1}^{n_i} x_{ij}^2 - CF$$

Step VI. Calculate the sum of squares due to treatments (SST):

$$SST = \sum_{i=1}^{k} \frac{T_i^2}{n_i} - CF$$

Step VII. Calculate the sum of squares due to error (SSE):

$$SSE = TSS - SST$$

Step VIII. Complete the ANOVA table:

Source	Degree of freedom (df)	Sum of Squares (SS)	Mean Sum of Squares (MS)	Compute F-ratio
Treatments	$(k-1)$	SST	$MST = \dfrac{SST}{k-1}$	$F_{cal} = \dfrac{MST}{MSE}$
Error	$(n-k)$	SSE	$MSE = \dfrac{SSE}{n-k}$	
Total	$(n-1)$	TSS		

Step IX. Critical value: From **Table V** given in the Appendix, for ANOVA, find only one critical F value given by:

$$F_c = F_\alpha \left(v_1 = k - 1, \ v_2 = n - k \right)$$

Step X. Decision: If $F_{cal} \geq F_c$, then reject the null hypothesis H_0 at a $\alpha\%$ level of significance.

Example 9.16. (COMPARING TREATMENTS) A scientist applied four treatments: I, II, III, and IV to find out if there is any difference between the strength of the smell of roses based on quantitative scores. Treatment I is applied to $n_1 = 5$ plants, treatment II is applied to $n_2 = 4$ plants, treatment III is applied to $n_3 = 6$ plants and treatment IV is applied to $n_4 = 3$ plants of roses as shown below:

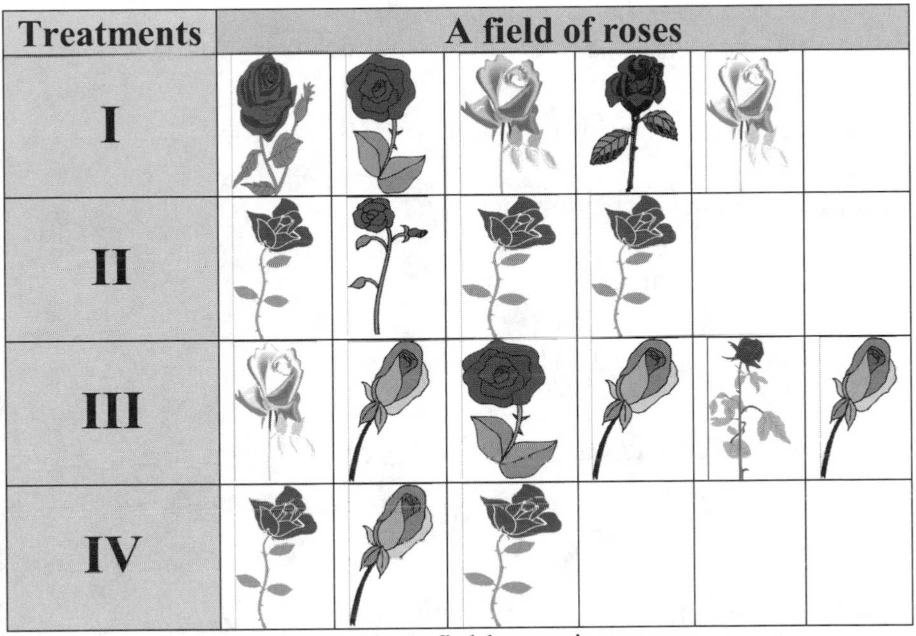

Fig. 9.83. A field experiment.

A special smell detector instrument was designed and used to quantify the strengths of the smell of different roses as given below:

Treatments	Observed data						Total
I	50	25	28	30	19		152
II	8	7	8	8			31
III	20	10	25	10	27	10	102
IV	12	10	14				36

Carry out the analysis of variance (ANOVA) to find if there is any significant difference between the four treatments while growing roses.

Solution. We apply the ten steps of the ANOVA table as follows:

Step I. Set the null hypothesis:

$$H_0 : \mu_1 = \mu_2 = \mu_3 = \mu_4 \text{ (All the treatment means are equal.)}$$

Step II. Set the alternative hypothesis:

$$H_a : \text{At least one of the treatment means is different.}$$

Step III. Level of significance: Let $\alpha = 0.05$.

Step IV. Find the correction factor (CF):

Here the grand total (GT) is given by:

$$GT = \sum_{i=1}^{k} T_i = 152 + 31 + 102 + 36 = 321$$

and, total number of observations (n) is given by:

$$n = \sum_{i=1}^{k} n_i = 5 + 4 + 6 + 3 = 18.$$

Thus, the correction factor (CF) is given by:

$$CF = \frac{(GT)^2}{n} = \frac{321^2}{18} = 5{,}724.50.$$

Step V. Find the corrected total sum of squares (TSS):

$$TSS = \sum_{i=1}^{k} \sum_{j=1}^{n_i} x_{ij}^2 - CF$$

$$= \left[50^2 + 25^2 + 28^2 + 30^2 + 19^2 + 8^2 + 7^2 + 8^2 + 8^2 + 20^2 \right. $$
$$\left. + 10^2 + 25^2 + 10^2 + 27^2 + 10^2 + 12^2 + 10^2 + 14^2 \right] - 5{,}724.50$$

$$= 2{,}180.50.$$

Step VI. Find the corrected sum of squares due to treatment totals (SST):

$$\text{SST} = \sum_{i=1}^{k} \frac{T_i^2}{n_i} - \text{CF} = \frac{152^2}{5} + \frac{31^2}{4} + \frac{102^2}{6} + \frac{36^2}{3} - 5,724.50 = 1,302.55.$$

Step VII. Find the sum of squares due to error (SSE):

$$\text{SSE} = \text{TSS} - \text{SST} = 2,180.50 - 1,302.55 = 877.95.$$

Step VIII. Complete the ANOVA table:

Sources of variation	Df	SS	MS	F-ratio
Between treatments	4–1 = 03	1,302.55	434.183	6.92
Within treatments	18–4 = 14	877.95	62.711	
Total	18–1 = 17	2,180.50		

Step IX. Critical value: Using **Table V** from the Appendix, the critical value for a right-tailed test is given by:

$$F_c = F_\alpha(k-1,\ n-k) = F_{0.05}(3,\ 14) = 3.34.$$

Step X. Decision: Note that $F_{cal} = 6.92 \geq F_c = 3.34$, so we reject the null hypothesis H_0 at a 5% level of significance.

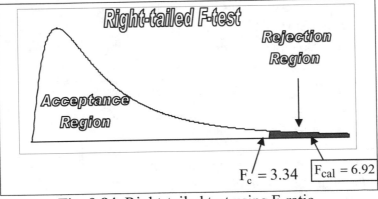

Fig. 9.84. Right-tailed test using F-ratio.

In other words, at least one of the treatment means is different from the others, that is, at least one treatment is better or worse than the others. To pick up the best or worst treatment among these treatments, please consult advanced books related to the design of experiments.

LUDI 9.1. (SOMETIMES CHOICES ARE DIFFICULT) Circle the correct answer:

Statement	(a)	(b)	(c)
(1) The t-score is called the Tiger score.	True	False	
(2) The Z-score is called the Zebra score.	True	False	
(3) The t-score is called the Student's t-score.	True	False	
(4) A F-test is used to compare two-----	Foxes	Variances	Means
(5) A Chi-square test is used for -----	Mean	Variance	Cheetah
(6) A t-test is used to compare two-----	Means	Variances	Tigers
(7) A Z-test is used to compare two-----	Means	Variances	Zebras
(8) If the p-value is 0.045, then the result is significant at 5%.	True	False	
(9) If a result is significant at a 5% level, then this result is ----- significant at a 1% level.	Always	Some-times	Never
(10) If a result is significant at a 1% level, then this result is ----- significant at a 5% level.	Always	Some-times	Never
(11) If a sample size increases, neither the chance of α nor of β will increase.	True	False	
(12) For a fixed sample size, the value of β can be reduced by reducing α.	True	False	
(13) The significance level is the chance that H_0 is true.	True	False	
(14) The p-value can be determined without observing data.	True	False	
(15) A new idea or change in a population is stated in the ----- hypothesis.	Null	Alternative	
(16) If we accept H_0 when in fact the H_a is true, we make a ----- error.	Type I	Type II	
(17) The chance of accepting H_0, when in fact it is false is called -----	α	β	p-value
(18) The sum of the probability of type I and type II errors is always -----	One	Between 0 and 1	Two

LUDI 9.2. (FAMILY PLANNING) In a survey of 440 parents with a total of 940 children younger than 8 years, 320 said that a family with two kids is considered to be an optimum family size.

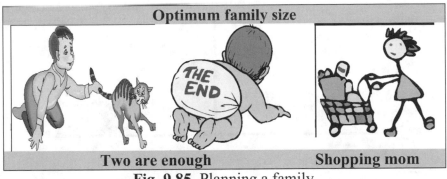

Fig. 9.85. Planning a family.

Does the survey support the conjecture that the proportion of parents who like two kids at home is greater than 0.70? Use a 5% level of significance to test.

Step I. Set the null hypothesis: H_0 :-------

Step II. Set the alternative hypothesis:

H_1 :------- (Is it a one-tailed or a two-tailed test?)

Step III. Level of significance: Given α = --------

Step IV. Compute the Z-statistic: Find x = ---- and n = ----,
Find the sample proportion:

$$\hat{p} = \frac{x}{n} = \frac{----}{----} = --------$$

Find P_0 = --------

Compute the Z-score:

$$Z_{cal} = \frac{\hat{p} - P_0}{\sqrt{\dfrac{P_0(1 - P_0)}{n}}} = --------$$

Step V. Find the critical value(s): Given that α = ---- and that we are using a - - - tailed test, then the critical value(s) are:

$$Z_c = -----$$

Step VI. Decision: Now Z_{cal} = -------- is inside (or outside) the critical region, so we should -------- the null hypothesis.

Based on the above steps, complete the following picture:

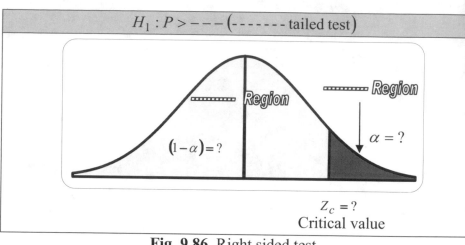

$$H_1 : P > --- \left(------ \text{tailed test} \right)$$

------- Region

------- Region

$(1-\alpha)=?$

$\alpha = ?$

$Z_c = ?$
Critical value

Fig. 9.86. Right sided test.

Step VII. Find the p-value:

Note that we are using a one-tailed test, so the p-value will be given by the area to the right of: $Z_{\text{cal}} = --------$

LUDI 9.3. (WATCH FOR THE POSTED SPEED LIMIT WHILE DRIVING) Due to a sharp turn in the road, the traffic administration decided to reduce the speed limit to 15 miles/hour. A random sample of 25 vehicles shows the following speed at the time of turning.

	15	18	16	21	20
	14	16	17	25	13
	22	28	20	16	18
	16	15	18	12	15
	18	20	21	20	12

Fig. 9.87. Traffic rules.

(a) Test at a 5% level of significance that the vehicles are following the traffic rules.
(b) Construct a 95% confidence interval estimate of the average speed of the vehicles. Does the recommended speed of 15 miles/hour fall into it?

LUDI 9.4. (CAR LOVERS: SPEED THRILLS, BUT KILLS)

Assume that on a three-lane highway there is a significant amount of snowfall during a winter season. It is recommend that the state driving speed should be 25 miles/hour. A random sample from the camera surveillance record of 30 cars driving on the highway is as follows:

Fig. 9.88. Highway look.

Does the survey support the conjecture that the people are following the traffic rules? Use a 5% level of significance to test.

Step I. Set the null hypothesis:

$$H_0 : \mu = - - - - - - -$$

Step II. Set the alternative hypothesis:

$$H_1 : \mu = - - - - - - - - \text{ (One or two-tailed tests)}$$

Step III. Level of significance: Given $\alpha = - - - - - - -$

Step IV. Calculate the Z-statistic: We are given $n = -------$.
Find the sample mean:

$$\bar{x} = \frac{\sum\limits_{i=1}^{n} x_i}{n} = --------$$

the sample variance:

$$s^2 = \frac{n\left(\sum\limits_{i=1}^{n} x_i^2\right) - \left(\sum\limits_{i=1}^{n} x_i\right)^2}{n(n-1)} = --------$$

and, the sample standard deviation:

$$s = \sqrt{s^2} = --------$$

Thus, the calculated Z-statistic is given by:

$$Z_{cal} = \frac{\bar{x} - \mu}{s/\sqrt{n}} = --------$$

Step V. Critical value(s): We are given $\alpha = 0.05$ and we are using a one (or two) tailed test, so the critical value(s) are given by:

$$Z_c = --------$$

Step VI. Decision: The calculated Z-value falls inside (or outside) the critical region, as shown below, and hence we reject (or accept) the null hypothesis. Complete the following figure.

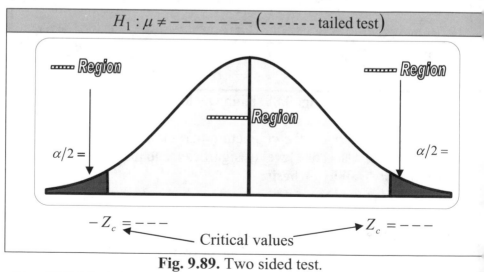

Fig. 9.89. Two sided test.

Step VII. Find the p-value.

LUDI 9.5. (MOTHER'S DAY) A study that followed 2600 children through age 18 has found that those who were breast-fed as children fared better in school, both in teacher ratings and in performance of other activities. The authors of the study claim that fatty acids found in breast milk, but not in formula, may be good for boosting the brain.

Fig. 9.90. Child and Mother.

(a) In contrast, a criticism of the study is that breast-fed children tended to have mothers who were older, educated, wealthier and/or healthier. A question that by itself could account for the differences in intelligence is called a - - - - - - -

      (i) response variable (ii) confounding variable
      (iii) continuous variable (iv) control variable.

(b) The study states that they adjusted for the question mentioned above, but still found small but consistent tendencies for the duration of breastfeeding have a positive correlation with intelligence. Suppose that the hypotheses being tested were as follows:

    H_0 : Breastfeeding duration and intelligence are related.

    H_1: Breastfeeding duration and intelligence are not related.

Assuming that the results were statistically significant at a 5% level, circle your choice:

(i)	Intelligence is an independent variable.	True	False
(ii)	A Type I error is 0.05.	True	False
(iii)	H_0 is rejected.	True	False
(iv)	The p-value is more than 0.05.	True	False
(v)	Type II error is 0.95.	True	False
(vi)	The p-value is less than 0.01.	True	False
(vii)	H_1 is accepted.	True	False

LUDI 9.6 (SHOE SIZE) It is claimed that 75% of women wear tight shoes. The *American Shoe Society,* in a study of 400 women, found that 320 women were wearing tight shoes. Use the study results to test the hypothesis at a 5% level of significance that the proportion of women wearing such shoes is greater than 0.75.

Fig. 9.91. Shoe store.

Step I. Set the null hypothesis:
$$H_0 :-------$$
Step II. Set the alternative hypothesis:
$$H_1 :-------\text{(Is it a one-tailed or a two-tailed test?)}$$
Step III. Level of significance: Given $\alpha = --------$
Step IV. Compute the Z-statistic:
Find $x = -------$ and $n = -------$
Find the sample proportion:
$$\hat{p} = \frac{x}{n} = \frac{----}{----} = -------$$
Find $P_0 = ---------$. Compute the Z-score:
$$Z_{cal} = \frac{\hat{p} - P_0}{\sqrt{\dfrac{P_0(1 - P_0)}{n}}} = -------$$

Step V. Critical value(s): Given that $\alpha = -------$ and that we are using a $-------$ tailed test, then the critical value(s) are:
$$Z_c = -------$$
Step VI. Decision: Now $Z_{cal} = -------$, falls outside (or inside) the critical region, thus we reject (or accept) the null hypothesis H_0.
Step VII. Find the p-value and interpret it.
Step VIII. Construct a 95% confidence interval estimate of the required parameter and interpret it.

LUDI 9.7. (SAVE YOUR MONEY FOR HIGHER EDUCATION)

Amy wants to compare the cost of one and two-bedroom apartments in the area of her campus. She took a random sample of five one-bedroom and eight two-bedroom rents from a local magazine, called the owners, and collected the following information:

| Two Bedroom ($) | 595 | 500 | 580 | 675 | 680 | 495 | 580 | 677 |
| One Bedroom ($) | 500 | 650 | 450 | 515 | 395 | | | |

Fig. 9.92. Amy searching an apartment.

Amy wonders if renting a two-bed room apartment is significantly more than a one-bedroom. To test her claim at a 5% level of significance, give the following information:

(a) Set $H_0 : - - - - - -$ against $H_a : - - - - - -$

(b) Find $\alpha = - - - - - -$

(c) To apply the appropriate test, complete the following table:

	Sample I (Two-bedroom)			Sample II (One-bedroom)		
	x_{1i}	$(x_{1i} - \bar{x}_1)$	$(x_{1i} - \bar{x}_1)^2$	x_{2i}	$(x_{2i} - \bar{x}_2)$	$(x_{2i} - \bar{x}_2)^2$
Sum						

From the above table, find:

$n_1 = - - -,$ $n_2 = - - -,$ $\bar{x}_1 = - - - - - - -,$ $\bar{x}_2 = - - - - - - -,$

$s_1^2 = - - - - - - -$ and $s_2^2 = - - - - - - -$

Then find the pooled sample variance as:

$$s_p^2 = - - - - - - -$$

Then compute the t_{cal} statistic as:

$$t_{cal} = \frac{(\bar{x}_1 - \bar{x}_2) - 0}{\sqrt{s_p^2\left(\dfrac{1}{n_1} + \dfrac{1}{n_2}\right)}} = --------$$

(d) Sketch the critical region.
(e) State your decision.
(f) Sketch and find the p-value.

LUDI 9.8. (PEAR LOVERS) At an orchard, the average weight of a pear was determined for a sample of size 36, resulting in a sample mean of 90 gm and a sample standard deviation of 12 gm.

(a) Construct a 75% confidence interval estimate.
(b) Construct a 85% confidence interval estimate.
(c) Construct a 90% confidence interval estimate.
(d) Construct a 95% confidence interval estimate.
(e) Construct a 98% confidence interval estimate.
Give your interpretation in each case.

Fig. 9.93. Pear.

LUDI 9.9. (HOT AIR BALOONS LOOK CUTE) The following table lists the results of an experiment in which two different methods of determining Helium content were used on samples of hot-air balloons. The observations are in mg/l.

	MSI-method	0.39	0.84	1.76	3.35	4.69	7.70
	SIB-method	0.36	1.35	2.56	3.92	5.35	8.33

Fig. 9.94. Experimental data.

(a) Does the true average content measured by one method appear to differ from that measured by another method?
State and test the appropriate hypothesis.
(b) Does the conclusion depend on whether a significance level of 0.05, 0.01, or 0.001 is used?
(c) What is the exact level of significance?

LUDI 9.10. (FRUITS ARE GOOD FOR YOUR HEALTH) There are two types of oranges, sweet and sour. A vendor claims that the average weight of sweet and sour oranges is the same.

Fig. 9.95. Vendor and customer.

A customer took two independent samples of sweet and sour oranges, and noted the information in the following table.

Information	Number of oranges selected	Average weight (gm)	Sample standard deviation
Sweet oranges	$n_1 = 12$	$\bar{x}_1 = 108$	$s_1 = 1.7$
Sour oranges	$n_2 = 9$	$\bar{x}_2 = 106$	$s_2 = 1.6$

Test the vendor's claim at a 95% level of confidence.

LUDI 9.11. (WHY DOES EXPRESS MAIL COST MORE?) The average delivery time (in hours) of two mailing systems, express mail and air mail, was studied based on two independent samples selected from the customers who used these mailing systems at a particular post office. The results after the delivery of all the mail are listed below.

	$n_1 = 60$	$\bar{x}_1 = 26.99$	$s_1 = 4.89$
 AIR MAIL	$n_2 = 60$	$\bar{x}_2 = 35.76$	$s_2 = 6.34$

Fig. 9.96. Comparing mail.

Test the claim that the average delivery time of express mail is less than air mail at a 99% confidence level.

LUDI 9.12. (KIDS R US) A medical student considers the study of iron deficiencies among kids due to different feeding methods. She took a random sample of breast-fed kids, while the kids in another group were fed a standard baby formula without any iron supplements. She then recorded the results of blood hemoglobin levels at certain ages as follows:

Group	n_i	$\bar{x}_i$	$\dfrac{s_i}{\sqrt{n_i}}$
Breast-fed	15	13.3	0.45
Formula	12	12.4	0.52

Fig. 9.97. Babies need attention.

Is there significant evidence that the mean hemoglobin level is higher among breast-fed babies? Test at a 5% level of significance.

(a) Carry out the six steps of the appropriate t test.

(i) State H_0:

(ii) State H_a:

(iii) Find $\alpha =$ --------.

(iv) Find $t_{cal} =$ --------.

(v) Sketch the critical region.

(vi) State your decision.

(b) Find the p-value.

LUDI 9.13. (LAB EXPERIMENTS NEED ATTENTION) In an inter-laboratory study, each of the four laboratories (designated Lab 1 - Lab 4) is asked to make three or four determinations of the percentage of Nitrogen in specimens of a certain compound. The specimens all came from the same batch of compound as shown below:

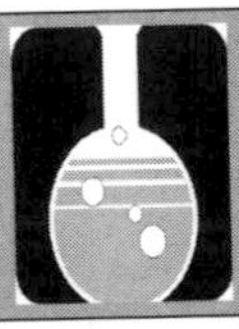

Lab 1	85.06	85.25	84.87	85.67
Lab 2	84.99	84.28	84.88	
Lab 3	84.48	84.72	85.10	85.20
Lab 4	84.10	84.55	84.05	

Fig. 9.98. Experimental data.

(a) State and test the relevant hypothesis with $\alpha = 0.05$.

(b) Can you conclude that there is a difference between the average percentages reported by the laboratories?

LUDI 9.14. (TALES OF TAILS) Find the p-value in each one of the following hypotheses:

(a)

H_0 : Cherries and tomatoes are the same size.

H_1 : Cherries are smaller than tomatoes (------- tailed test).

Let

$$Z_{cal} = -3.08$$

Fig. 9.99. Cherry and tomato.

Then the p-value is the area to the ------- of the calculated Z-score as shown:

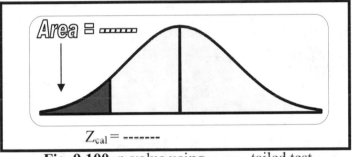

Fig. 9.100. p-value using ------- tailed test.

Thus, using **Table II** from the Appendix, the p-value is given by:

p-value ≡ ------------

Here, the p-value ------- means, we are ------- sure we should reject the null hypothesis.

(b)

H_0 : Oranges and apples are the same size.

H_1 : Oranges are bigger than apples (------- tailed test).

Let

$$Z_{cal} = +1.68$$

Fig. 9.101. Apple and Orange.

Then, the p-value is the area to the ------- of the calculated Z-score as:

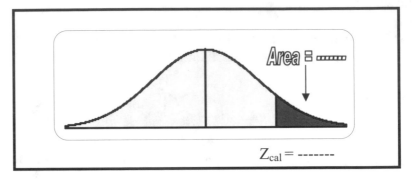

Fig. 9.102. p-value using ------- tailed test.

Thus, using **Table II** from the Appendix, the p-value is given by:

Here the p-value ------- means, we are ------- sure we should reject the null hypothesis.

(c)

H_0 : Pumpkins and watermelons are the same size.

H_1 : Pumpkins and watermelons are not the same size (---tailed test).

Let

$$Z_{cal} = +1.22$$

Fig. 9.103. Pumpkin and Watermelon.

Then, the p-value is the total area to the ------- of the calculated Z-score and its image on the ------- side.

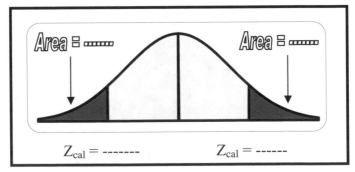

Fig. 9.104. p-value using ------- tailed test.

Thus, using **Table II** from the Appendix, the p-value is given by:

$$p\text{-value} \equiv \text{------}$$

Here the p-value ------- means that we are ------- sure we should reject the null hypothesis.

LUDI 9.15. (NO FAIR DIE ON ANY LOTTERY FAIR) To assess whether a die is a "fair" die, Michael rolled a die 10,000 times. He observed that 5080 times an "odd number" appeared.

(a) Find the p-value while testing:
   $H_0 : P = 0.50$ against $H_1 : P \neq 0.50$
(b) Find the p-value while testing:
   $H_0 : P = 0.50$ against $H_1 : P > 0.50$

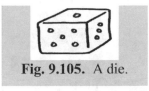

Fig. 9.105. A die.

LUDI 9.16. (GROW MORE TREES) Due to some special type of soil and environment, a garden grows apples very well, and everybody likes to have apples from this particular garden. The gardener packs apples in boxes each consisting of 12 lbs and sells them to shopkeepers in the market.

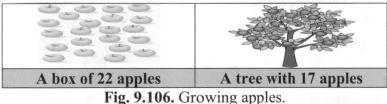

A box of 22 apples	A tree with 17 apples

Fig. 9.106. Growing apples.

One shopkeeper had a complaint from a customer that their apples were not the right weight as expected from that garden. The shopkeeper doubted the gardener. He thought the gardener may be mixing apples in the boxes from another orchard having low quality apples. Then the shopkeeper visited the gardener and selected one tree to weigh all the apples on it, and found that the total weight of all the apples on the selected tree was 10 lbs. Then the shopkeeper claimed that the average weight of an apple in the boxes was less than that in the orchard. The experience shows that the population standard deviation of weight of the apples in the boxes is 2.2 and that in the orchards is 2.5.

(a) Test the shopkeeper's claim at a 5% level of significance.
(b) Find the p-value.

LUDI 9.17. (GOOD GRADES NEED HARD WORK) Generally, old instructors are stricter than young instructors. Amy wants to register for statistics courses, so she contacted 6 students from sections taught by an old instructor, 8 students from sections taught by a young instructor, and collected the following information about their mid-term scores:

							Amy
 **Old**	75	85	65	35	92	68	 Back to School
 **Young**	88	90	95	77	73	68	85	72

Fig. 9.107. How does Amy compare the two instructors?

Amy wonders if the average score of the students registered with the old instructor is significantly lower than those registered with the young instructor. To test her claim at a 5% level of significance, give the following information:

(i) Set the null hypothesis:

$$H_0 : - - - - - - -$$

(ii) Set the alternative hypothesis:

$$H_a : - - - - - - - -$$

(iii) Find $\alpha = - - - - - - -$

(iv) To calculate the appropriate test statistic, complete the table:

	Sample I (Old Instructor)			Sample II (Young Instructor)		
	x_{1i}	$(x_{1i} - \bar{x}_1)$	$(x_{1i} - \bar{x}_1)^2$	x_{2i}	$(x_{2i} - \bar{x}_2)$	$(x_{2i} - \bar{x}_2)^2$
Sum						

From the above table, find:

$n_1 = ---$, $n_2 = ---$, $\bar{x}_1 = -------$, $\bar{x}_2 = -------$,

$s_1^2 = --------$, and $s_2^2 = --------$

Then, find the pooled sample variance:

$$s_p^2 = -------$$

Then, compute the t_{cal} statistic as:

$$t_{cal} = \frac{(\bar{x}_1 - \bar{x}_2) - 0}{\sqrt{s_p^2\left(\dfrac{1}{n_1} + \dfrac{1}{n_2}\right)}} = -------$$

(v) Sketch the critical region.
(vi) State your decision.
(vii) Sketch and find the p-value.
(viii) Construct a 95% confidence interval estimate for the difference between the average scores of the old and the young instructor.

LUDI 9.18. (GREENHOUSE EXPERIMENTS ARE USEFUL)
Michael conducted an experiment in a greenhouse to study the effect of five different treatments for the number of days it takes to get flowers from special types of roses:

Fig. 9.108. Greenhouse.

Michael noted the number of days it took for each plant to show the first flower as given below:

Treatments	Data values						
I	160	172	170	180	189	200	150
II	140	170	156	178	187	201	154
III	201	180	156	160	170	145	176
IV	155	167	176	189	202	146	187
V	190	180	156	187	174	201	156

Apply the ANOVA approach to find if there is any significant difference among the treatments in growing these roses.

LUDI 9.19. (GOOD FEEDS HELP) Michael designed a special cage for four hens to count the number of eggs per month laid by each hen. He wants to know if the number of eggs depends upon the type of feed. In the market, he found two types of feed, feed A and feed B. Feed A has special kinds of stones and grains, while feed B has fish and grains.

Michael's Poultry Farm Experiment		Feed-A	
		Stones	Grains
Feed-B	Fish		
	Grains		

Fig. 9.109. Michael's hens.

After one month, Michael counted the following number of eggs by each one of the four hens:

Michael's eggs basket		Feed-A	
		Stones	Grains
Feed-B	Fish	30	25
	Grains	28	22

Apply the chi-square test to see whether there is any association between the two types of feed and state your views.

LUDI 9.20. (QUALITY CONTROL) It is believed that the precision (as measured by the variance) of a drill is no more than 10. Test at a 5% level of significance, given that the 10 observations taken from the drill have the values: 25, 24, 23, 27, 30, 32, 30, 20, 21 and 28.

Fig. 9.110. Drill.

LUDI 9.21. (FEMALES ARE GOOD GROCERY SHOPPERS)
Angie and Bob independently bought two baskets of apples, each
weighing 10 lbs and costing $7.00.

Fig. 9.111. Angie and Bob's conflict.

In the evening at home, they found that Angie's basket has 20 apples
whereas Bob's basket has 25 apples.

Weights (lbs) of Angie's apples					
	0.40	0.54	0.50	0.48	0.45
	0.60	0.53	0.54	0.43	0.56
	0.59	0.42	0.52	0.64	0.50
Angie's basket	0.50	0.53	0.42	0.53	0.32
Weights (lbs) of Bob's apples					
	0.35	0.25	0.50	0.35	0.25
	0.60	0.53	0.54	0.43	0.30
	0.45	0.42	0.30	0.35	0.25
	0.25	0.53	0.42	0.53	0.25
Bob's basket	0.20	0.53	0.35	0.62	0.45

Fig. 9.112. Angie and Bob's shopping.

Angie claims that on average her apples are better than Bob's apples,
making Bob's shopping not as good.
(a) Test her claim at a 5% level of significance and comment.
(b) Construct a 95% confidence interval estimate for the difference
between the average weights of two types of apples.

LUDI 9.22. (ANOVA IS VERY USEFUL IN AGRICULTURAL FIELD EXPERIMENTS) Bob bought 20 plots of land each of equal size. He used two plots to make his big farm house, one for a fishery farm, one for his machinery, one for having barbeques, one for animals, and the rest of the 14 plots he used for growing wheat grains.

N-10 kg	N-20	N-30	N-40
112	128	118	125
119	121	108	115
176	106		125
148	101		
126			

Fig. 9.113. Bob as a farmer.

Bob bought a new farm and wants to know if there is any difference if he uses different levels of nitrogen in the fertilizer to grow wheat grains. He decided to use 10 kg of nitrogen (N) in five plots, 20 kg of N in four plots, 30 kg of N in two plots, and 40 kg of N in three plots. At the end of the season, the yield from each plot was recorded as shown above. Carry out the analysis of variance and point out if there is any difference between the uses of different levels of nitrogen. Which principles of design of experiments seem to be violated by Bob?

LUDI 9.23. (CHICKEN LOVERS) Angie went to the market and bought 3 chickens. She noted their weights, and gave them a special feed for a period of 2 weeks and recorded the change in weights:

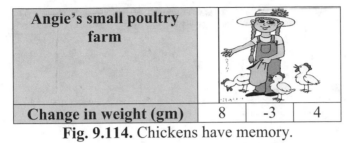

Angie's small poultry farm			
Change in weight (gm)	8	-3	4

Fig. 9.114. Chickens have memory.

Apply the paired t-test to find out if the gain in weight by the chickens after 2 weeks is significant at a 5% level. Also, find the p-value and interpret it.

LUDI 9.24. (COW MILK IS BEST FOR YOUR HEALTH) Angie and Bob bought two cows and recorded the milk in liters for eight days as shown below:

 Angie's Cow					 Bob's Cow			
4.3	3.6	3.2	4.7		3.5	4.5	4.9	3.8
3.2	3.8	4.5	3.7		3.9	4.6	4.7	3.7

Fig. 9.115. Angie and Bob's cows.

Angie wonders if her cow is as good as Bob's cow.

(a) Test her doubt at a 5% level of significance by comparing the average milk per day by both cows.
(b) Construct a 95% confidence interval estimate for the difference between the average milk per day.

LUDI 9.25. (USE OF STATISTICS IN SCIENCE) Consider 5 kinds of robots designed to plug certain kinds of screws into a satellite.

	 Robot-1	 Robot-2	 Robot-3	 Robot-4	 Robot-5
1	763	1335	596	3742	1632
2	4365	1262	1448	1833	5078
3	2144	217	1183	375	3010
4	1998	4100	3200	2010	671
5	5412	2948	630	743	2145

Fig. 9.116. Comparing robots.

Five robots of each model were randomly selected to test their efficiency of plugging screws into a satellite. The numbers of screws plugged in by different models of robots during an hour were counted. Test at a 5% level of significance if the average efficiency of all the robots is the same.

LUDI 9.26. (STUDENT LIFE NEVER COMES BACK) At a university, the students were classified in a large random sample according to their intelligence and economic conditions.

	 Rich	 Poor	Total
Excellent	50	80	
Good	200	180	
Mediocre	180	200	
Dumb	90	100	
Total			

Fig. 9.117. Student life is the best.

Apply the chi-square test to see whether or not there is any association between economic conditions and intelligence and state your views.

LUDI 9.27. (PEPPERONI PIZZA LOVERS) Bob went to market and bought 5 pigs. He recorded their weights and then fed them with a special diet for 6 months and recorded the change in weight:

Bob's small Pig farm					
Change in weight (lbs)	15	14	12	16	10

Fig. 9.118. Bob's pig farm.

Apply the paired t-test to find out if the gain in weight by the pigs after six months is significant at a 5% level. Also find the p-value and interpret it.

LUDI 9.28. (PLUMBERS NEED GOOD DRILLS) The precision of a drill is measured by the variance of measurements of a certain subject. For a hand drill in a sample of 11 data values of such subjects the sum of the

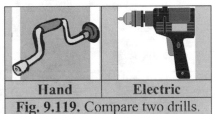

Hand	Electric

Fig. 9.119. Compare two drills.

square of deviations from the sample mean was 900. For an electric drill it was 2000 based on another random sample of 21 such subjects. Test whether or not both drills have the same precision at a 5% level of significance.

LUDI 9.29. (SPICY QUESTION) Consider the effects of four feeds on the quality of eggs laid by a sample of 48 chickens. Assuming that each chicken received only one kind of feed, complete the following ANOVA table:

Source	df	SS	MS	F
Feeds				14.71
Within Feeds			12.84	
Total				

Fig. 9.120. Chicken lovers.

LUDI 9.30. (ORCHARDS NEED ATTENTION) One day the gardener Monkey was away from the orchard and an oriole entered and dropped a huge number of apples from the trees.

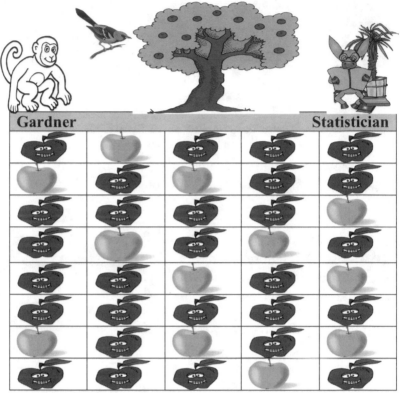

Fig. 9.121. A sample of apples dropped by Oriole.

Assume that a smiley face apple dropped by the oriole is not waste and can be easily sold in the market. The monkey asked the statistician if the damage is too much. The statistician reported to the monkey that the proportion of all the non-smiling apples in the whole orchard dropped by the oriole is less than 0.50, thus at this time he should not be worried. Based on a random sample of 40 apples shown above, use the 5% level of significance to test the statistician's claim by performing the appropriate test.

10. ANALYZING BIVARIATE DATA

10.1 INTRODUCTION

In this chapter, we introduce the concept of bivariate data, correlation, and regression analysis.

10.2 BIVARIATE DATA

If two variables are measured on each experimental unit, such data is called bivariate data. For example, consider that we select a sample of five students and record their GPA and the number of classes they missed.

Classes missed (x)	1	6	3	8	2	5
GPA (y)	3.80	2.00	3.00	2.00	3.50	2.50

Fig. 10.1. Missing classes is not fun.

Note that if the number of classes missed by a student increases then his/her GPA decreases. Also note that for each student we are measuring two variables: GPA and the number of missed classes. Such a data set is called bivariate data.

10.3 CORRELATION ANALYSIS

We consider here two methods to study the correlation between two quantitative variables:

(a) Graphical method: Scatter plot
(b) Numerical method: Correlation coefficient

Let us discuss each one of these methods as follows:

10.3.1 GRAPHICAL METHOD: SCATTER PLOT

Let x and y be the two quantitative variables. When each pair (x, y) of data values is plotted as a point on a two-dimensional graph the resulting graph is called a scatter plot.

Example 10.1.(GOOD COUPLES ARE RARE) The following data shows the number of household members, x, and the amount spent (in dollars) on groceries per week, y, measured for six households.

 x	2	2	3	4	1	5
y	40	60	60	100	35	130

Fig. 10.2. Bivariate data.

Draw a scatter plot of these data values.
Solution. The following graph shows the scatter plot of the number of household members and expenditures.

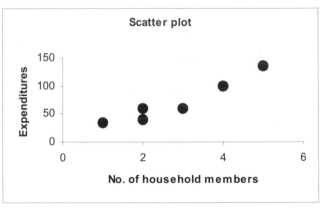

Fig. 10.3. Scatter plot.

The above scatter plot shows that as the number of household members increases, expenditures also increase. Thus, the relationship between the number of household members and expenditures can be considered positive.

A scatter plot can be used to find the nature of the linear relationship between two variables as follows:

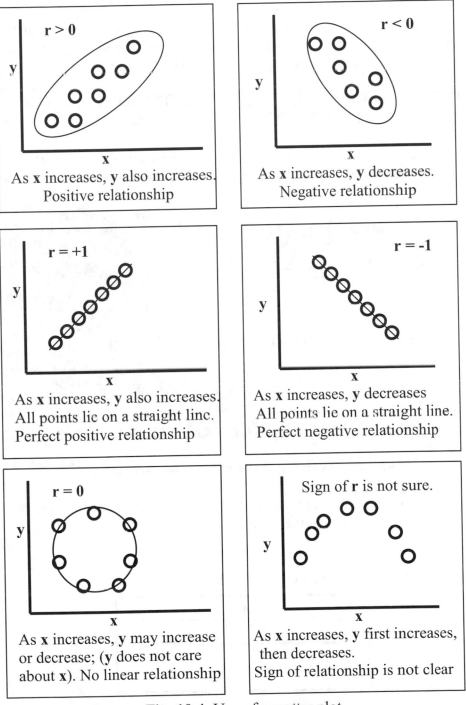

Fig. 10.4. Use of a scatter plot.

Thus, a scatter plot can be used to guess the sign of the linear relationship between two quantitative variables, say **x** and **y**, which may be negative, zero or positive. The relationship can be quantified with a measure called Pearson's correlation coefficient.

10.3.2 NUMERICAL METHOD: PEARSON'S CORRELATION COEFFICIENT

The linear relation between two variables can be quantified with the help of the Pearson's correlation coefficient, r, defined as:

$$r = \frac{n\left(\sum x_i y_i\right) - \left(\sum x_i\right)\left(\sum y_i\right)}{\sqrt{n\left(\sum x_i^2\right) - \left(\sum x_i\right)^2}\sqrt{n\left(\sum y_i^2\right) - \left(\sum y_i\right)^2}}$$

where

n = number of pairs (**x**, **y**) in the data set.

Fig. 10.5.
Prof. Karl Pearson
(1857-1936).
Source: Refer to Bibliography

10.3.3 PROPERTIES OF THE CORRELATION COEFFICIENT

The correlation coefficient, r, has the following characteristics:

(i) The value of the correlation coefficient, r, lies between negative one and positive one. Mathematically,

$$-1 \leq r \leq +1$$

or its pictorial representation is given by:

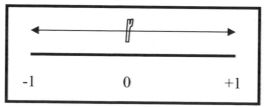

Fig. 10.6. Range of the value of the correlation coefficient.

(ii) The value of the correlation coefficient, r , is a pure number. It is free from the unit of measurement of the variables x and y.
(iii) The value of the correlation coefficient , r , is independent of the change of origin and of the scale of the original variables x and y.
(iv) The value of the correlation coefficient between x and y is the same as that between y and x.

Example 10.2. (POSITIVE CORRELATION) Consider 5 pairs of observations as given below:

x	10	20	30	40	50
y	5	25	30	35	45

(a) Construct a scatter plot and guess the sign of the Pearson's correlation coefficient
(b) Find the value of the correlation coefficient.
(c) Was your guess correct about the sign of the value of the correlation coefficient?
Solution. (a) Scatter plot and guess:

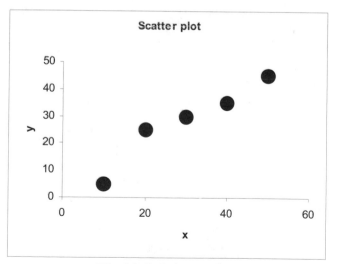

Fig.10.7. Scatter plot.

Guess: As the value of x increases the value of y also increases, so the relationship between them is positive. Thus, the value of the Pearson's correlation coefficient will be positive.

(b) Correlation coefficient: Let us make a five column table:

x_i	y_i	x_i^2	y_i^2	$x_i y_i$
10	5	100	25	50
20	25	400	625	500
30	30	900	900	900
40	35	1,600	1,225	1,400
50	45	2,500	2,025	2,250
Sum	**150**	**140**	**5,500**	**4,800**	**5,100**

From the above table:

$n = 5$, $\Sigma x_i = 150$, $\Sigma y_i = 140$, $\Sigma x_i^2 = 5,500$, $\Sigma y_i^2 = 4,800$, and $\Sigma x_i y_i = 5,100$.

Thus, the value of the correlation coefficient r is given by:

$$r = \frac{n(\Sigma x_i y_i) - (\Sigma x_i)(\Sigma y_i)}{\sqrt{n(\Sigma x_i^2) - (\Sigma x_i)^2}\sqrt{n(\Sigma y_i^2) - (\Sigma y_i)^2}}$$

$$= \frac{5 \times 5,100 - 150 \times 140}{\sqrt{5 \times 5,500 - 150^2}\sqrt{5 \times 4,800 - 140^2}}$$

$$= \frac{25,500 - 21,000}{\sqrt{27,500 - 22,500}\sqrt{24,000 - 19,600}}$$

$$= \frac{4,500}{\sqrt{5,000}\sqrt{4,400}} = \frac{4,500}{70.71 \times 66.33} = \frac{4,500}{4,690.19} = +0.959.$$

(c) Yes!, our guess of a positive relationship (or correlation coefficient) was correct.

In the next example, let us reverse one of the variables as follows:

Example 10.3. (NEGATIVE CORRELATION) Consider the 5 pairs of observations given below:

x	50	40	30	20	10
y	5	25	30	35	45

(a) Construct a scatter plot and guess the sign of the value of the correlation coefficient.
(b) Find the value of the correlation coefficient.
(c) Was your guess correct about the sign of the correlation coefficient?

Solution. (a) Scatter plot and guess:

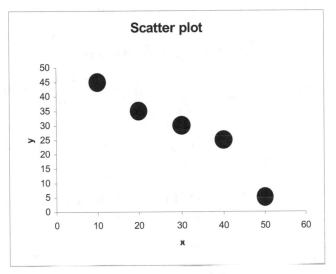

Fig. 10.8. Scatter plot.

Guess: Correlation coefficient is negative.

(b) Correlation coefficient: Let us make a five column table as:

x_i	y_i	x_i^2	y_i^2	$x_i y_i$
50	5	2,500	25	250
40	25	1,600	625	1,000
30	30	900	900	900
20	35	400	1,225	700
10	45	100	2,025	450
Sum 150	140	5,500	4,800	3,300

From the above table:

$n = 5,\ \Sigma x_i = 150,\ \Sigma y_i = 140,\ \Sigma x_i^2 = 5{,}500,\ \Sigma y_i^2 = 4{,}800,$ and $\Sigma x_i y_i = 3{,}300.$

Thus, the value of the correlation coefficient r is given by:

$$r = \frac{n(\Sigma x_i y_i) - (\Sigma x_i)(\Sigma y_i)}{\sqrt{n(\Sigma x_i^2) - (\Sigma x_i)^2}\sqrt{n(\Sigma y_i^2) - (\Sigma y_i)^2}}$$

$$= \frac{5 \times 3,300 - 150 \times 140}{\sqrt{5 \times 5,500 - 150^2}\sqrt{5 \times 4,800 - 140^2}}$$

$$= \frac{16,500 - 21,000}{\sqrt{27,500 - 22,500}\sqrt{24,000 - 19,600}}$$

$$= \frac{-4,500}{\sqrt{5,000}\sqrt{4,400}} = \frac{-4,500}{70.71 \times 66.33} = \frac{-4,500}{4,690.19} = -0.959.$$

(c) Yes!, our guess of a negative relationship (or correlation coefficient) was correct.

Example 10.4. (PERFECT POSITIVE CORRELATION)
Consider the 5 pairs of observations given below:

x	10	20	30	40	50
y	5	10	15	20	25

(a) Construct a scatter plot and guess the sign of the value of the correlation coefficient.
(b) Find the value of the correlation coefficient.
(c) Was your guess correct?
Solution. (a) Scatter plot and guess:

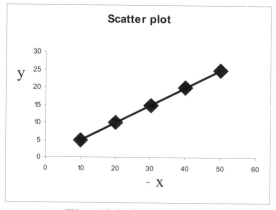

Fig. 10.9. Scatter plot.

Guess from the scatter plot: As x increases, y also increases and all points lie on a straight line, thus a perfect positive relationship is expected.

(b) Correlation coefficient: Let us make a five column table:

x_i	y_i	x_i^2	y_i^2	$x_i y_i$
10	5	100	25	50
20	10	400	100	200
30	15	900	225	450
40	20	1,600	400	800
50	25	2,500	625	1,250
Sum **150**	**75**	**5,500**	**1,375**	**2,750**

From the above table:

$n = 5$, $\Sigma x_i = 150$, $\Sigma y_i = 75$, $\Sigma x_i^2 = 5{,}500$, $\Sigma y_i^2 = 1{,}375$, and $\Sigma x_i y_i = 2{,}750$.

Thus, the value of the correlation coefficient, r, is given by:

$$r = \frac{n\left(\Sigma x_i y_i\right) - \left(\Sigma x_i\right)\left(\Sigma y_i\right)}{\sqrt{n\left(\Sigma x_i^2\right) - \left(\Sigma x_i\right)^2}\,\sqrt{n\left(\Sigma y_i^2\right) - \left(\Sigma y_i\right)^2}}$$

$$= \frac{5 \times 2{,}750 - 150 \times 75}{\sqrt{5 \times 5{,}500 - 150^2}\,\sqrt{5 \times 1{,}375 - 75^2}}$$

$$= \frac{13{,}750 - 11{,}250}{\sqrt{27{,}500 - 22{,}500}\,\sqrt{6{,}875 - 5{,}625}}$$

$$= \frac{2{,}500}{\sqrt{5{,}000}\,\sqrt{1{,}250}} = \frac{2{,}500}{\sqrt{6{,}250{,}000}} = \frac{2{,}500}{2{,}500} = +1.00 .$$

(c) Yes!, our guess of a positive perfect relationship (or correlation coefficient, $r = +1.00$) was correct.

In the next example let us again reverse one of the variables as follows:

Example 10.5. (PERFECT NEGATIVE CORRELATION)
Consider the 5 pairs of observations as given below:

x	50	40	30	20	10
y	5	10	15	20	25

(a) Construct a scatter plot and guess the sign of the value of the correlation coefficient.
(b) Find the value of the correlation coefficient.
(c) Was your guess correct about the sign of the correlation coefficient?
Solution. (a) Scatter diagram and guess:

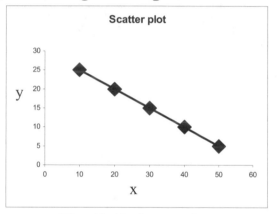

Fig. 10.10. Scatter plot.

Guess from the scatter plot: As x increases, y decreases and all points lie on a straight line, therefore a perfect negative relationship is expected.
(b) **Correlation coefficient:** Let us make a five column table:

x_i	y_i	x_i^2	y_i^2	$x_i y_i$
50	5	2,500	25	250
40	10	1,600	100	400
30	15	900	225	450
20	20	400	400	400
10	25	100	625	250
Sum 150	75	5,500	1,375	1,750

From the above table:

$n = 5, \ \Sigma x_i = 150, \ \Sigma y_i = 75, \ \Sigma x_i^2 = 5,500, \ \Sigma y_i^2 = 1,375, \ \text{and} \ \Sigma x_i y_i = 1,750.$

Thus, the value of the correlation coefficient, r, is given by:

$$r = \frac{n\left(\Sigma x_i y_i\right) - \left(\Sigma x_i\right)\left(\Sigma y_i\right)}{\sqrt{n\left(\Sigma x_i^2\right) - \left(\Sigma x_i\right)^2} \ \sqrt{n\left(\Sigma y_i^2\right) - \left(\Sigma y_i\right)^2}}$$

$$= \frac{5 \times 1,750 - 150 \times 75}{\sqrt{5 \times 5,500 - 150^2} \ \sqrt{5 \times 1,375 - 75^2}}$$

$$= \frac{8,750 - 11,250}{\sqrt{27,500 - 22,500} \ \sqrt{6,875 - 5,625}}$$

$$= \frac{-2,500}{\sqrt{5,000} \ \sqrt{1,250}} = \frac{-2,500}{\sqrt{6,250,000}} = \frac{-2,500}{2,500} = -1.00.$$

(c) Yes!, our guess of a perfect negative relationship (or correlation coefficient, $r = -1.00$) was correct.

Remark: Note that if we reverse the direction of one of the variables, then the sign of the correlation coefficient also reverses, but the magnitude of the correlation coefficient may increase, decrease or remain the same. To confirm this, solve the following question: Consider the 5 pairs of observations given below:

x	0	10	60	30	50
y	40	30	0	20	10

(a) Construct a scatter plot and guess the sign of the correlation coefficient. Find the value of the correlation coefficient.

(b) Now reverse one of the variables as given below:

x	50	30	60	10	0
y	40	30	0	20	10

Compare the sign and magnitude of the correlation coefficient in both cases. Do it yourself!

Example 10.6. (ZERO CORRELATION) Consider the 5 pairs of observations given below:

x	10	10	30	50	50
y	5	25	15	5	25

(a) Construct a scatter plot and guess the sign of the value of the correlation coefficient.
(b) Find the value of the correlation coefficient.
(c) Was your guess correct?
Solution. (a) Scatter plot and guess:

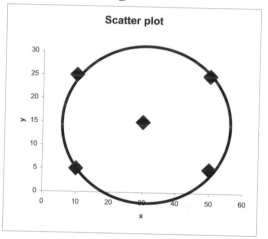

Fig. 10.11. Scatter plot.

Guess from the scatter plot: As x increases, y sometime increases, sometime decreases and the points form a circle instead of a straight line, so the linear relationship seems to be zero. Note that there may be a circular relationship, but not a linear relationship.

(b) Correlation coefficient: Let us make a five column table:

x_i	y_i	x_i^2	y_i^2	$x_i y_i$
10	5	100	25	50
10	25	100	625	250
30	15	900	225	450
50	5	2,500	25	250
50	25	2,500	625	1,250
Sum 150	75	6,100	1,525	2,250

From the above table:

$n = 5$, $\Sigma x_i = 150$, $\Sigma y_i = 75$, $\Sigma x_i^2 = 6{,}100$, $\Sigma y_i^2 = 1{,}525$, and $\Sigma x_i y_i = 2{,}250$.

Thus, the value of the correlation coefficient r is given by:

$$
\begin{aligned}
r &= \frac{n\left(\Sigma x_i y_i\right) - \left(\Sigma x_i\right)\left(\Sigma y_i\right)}{\sqrt{n\left(\Sigma x_i^2\right) - \left(\Sigma x_i\right)^2}\ \sqrt{n\left(\Sigma y_i^2\right) - \left(\Sigma y_i\right)^2}} \\[2mm]
&= \frac{5 \times 2{,}250 - 150 \times 75}{\sqrt{5 \times 6{,}100 - 150^2}\ \sqrt{5 \times 1{,}525 - 75^2}} \\[2mm]
&= \frac{11{,}250 - 11{,}250}{\sqrt{30{,}500 - 22{,}500}\ \sqrt{7{,}625 - 5{,}625}} \\[2mm]
&= \frac{0}{\sqrt{8{,}000}\ \sqrt{2{,}000}} = \frac{0}{89.44 \times 44.72} = \frac{0.0}{3{,}999.76} = 0.0.
\end{aligned}
$$

(c) Yes!, our guess of no linear relationship (or the correlation coefficient, $r = 0.0$) was correct.

10.4 REGRESSION ANALYSIS

Sometimes the variables x and y are related in a particular way. It may be that the value of y depends on the value of x.

For example:

(a) The yield (y) of a crop may depend upon the amount of a particular type of fertilizer used (x).

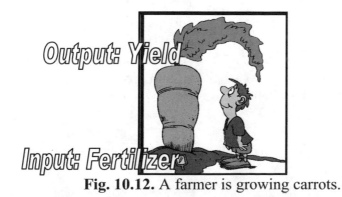

Fig. 10.12. A farmer is growing carrots.

Note that the use of fertilizer is an input and yield is an output, and both fertilizer and yield vary from plot to plot. The use of less or more fertilizer is in our hands, but the output yield is not in our hands. Fertilizer is called an input variable and the yield is called an output variable.

(**b**) Grade point average (y) of a student may depend on his/her number of study hours/week.

Fig. 10.13. Apple is studying to get an A^+ grade.

Note that the number of study hours is an input and the grade point average (GPA) is an output, and both study hours and GPA vary from student to student. The use of less or more study hours is in the student's hands, but the output GPA is not in his/her hands.
In such situations:

> y is called an output variable or dependent variable,
> and
> x is called an input variable or independent variable.

Note that the input variable is also called the auxiliary variable, explanatory variable, cause variable, treatment, or factor. Further, note that the output variable is also called the effect variable, response variable or study variable.

The input variable is in our hands and can be manipulated, whereas the output variable cannot be directly manipulated. For example, a student can adjust study hours/week, but cannot adjust the GPA.

The following flow chart will be helpful in remembering the names of input and output variables in the regression analysis.

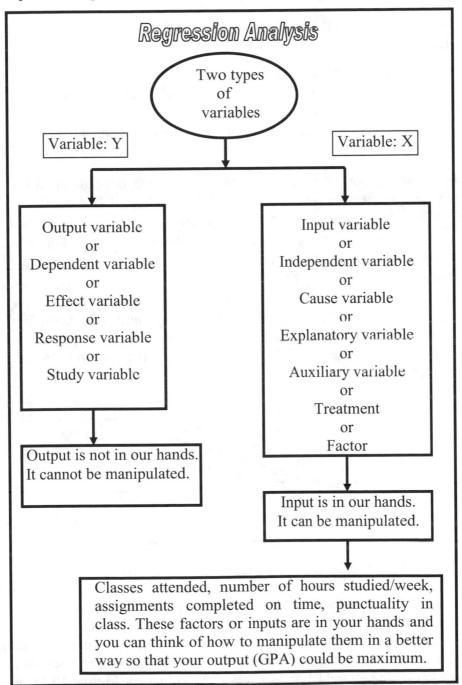

Fig. 10.14. Names of input and output variables.

The dependent variable y and independent variable x are related to each other through a linear regression model given by:

$$y_i = a + bx_i + e_i$$

where

$a =$ intercept with the y-axis
    $=$ value of output (y) when the value of input (x) is zero.
$b =$ slope $=$ regression coefficient
    $=$ change in the output (y) for a unit change in the input (x)
$e_i =$ error term

For example, consider that the GPA of a student depends upon the number of hours he/she spends studying at home. If the student spends no hours studying at home, then the input is $x = 0$ hours, and the output GPA is $y_i = a =$ Intercept by assuming $e_i = 0$. This is a reasonable assumption because the student might have learned something on campus (Assuming the intercept a is positive). Sometimes, the value of a is also called a *constant*. Assume the relationship is positive between GPA and the number of hours per week spent studying at home. Then, if a student starts spending 6 hours per week instead of 5 hours per week, the increase in GPA will be given by the value of the slope, b.

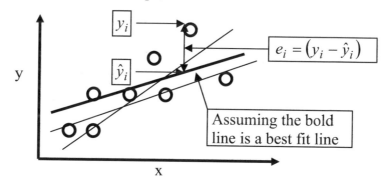

Fig. 10.15. Which line is the best fit?

Thus, note that the value of slope b can be positive, negative or zero. It will be zero if there is no change in GPA after adding or subtracting an hour of study time per week at home. Clearly, the graphical method can be used to find the intercept a and slope b if the exact position of the straight line is known. As shown in **Figure 10.15**, a large number of lines can be passed through the scatter plot, and it is difficult to find which one is the best fit. So the graphical method is not accurate. The vertical distance between a real point y_i and a point on the best fit line $\hat{y}_i$ for a given value of x_i is called an error term. Note that some points are above, some are below, and others fall on the best fit line. Thus, the error term can be positive, negative or zero. **The best fit or least squares line can be obtained by passing a line across the scatter plot such that the sum of all the error terms is zero and the sum of the squares of the error terms is minimum.** Such a **best fit** line can be found by calculating the slope (or regression coefficient) and intercept with the formulae given by:

$$\text{Regression Coefficient} = \text{Slope} = b = \frac{n(\Sigma x_i y_i) - (\Sigma x_i)(\Sigma y_i)}{n(\Sigma x_i^2) - (\Sigma x_i)^2}$$

and

$$\text{Constant} = \text{Intercept} = a = \bar{y} - b\bar{x}$$

where

$$\bar{y} = \frac{1}{n}\sum_{i=1}^{n} y_i \text{ and } \bar{x} = \frac{1}{n}\sum_{i=1}^{n} x_i$$

Therefore, the best fit line is given by:

$$\boxed{\text{BEST FIT LINE: } \hat{y}_i = a + bx_i}$$

The predicted value of the dependent or output variable y_i for a given value of input variable x_i^* is given by:

$$\hat{y}_i = a + bx_i^*$$

The error term is the difference between the real value and predicted value of the dependent variable and is given by:

$$e_i = (y_i - \hat{y}_i)$$

10.4.1 SOME FACTS ABOUT THE BEST FIT LINE

(a) If x increases and y also increases, the slope b will be positive.
(b) If x increases and y decreases, the slope b will be negative.

(c) If x changes (increases or decreases) and the value of y remains the same, the value of slope b will be zero.

(d) The value of slope b is a real number from minus infinity to plus infinity, that is, it can be positive, negative, or zero.

(e) The value of intercept a is a real number from minus infinity to plus infinity, that is, it can be positive, negative or zero.

(f) The regression line always passes through the point $(\bar{x}, \bar{y})$.

(g) $\Sigma e_i = 0$, Σe_i^2 is minimum and $\Sigma y_i = \Sigma \hat{y}_i$.

A pictorial representation of the above linear model is shown below:

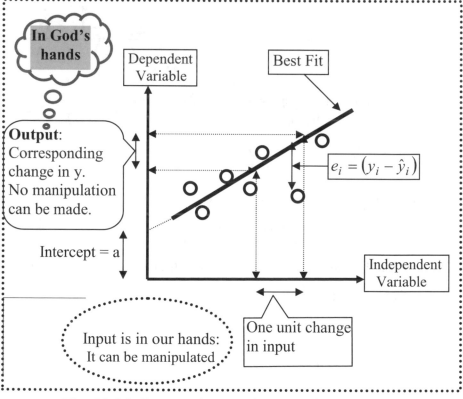

Fig. 10.16. Input and output in regression analysis.

From the above figure, the slope of the regression line is given by:

$$\text{Regression coefficient} = \text{Slope} = b = \frac{\text{Change in } y}{\text{Change in } x} = \frac{\text{rise}}{\text{run}}$$

Example 10.7. (SCHEDULE YOUR STUDY HOURS EVERY DAY) Consider the following data of GPA and the number of study hours/week a student spends at home:

Do your home work						
GPA	2.00	2.30	2.50	3.00	3.50	3.90
Hours	0	1	2	3	5	6

Fig. 10.17. Study hours versus GPA.

(a) Draw a scatter plot, and denote the dependent variable with y and the independent variable with x. Guess the sign of the slope.
(b) Find the best fit line.
(c) Interpret the intercept and regression coefficient (or slope).
(d) Predict the GPA if a student spends 6.25 hours per week at home (**Extrapolation**: Prediction outside the given data set, as we forecast weather).
(e) Predict the GPA if a student spends 4 hours per week at home (**Interpolation**: Prediction within the given data set).
(f) Draw the predicted line on the scatter plot.
Solution. (a) Scatter plot: Here the dependent variable is GPA and the independent variable is hours, so $y = GPA$ and $x = Hours$.

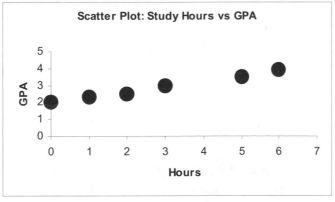

Fig. 10.18. Make a study group of good friends.

Note that as x (hours) increases the value of y (GPA) also increases. Thus, the sign of slope b will be positive.

(**b**) To find the **best fit** line, we have to make a five column table as:

x_i	y_i	x_i^2	y_i^2	$x_i y_i$
0	2.00	0	4.00	0.00
1	2.30	1	5.29	2.30
2	2.50	4	6.25	5.00
3	3.00	9	9.00	9.00
5	3.50	25	12.25	17.50
6	3.90	36	15.21	23.40
Sum **17**	**17.20**	**75**	**52.00**	**57.20**

From the above table, we have:

$n = 6$, $\Sigma x_i = 17$, $\Sigma y_i = 17.20$, $\Sigma x_i^2 = 75$, $\Sigma y_i^2 = 52.00$, and $\Sigma x_i y_i = 57.20$

Thus, the value of slope is given by:

$$\text{Regression coefficient} = \text{Slope} = b = \frac{n\left(\Sigma x_i y_i\right) - \left(\Sigma x_i\right)\left(\Sigma y_i\right)}{n\left(\Sigma x_i^2\right) - \left(\Sigma x_i\right)^2}$$

$$= \frac{6 \times 57.20 - 17 \times 17.20}{6 \times 75 - 17^2}$$

$$= \frac{343.20 - 292.40}{450 - 289} = \frac{50.80}{161} = 0.32.$$

Now

$$\bar{y} = \frac{1}{n}\sum_{i=1}^{n} y_i = \frac{1}{6} \times 17.20 = 2.87 \text{ and } \bar{x} = \frac{1}{n}\sum_{i=1}^{n} x_i = \frac{1}{6} \times 17 = 2.83.$$

Thus,

$$\text{Intercept} = a = \bar{y} - b\bar{x}$$

$$= 2.87 - 0.32 \times 2.83$$

$$= 2.87 - 0.91$$

$$= 1.96.$$

Therefore,

$$\boxed{\textbf{BEST FIT: } \hat{y}_i = a + bx_i = 1.96 + 0.32\, x_i}$$

(**c**) **Interpretation of the intercept:** If a student spends zero study hours per week at home, then his/her GPA will be: $a = 1.96$.

Interpretation of the slope: If a student increases one study hour per week, then the change in GPA will be: $+0.32$.

(d) The predicted value of the dependent or output variable y_i for a given value of the input variable x_i^* is given by:

$$\hat{y}_i = 1.96 + 0.32\, x_i^*.$$

Extrapolation: If a student uses $x_i^* = 6.25$ hours, then his/her predicted GPA will be:

$$\hat{y}_i = 1.96 + 0.32 \times 6.25 = 1.96 + 2.00 = 3.96.$$

Caution! Sometimes too much extrapolation is dangerous. For example, if a student starts spending, $x_i^* = 10$ hours/week, then the predicted GPA will be:

$$\hat{y}_i = 1.96 + 0.32 \times 10 = 1.96 + 3.20 = 5.16$$

We know that GPA cannot be more than 4.00. So, be careful while doing extrapolation. This is the reason, that it is difficult to predict our daily weather for a period of more than a week or a month etc.!!

(e) **Interpolation:** If a student uses $x_i^* = 4$ hours, then his/her predicted GPA will be:

$$\hat{y}_i = 1.96 + 0.32 \times 4 = 1.96 + 1.28 = 3.24.$$

Note that interpolation is not as dangerous as extrapolation. Most of the time these models work well for interpolation.

(f) **It is tricky:** To plot the predicted line, pick up any two values of the independent variable (x) from the given data set, say minimum value (x_{min}) and maximum value (x_{max}). Then predict the corresponding values of the dependent variable as:

$$\hat{y}_{min} = a + bx_{min} \quad \text{and} \quad \hat{y}_{max} = a + bx_{max}$$

Mark the points $P_1 \leftrightarrow (x_{min}, \hat{y}_{min})$ and $P_2 \leftrightarrow (x_{max}, \hat{y}_{max})$ with square boxes on the scatter plot and join with a straight line.
In our data set:

$$x_{min} = 0, \text{ so } \hat{y}_{min} = a + bx_{min} = 1.96 + 0.32 \times 0 = 1.96$$
$$x_{max} = 6, \text{ so } \hat{y}_{max} = a + bx_{max} = 1.96 + 0.32 \times 6 = 3.88$$

So, to draw the best fit or the predicted line, join the points $P_1 \leftrightarrow (x_{min}, \hat{y}_{min}) = (0, 1.96)$ and $P_2 \leftrightarrow (x_{max}, \hat{y}_{max}) = (6, 3.88)$ with a straight line on the scatter plot as shown in **Figure 10.19**:

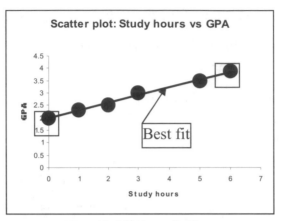

Fig. 10.19. Best fit line.

Example 10.8. (NEVER MISS YOUR CLASS, BECAUSE YOU PAY FOR IT) Consider the following data of GPA and the number of classes missed by the students during a semester:

Classes missed (x)	1	6	3	8	2	5
GPA (y)	3.80	2.00	3.00	2.00	3.50	2.50

Fig. 10.20. Missing classes is not fun.

(a) Draw a scatter plot, and denote the dependent (or output) variable with y and the independent (or input) variable with x. Guess the sign of the slope.
(b) Find the best fit line.
(c) Interpret intercept and slope.
(d) Predict the GPA if a student will miss 9 classes per semester. (**Extrapolation**: Prediction outside the given data set).
(e) Predict the GPA if a student will miss 7 classes per semester. (**Interpolation**: Prediction within the given data set).
(f) Plot the predicted line.
Solution. (a) We know that a grade depends upon the number of classes attended, so the dependent variable is GPA $= y$ and the independent variable is number of classes missed $= x$.

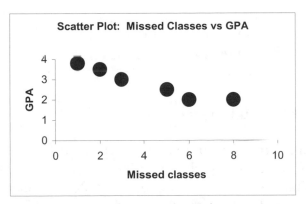

Fig. 10.21. Scatter plot.

From the scatter plot, we found that as the independent variable, number of missed classes = x increases, the dependent variable GPA = y decreases, therefore the sign of slope b will be negative.

(b) To find the best fit line, we have to make a five column table as:

x_i	y_i	x_i^2	y_i^2	$x_i y_i$
1	3.80	1	14.44	3.80
6	2.00	36	4.00	12.00
3	3.00	9	9.00	9.00
8	2.00	64	4.00	16.00
2	3.50	4	12.25	7.00
5	2.50	25	6.25	12.50
Sum	**25**	**16.80**	**139**	**49.94**	**60.30**

From the above table, we have:

$n = 6$, $\Sigma x_i = 25$, $\Sigma y_i = 16.80$, $\Sigma x_i^2 = 139$, $\Sigma y_i^2 = 49.94$, and $\Sigma x_i y_i = 60.30$

Thus, the value of the regression coefficient (or slope) is given by:

$$\text{Slope} = b = \frac{n\left(\Sigma x_i y_i\right) - \left(\Sigma x_i\right)\left(\Sigma y_i\right)}{n\left(\Sigma x_i^2\right) - \left(\Sigma x_i\right)^2} = \frac{6 \times 60.30 - 25 \times 16.80}{6 \times 139 - 25^2}$$

$$= \frac{361.80 - 420.00}{834 - 625} = \frac{-58.20}{209} = -0.28.$$

Now

$$\bar{y} = \frac{1}{n}\sum_{i=1}^{n} y_i = \frac{1}{6} \times 16.80 = 2.80 \text{ and } \bar{x} = \frac{1}{n}\sum_{i=1}^{n} x_i = \frac{1}{6} \times 25 = 4.17.$$

Thus,

$$\text{Intercept} = a = \bar{y} - b\bar{x}$$
$$= 2.80 - (-0.28) \times 4.17$$
$$= 2.80 + 1.17$$
$$= 3.97.$$

Therefore,

BEST FIT: $\hat{y}_i = a + bx_i = 3.97 - 0.28 \, x_i$

(c) **Interpret intercept and slope:** Intercept $a = 3.97$ says that if a student does not miss any class then his/her predicted GPA is 3.97. The regression coefficient (or slope) $b = -0.28$ says that if a student misses one class, then his/her GPA will decrease by 0.28.

(d) **Extrapolation:** The predicted value of the dependent or output variable y_i for a given value of input variable x_i^* is:

$$\hat{y}_i = 3.97 - 0.28 \, x_i^*$$

If a student will miss $x_i^* = 9$ classes, his/her predicted GPA will be:

$$\hat{y}_i = 3.97 - 0.28 \times 9 = 3.97 - 2.52 = 1.45.$$

(this student may be dropped!)

(e) **Interpolation:** If a student misses $x_i^* = 7$ classes, then his/her predicted GPA will be:

$$\hat{y}_i = 3.97 - 0.28 \times 7 = 3.97 - 1.96 = 2.01.$$

Thus, this regression analysis recommends that do not miss any class.

(f) **It is tricky:** To plot the predicted line, pick up any two values of the independent variable (x) from the given data set, say minimum value (x_{min}) and maximum value (x_{max}). Note the slope is negative, thus predict the corresponding values of the dependent variable as:

$$\hat{y}_{max} = a + bx_{min} \quad \text{and} \quad \hat{y}_{min} = a + bx_{max}$$

Mark the points $P_1 \leftrightarrow (x_{min}, \hat{y}_{max})$ and $P_2 \leftrightarrow (x_{max}, \hat{y}_{min})$ with square boxes on the scatter plot and join them with a straight line. In our data set:

$$x_{min} = 1, \text{ so } \hat{y}_{max} = a + bx_{min} = 3.97 - 0.28 \times 1 = 3.69$$
$$x_{max} = 8, \text{ so } \hat{y}_{min} = a + bx_{max} = 3.97 - 0.28 \times 8 = 1.73$$

So to plot the best fit line join the points $P_1 \leftrightarrow (x_{min}, \hat{y}_{max}) = (1, 3.69)$ and $P_2 \leftrightarrow (x_{max}, \hat{y}_{min}) = (8, 1.73)$ on the scatter plot as shown in **Figure 10.22**:

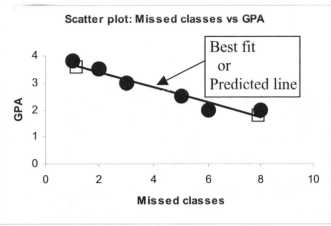

Fig. 10.22. Best fit line.

10.5 RELATION BETWEEN Z-SCORE AND THE CORRELATION COEFFICIENT

Let x_i and y_i, $i = 1,2,...,n$ be the two variables measured on the ith experimental unit. Obviously, their respective sample means are:

$$\bar{x} = \frac{1}{n} \sum_{i=1}^{n} x_i, \text{ and } \bar{y} = \frac{1}{n} \sum_{i=1}^{n} y_i$$

and the respective sample variances are given by:

$$s_x^2 = \frac{\sum_{i=1}^{n}(x_i - \bar{x})^2}{n-1} \text{ and } s_y^2 = \frac{\sum_{i=1}^{n}(y_i - \bar{y})^2}{n-1}$$

Thus, the sample standard deviations of the variables x and y respectively, are given by:

$$s_x = \sqrt{s_x^2} \text{ and } s_y = \sqrt{s_y^2}$$

The sample Z-scores or sample standardized Z-scores of the ith unit for the variables x and y are, respectively, given by:

$$Z_{x_i} = \frac{(x_i - \bar{x})}{s_x} \text{ and } Z_{y_i} = \frac{(y_i - \bar{y})}{s_y}$$

Then the correlation coefficient r can be obtained by:

$$r = \frac{\sum\limits_{i=1}^{n} Z_{x_i} Z_{y_i}}{n-1}$$

or equivalently:

$$r = \frac{1}{n-1} \sum\limits_{i=1}^{n} \left(\frac{x_i - \bar{x}}{s_x} \right) \left(\frac{y_i - \bar{y}}{s_y} \right)$$

In other words, if values of the Z-scores for all units in a sample are known about two characteristics, the value of the correlation coefficient can be obtained by dividing the sum of the product of these Z-scores with $(n-1)$.

10.6 RELATION BETWEEN CORRELATION AND REGRESSION COEFFICIENTS

If the value of the correlation coefficient is known, then it can be used to find the value of slope (or regression coefficient) as:

$$b = r \frac{s_y}{s_x}$$

and intercept is given by:

$$a = \bar{y} - b\bar{x}$$

The best fit line or predicted line is given by:

$$\hat{y}_i = a + bx_i$$

Thus, the error term in the ith observed value and predicted value is given by:

$$e_i = y_i - \hat{y}_i$$

These error terms are useful in determining a very useful measure or tool in regression analysis called the coefficient of determination given by:

$$r^2 = 1 - \frac{\sum\limits_{i=1}^{n} e_i^2}{\sum\limits_{i=1}^{n} (y_i - \bar{y})^2}$$

More discussion on the coefficient of determination is beyond the scope of this book.

Example 10.9. (COMPARE YOUR HEIGHT WITH YOUR MOM'S HEIGHT) The following table gives the age (in months) and height (in cms) for a sample of five children.

Age (x)	18	20	25	28	29
Height (y)	76	78	81	82	83

Fig. 10.23. Measuring height.

(a) Draw a scatter plot. Guess the sign of the correlation coefficient, r, and the regression coefficient b.
(b) Find the Z-score of each child based on age.
(c) Find the Z-score for each child based on height.
(d) Find the value of the correlation coefficient and compare its sign with your guess in (a).
(e) Find the best fit line. Interpret intercept and slope.
(f) Predict the height of a child of 32 months. Is it interpolation or extrapolation? Can we predict the height of a child after 1,200 months or say 100 years? Comment on such a prediction.
(g) Predict the height of a child of 26 months. Is it interpolation or extrapolation?
(h) Draw a predicted line on the scatter plot.
(i) Find the error term corresponding to each child in the sample. Use these error terms to find the value of the coefficient of determination, and interpret it.

Solution. (a) Scatter plot:

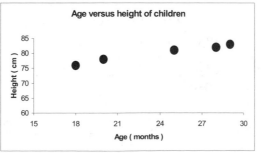

Fig. 10.24. Guessing signs.

As age increases, the height of a child increases. Thus, the signs of both the correlation coefficient and the regression coefficient (or slope) will be positive.

(b) Z-score of each child based on age (x):

Age (x_i) months	$(x_i - \bar{x})$ months	$(x_i - \bar{x})^2$ months^2	$Z_{x_i} = \dfrac{(x_i - \bar{x})}{s_x}$ no unit
18	$(18 - 24) = -6$	$(-6)^2 = 36$	$-6/4.85 = -1.24$
20	$(20 - 24) = -4$	$(-4)^2 = 16$	$-4/4.85 = -0.82$
25	$(25 - 24) = +1$	$(+1)^2 = 1$	$+1/4.85 = +0.21$
28	$(28 - 24) = +4$	$(+4)^2 = 16$	$+4/4.85 = +0.82$
29	$(29 - 24) = +5$	$(+5)^2 = 25$	$+5/4.85 = +1.03$
Sum 120	0	94	0.00

In the above table, we used:

$$\bar{x} = \frac{\sum\limits_{i=1}^{n} x_i}{n} = \frac{120}{5} = 24 \text{ months}, \quad s_x^2 = \frac{\sum\limits_{i=1}^{n}(x_i - \bar{x})^2}{n-1} = \frac{94}{4} = 23.5 \text{ months}^2, \text{ and}$$

$s_x = \sqrt{23.5} = 4.85 \text{ months}$.

(c) Z-score of each child based on height (y):

Height (y_i) cms	$(y_i - \bar{y})$ cms	$(y_i - \bar{y})^2$ cms^2	$Z_{y_i} = \dfrac{(y_i - \bar{y})}{s_y}$ no unit
76	$(76 - 80) = -4$	$(-4)^2 = 16$	$-4/2.92 = -1.37$
78	$(78 - 80) = -2$	$(-2)^2 = 4$	$-2/2.92 = -0.68$
81	$(81 - 80) = +1$	$(+1)^2 = 1$	$+1/2.92 = +0.34$
82	$(82 - 80) = +2$	$(+2)^2 = 4$	$+2/2.92 = +0.68$
83	$(83 - 80) = +3$	$(+3)^2 = 9$	$+3/2.92 = +1.03$
Sum 400	0	34	0.00

In the above table, we used:

$$\bar{y} = \frac{\sum\limits_{i=1}^{n} y_i}{n} = \frac{400}{5} = 80 \text{ cms}, \quad s_y^2 = \frac{\sum\limits_{i=1}^{n}(y_i - \bar{y})^2}{n-1} = \frac{34}{4} = 8.5 \text{ cms}^2,$$

and $s_y = \sqrt{8.5} = 2.92 \text{ cms}$.

(**d**) The value of the correlation coefficient (r):

	Z_{x_i}	Z_{y_i}	$Z_{x_i} \times Z_{y_i}$
	-1.24	-1.37	$(-1.24)(-1.37) = 1.70$
	-0.82	-0.68	$(-0.82)(-0.68) = 0.56$
	+0.21	+0.34	$(+0.21)(+0.34) = 0.07$
	+0.82	+0.68	$(+0.82)(+0.68) = 0.56$
	+1.03	+1.03	$(+1.03)(+1.03) = 1.06$
Sum	**0.00**	**0.00**	**3.95**

Thus, the value of the correlation coefficient (r) is given by:

$$r = \frac{1}{n-1} \sum_{i=1}^{n} Z_{x_i} Z_{y_i} = \frac{1}{5-1} \times 3.95 \approx 0.99 \quad \textbf{(no unit)}$$

Thus, the sign of the correlation coefficient (r) is the same as the guess from the scatter plot.

(**e**) **Best fit line:** The value of the slope (or regression coefficient) is:

$$b = r\frac{S_y}{S_x} = 0.99 \times \frac{2.92}{4.85} \approx +0.59 \text{ cms/month}$$

and, the value of the intercept is given by:

$$a = \bar{y} - bx = 80 - (0.59)(24) = 65.84 \text{ cms}.$$

Note that the sign of the slope is the same as the guess from the scatter plot. Now the best fit line is given by:

$$\hat{y}_i = a + bx_i = 65.84 + 0.59x_i$$

or equivalently we model height on age as:

$$\text{Height} = 65.84 + 0.59(\text{Age})$$

Here $a = 65.84$ cms means that if age $= 0$ (*i.e.,* at birth time), the expected height of a child is 65.84 cms, and $b = 0.59$ cms/month means the height of a child increases by 0.59 cms each month.

(**f**) If age $= 32$ months, then the expected height of the child is:

$$\text{Height} = 65.84 + 0.59(32) = 84.72 \text{ cms}.$$

This is called extrapolation.

Note that if a child lives for 100 years, or 1,200 months, then the predicted heights is:

$$\text{Height} = 65.84 + 0.59(1,200) = 773.84 \text{ cms} \approx 25.38 \text{ feet}$$

which is not feasible. It looks like this model cannot be used for predicting the age of a child after 100 years.

Moreover, a child will not be a child after 100 years. We say, "Statistics is a collection of tools used for analyzing data." Experience and practice are important in making a decision to choose a statistical tool to analyze data in real life. For example, an axe may not cut an iron rod. Choosing an axe as a tool to cut an iron rod is a wrong decision.

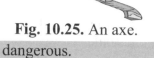

Fig. 10.25. An axe.

Caution! Too much extrapolation is dangerous.

(**g**) If age = 26 months, the expected height of the child is:

$$\text{Height} = 65.84 + 0.59(26) = 81.18 \text{ cms}.$$

This is called interpolation and this model works well for interpolation.

(**h**) Here $x_{\min} = 18$ months implies:

$$\hat{y}_{\min} = a + bx_{\min} = 65.84 + 0.59(18) = 76.46 \text{ cms},$$
thus, we have:

$$P_1 \leftrightarrow (x_{\min}, \hat{y}_{\min}) = (18, 76.46)$$

Here $x_{\max} = 29$ months implies:

$$\hat{y}_{\max} = a + bx_{\max} = 65.84 + 0.59(29) = 82.95 \text{ cms},$$

thus, we have:
$$P_2 \leftrightarrow (x_{\max}, \hat{y}_{\max}) = (29, 82.95)$$

Mark the points P_1 and P_2 on the scatter plots with square boxes, and join them with a straight line and it will give the **predicted line** as shown in **Figure 10.27.**

Fig. 10.26.
A tall boy

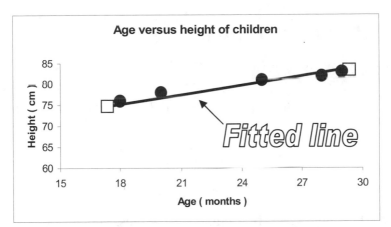

Fig. 10.27. Fitted line or predicted line.

(i) The error term $e_i = (y_i - \hat{y}_i)$ for each child is given below:

x_i	y_i	$\hat{y}_i = a + bx_i$	$(y_i - \hat{y}_i) = e_i$	e_i^2
18	76	$65.84 + 0.59(18) = 76.46$	$(76 - 76.46) = -0.46$	0.2116
20	78	$65.84 + 0.59(20) = 77.64$	$(78 - 77.64) = +0.36$	0.1296
25	81	$65.84 + 0.59(25) = 80.59$	$(81 - 80.59) = +0.41$	0.1681
28	82	$65.84 + 0.59(28) = 82.36$	$(82 - 82.36) = -0.36$	0.1296
29	83	$65.84 + 0.59(29) = 82.95.$	$(83 - 82.95) = +0.05$	0.0025
Sum	**400**	**400**	**0.00**	**0.6414**

Note that: $\sum\limits_{i=1}^{n} y_i = \sum\limits_{i=1}^{n} \hat{y}_i = 400$.

Thus, the value of the coefficient of determination is given by:

$$r^2 = 1 - \frac{\sum\limits_{i=1}^{n} e_i^2}{\sum\limits_{i=1}^{n} (y_i - \bar{y})^2} = 1 - \frac{0.6414}{34} = 0.98.$$

The meaning or the interpretation of the value of the coefficient of determination, $r^2 = 0.98$, is that 98% of the variation in the height of the children depends upon their age. The rest of the 2% variation depends on other factors such as diet, genes etc. which affect the height of the children, but have not been considered in our regression analysis. When two or more factors are considered in a regression analysis, it is called multiple regression analysis, which is beyond the scope of this book.

10.7 ADVANCED RULES OF MEAN, VARIANCE, AND STANDARD DEVIATION

10.7.1 RULES OF MEAN

Rule 1. If X is a random variable; a and b are real constants, then:

$$\mu_{(a+bX)} = a + b\mu_x$$

Mean of $(a + bX) = a + \{b \times \text{Mean of X}\}$

where μ_x is the mean of the random variable X.

Rule 2. If X and Y are two random variables, and a and b are real constants, then:

$$\mu_{(aX+bY)} = a\mu_x + b\mu_y$$

Mean of $(aX + bY) = \{a \times \text{Mean of X}\} + \{b \times \text{Mean of Y}\}$

where μ_x and μ_y are the means of X and Y respectively.

10.7.2 RULES OF VARIANCE

Rule 1. If X is a random variable; a and b are real constants, then:

$$\sigma^2_{(a+bX)} = b^2 \sigma^2_x$$

Variance of $(a + bX) = b^2 \times \text{Variance of X}$

where σ^2_x stands for the variance of X.

Rule 2. If X and Y are two random variables. Let ρ (rho) be the correlation between them. If a and b are real constants, then:

$$\sigma^2_{(aX+bY)} = a^2 \sigma^2_x + b^2 \sigma^2_y + 2ab\rho\sigma_x\sigma_y$$

Variance of $(aX + bY) = \{a^2 \times \text{Variance of X}\} + \{b^2 \times \text{Variance of Y}\}$
$$+ \{2 \times a \times b \times (\text{Correlation between X and Y})$$
$$\times (\text{Standard deviation of X})$$
$$\times (\text{Standard deviation of Y})\}$$

where σ^2_x and σ^2_y are the variances of X and Y respectively, and σ_x and σ_y are the standard deviations of X and Y respectively. If X and Y are independent, then $\rho = 0$. This means:

$$\sigma^2_{(aX+bY)} = a^2 \sigma^2_x + b^2 \sigma^2_y$$

10.7.3 RULES OF STANDARD DEVIATION

Rule 1. If X is a random variable; a and b are real constants, then:
$$\sigma_{(a+bX)} = b\sigma_Y$$
which implies:
SD of $(a + bX)=$ b $\times$ SD of (X)
where σ_x stands for the standard deviation (SD) of X.

Rule 2. If X and Y arc two random variables, let ρ (Rho) be the correlation between them. If a and b are real constants, then:

$$\sigma_{(aX+bY)} = \sqrt{\sigma^2_{(aX+bY)}}$$

where $\sigma_{(aX+bY)}$ stands for the standard deviation of $(aX+bY)$.

Example 10.10. (BUSINESSES NEED WISE PEOPLE) Amy receives a 20% return from her dairy farm and 80% from her poultry farm. If X is the annual return from her dairy farm and Y is the return from her poultry farm, then the total return R is given by:

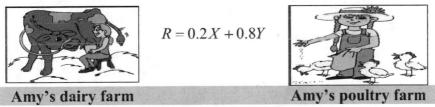

$$R = 0.2X + 0.8Y$$

Amy's dairy farm   Amy's poultry farm

Fig. 10.28. Return from two businesses.

The returns X and Y are random variables because they vary from time to time. Based on several annual returns, we have:
$\mu_X = 5.2\%$, $\mu_y = 13.3\%$, $\sigma_X = 2.9\%$, $\sigma_y = 17.0\%$, and $\rho = -0.1$.

Find (i) μ_R, (ii) σ^2_R, and (iii) σ_R.

Solution. Here $a = 0.2$ and $b = 0.8$.

(i) $\mu_R = \mu_{(0.2X+0.8Y)} = 0.2\mu_x + 0.8\mu_y = 0.2 \times 5.2 + 0.8 \times 13.3 = 11.68\%$

(ii) $\sigma^2_R = \sigma^2_{(0.2X+0.8Y)} = 0.2^2\sigma^2_x + 0.8^2\sigma^2_y + 2 \times 0.2 \times 0.8 \times \rho\sigma_x\sigma_y$

$= 0.2^2(2.9)^2 + 0.8^2(17.0)^2 + 2 \times 0.2 \times 0.8 \times (-0.1)(2.9)(17.0)$

$= 0.3364 + 184.96 - 1.5776$

$= 183.719 \ \%^2$.

(iii) $\sigma_R = \sqrt{\sigma^2_R} = \sqrt{183.719} = 13.55\%$.

LUDI 10.1. (A CIRCLE OF BAD FRIENDS IS NOT GOOD)
State if the statement is true or false.
"When the correlation coefficient is zero, the two variables are not related in any way."
(a) True
(b) False
Hint: All points on the circle show that there is a circular relationship. Thus, there is a relationship, but the value of the linear correlation coefficient may be zero.

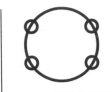

Fig. 10.29. Scatter plot.

LUDI 10.2. (BE CAREFUL ABOUT STRANGERS) Consider the
following scatter plot for a set of exam scores:

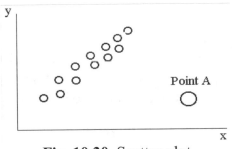

Fig. 10.30. Scatter plot.

The observation marked "Point A" is called - - - - - - -

(a) an outlier with respect to the regression line.
(b) an influential point.
(c) a residual.
(d) if the others are eggs of a hen, then it is an egg of a duck because of its large size.
(e) cannot say anything because I missed my classes.

LUDI 10.3. (USE TIGHT CLOTHES IN LABS) In the hope of preventing ecological damage from oil spills, a biochemical company is developing an enzyme to break up oil into less harmful chemicals. The table below shows the time it took for the enzyme to break up oil samples at different temperatures. The researcher plans to use this data set in correlation and regression analysis.

LAB EXPERIMENT	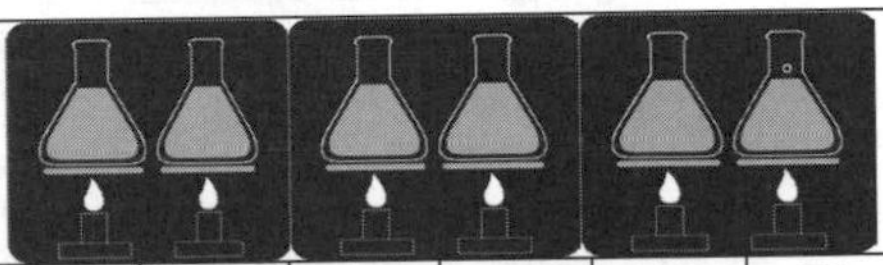					
Oil temperature (^{0}C) (x)	10	20	25	40	60	70
Time to break up (sec.) (y)	150	100	80	60	40	20

Fig. 10.31. Lab experiment.

(a) Construct a scatter diagram.
(b) Guess the signs of the correlation coefficient and slope.
(c) Find the value of the correlation coefficient.
(d) Fit the regression line.
(e) Interpret the slope and intercept.
(f) Sketch the predicted line.
(g) Predict the time to break up if the temp is 100 ^{0}C (Extrapolation).
(h) Predict the time to break up if the temp is 35 ^{0}C (Interpolation).

LUDI 10.4. (SMOKING PUTS YOU DOWN AMONG FRIENDS)
The following data set represents a trend in cigarette consumption per capita (in hundreds) and lung cancer mortality (per 100,000) for males of a particular country.

 Cigarette consumption	11	12	15	20	21	23
Mortality rate	10	16	22	26	34	42

Fig. 10.32. Better stop smoking.

(a) Construct a scatter plot and denote the dependent variable with y and the independent variable with x.
(b) Guess the sign of the correlation coefficient.

(c) Guess the sign of the regression coefficient.

(d) Complete the following table:

	x_i	y_i	x_i^2	y_i^2	$x_i y_i$
Sum					

Caution: Be careful while choosing the x and y variable.
From the above table, find:

$$n = ---, \quad \Sigma x_i = ---, \quad \Sigma y_i = ---, \Sigma x_i^2 = ---, \quad \Sigma y_i^2 = ---,$$

and $\Sigma x_i y_i = ---$

(e) Find the value of the correlation coefficient:

$$r = \frac{n(\Sigma x_i y_i) - (\Sigma x_i)(\Sigma y_i)}{\sqrt{n(\Sigma x_i^2) - (\Sigma x_i)^2} \sqrt{n(\Sigma y_i^2) - (\Sigma y_i)^2}} = --------$$

(f) Find the best fit line by computing:

$$\bar{y} = \frac{\Sigma y_i}{n} = --------, \quad \bar{x} = \frac{\Sigma x_i}{n} = --------$$

$$b = \frac{n(\Sigma x_i y_i) - (\Sigma x_i)(\Sigma y_i)}{n(\Sigma x_i^2) - (\Sigma x_i)^2} = -------- \quad \text{and} \quad a = \bar{y} - b\bar{x} = --------$$

BEST FIT: $\hat{y}_i = a + bx_i = --------$

(g) Use the least squares regression equation to predict the lung cancer mortality rate when the cigarette consumption per capita is 20. Is it interpolation or extrapolation?

(h) Interpret intercept:

(i) Interpret slope:

(j) Draw the predicted line on a scatter plot:

Find $x_{min} = --------$ and $x_{max} = --------$. Then predict the corresponding values of the dependent variable as:

$$\hat{y}_{min} = a + bx_{min} = -------- \quad \text{and} \quad \hat{y}_{max} = a + bx_{max} = --------$$

Show the pairs $(x_{min}, \hat{y}_{min}) = (---, ---)$ and $(x_{max}, \hat{y}_{max}) = (---, ---)$ on the scatter plot in square boxes and join them with a straight line.

LUDI 10.5. (SOMETIMES A GOOD MATCH IS DIFFICULT)
Bob is now 18 years old and planning to date. He wonders whether people of similar heights tend to date each other. He measures his mom and dad's heights, and his four friend's heights along with their girlfriend's heights in inches as shown below:

 Bob					
Height of woman (x)	65	66	66	65	70
Height of man (y)	65	72	70	68	71

Fig. 10.33. Dating may help.

(a) Make a scatter plot.
(b) Based on the scatter plot, do you expect the correlation to be positive or negative?
(c) Complete the following table:

x_i	y_i	x_i^2	y_i^2	$x_i y_i$
Sum				

From the above table, find:

$n = --$, $\Sigma x_i = --$, $\Sigma y_i = --$, $\Sigma x_i^2 = --$, $\Sigma y_i^2 = --$, $\Sigma x_i y_i = --$.

Find the value of the correlation coefficient r given by:

$$r = \frac{n(\Sigma x_i y_i) - (\Sigma x_i)(\Sigma y_i)}{\sqrt{n(\Sigma x_i^2) - (\Sigma x_i)^2}\sqrt{n(\Sigma y_i^2) - (\Sigma y_i)^2}} = -------$$

(d) Michael tells Bob that a woman likes to date a man exactly 2 inches taller than her. If Michael's observation is correct, Bob wonders what the correlation would be between male and female heights.

LUDI 10.6. (ALWAYS ATTEND YOUR CLASS, BECAUSE YOU PAY FOR IT) Professor Forgetful (*e.g.*, refer to the film 'The Nutty Professor' directed by Jerry Lewis) believes that the percentage of marks of students on an examination depends upon the number of classes attended by them.

MOM GIVING LUNCH BOX						
Percentage of marks	30	57	77	50	80	96
Number of classes attended	10	20	30	18	35	43

Fig. 10.34. Ready to go to school.

(a) Construct a scatter plot and denote the dependent variable with y and the independent variable with x.
(b) Guess the sign of the correlation coefficient.
(c) Guess the sign of the regression coefficient.
(d) Complete the following table (**Caution:** Be careful in choosing the x and y variable).

	x_i	y_i	x_i^2	y_i^2	$x_i y_i$
Sum					

From the above table, find: $n = ---$, $\Sigma x_i = ---$, $\Sigma y_i = ---$,
$\Sigma x_i^2 = ---$, $\Sigma y_i^2 = ---$, $\Sigma x_i y_i = ---$.

(e) Find the value of the correlation coefficient:
$$r = \frac{n(\Sigma x_i y_i) - (\Sigma x_i)(\Sigma y_i)}{\sqrt{n(\Sigma x_i^2) - (\Sigma x_i)^2} \sqrt{n(\Sigma y_i^2) - (\Sigma y_i)^2}} = -------$$

(f) Find the best fit line by computing:
$$\bar{y} = \frac{\Sigma y_i}{n} = --------, \quad \bar{x} = \frac{\Sigma x_i}{n} = --------,$$

$$b = \frac{n(\Sigma x_i y_i) - (\Sigma x_i)(\Sigma y_i)}{n(\Sigma x_i^2) - (\Sigma x_i)^2} = \text{--------} \quad \text{and} \quad a = \bar{y} - b\bar{x} = \text{-------}$$

BEST FIT: $\hat{y}_i = a + bx_i = \text{-------}$

(g) Professor Forgetful misplaced the mid-term exam of one student, but has information about the number of classes attended by the student. Find Professor Forgetful's prediction using the above fitted line if this student has attended 32 classes:

$$\hat{y}_i = a + bx_i^* = \text{-------}$$

(h) Interpret the intercept.
(i) Interpret the slope.
(j) Draw the predicted line on the scatter plot:
Find $x_{min} = \text{--------}$ and $x_{max} = \text{-------}$
Then predict the corresponding values of the dependent variable:
$\hat{y}_{min} = a + bx_{min} = \text{--------}$ and $\hat{y}_{max} = a + bx_{max} = \text{-------}$
Show the pairs:
$(x_{min}, \hat{y}_{min}) = (\text{--}, \text{---})$ and $(x_{max}, \hat{y}_{max}) = (\text{--}, \text{---})$
on the scatter plot in square boxes and join them with a straight line.

LUDI 10.7. (PREDICTING YOUR OWN MATCH) Amy is now 18 years old and wonders if she can date a man of similar height. She measures her mom and dad's heights, and then she measured her four friend's and their boyfriend's heights (in inches) as shown below:

 Amy's dating	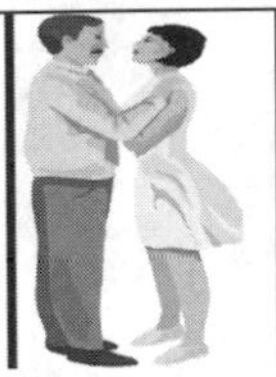				
Men	72	68	68	71	66
Women	66	64	65	70	65

Fig. 10.35. Understanding each other is important.

(a) Find the Z-score for each woman.
(b) Find the Z-score for each man.
(c) Multiply the Z-scores of each pair. Sum the product of the Z-scores and divide by $(n-1)$.
(d) Predict Amy's match's height if her height is 72 inches.

LUDI 10.8. (**DRIVE CAREFULLY**) The number of car accidents at a main intersection of a particular city is recorded per year from 1996 to 2003. The following table gives the number of cars passing through the intersection, x (in thousands), and the number of accidents, y, at the intersection for each year.

	Year	1996	1997	1998	1999	2001	2002	2003
DON'T DRINK & DRIVE... Enjoy the prom!	x	3.4	3.5	3.8	3.7	3.9	4.0	4.3
	y	22	22	27	23	32	34	36

Fig. 10.36. Obey the traffic rules.

(a) Construct a scatter plot.
(b) Guess: The sign of the correlation coefficient will be - - - - - - -
(c) Complete the following table:

x_i	y_i	x_i^2	y_i^2	$x_i y_i$
Sum				

From the above table, find:

$$n = ---, \ \Sigma x_i = ---, \ \Sigma y_i = ---, \ \Sigma x_i^2 = ---, \ \Sigma y_i^2 = ---$$

and $\Sigma x_i y_i = ---$.

Thus, the value of the correlation coefficient r is given by:

$$r = \frac{n(\Sigma x_i y_i) - (\Sigma x_i)(\Sigma y_i)}{\sqrt{n(\Sigma x_i^2) - (\Sigma x_i)^2} \sqrt{n(\Sigma y_i^2) - (\Sigma y_i)^2}} = -------$$

(d) Find the best fit line by computing:

$$\bar{y} = \frac{\Sigma y_i}{n} = -------, \quad \bar{x} = \frac{\Sigma x_i}{n} = -------,$$

$$b = \frac{n(\Sigma x_i y_i) - (\Sigma x_i)(\Sigma y_i)}{n(\Sigma x_i^2) - (\Sigma x_i)^2} = -------- \quad \text{and} \quad a = \bar{y} - b\bar{x} = --------.$$

Thus,

BEST FIT: $\hat{y}_i = a + bx_i = \text{-------}$

(e) Interpret the slope:

(f) Interpret the intercept:

(g) Find $x_{min} = \text{----}$ and $x_{max} = \text{----}$. Then predict the corresponding values of the dependent variable as:

$\hat{y}_{min} = a + bx_{min} = \text{---------}$, and $\hat{y}_{max} = a + bx_{max} = \text{---------}$

Show the pairs $(x_{min}, \hat{y}_{min}) = (\text{---}, \text{---})$ and $(x_{max}, \hat{y}_{max}) = (\text{---}, \text{---})$ on the scatter plot in square boxes and join them with a straight line.

(h) Predict the number of car accidents during 2005 if 4,500 cars will pass through the intersection.

$$\hat{y}_i = a + bx_i^* = \text{--------}$$

Is it interpolation or extrapolation?

(i) Predict the number of car accidents that occurred during 2000 if 3,780 cars crossed the intersection.

$$\hat{y}_i = a + bx_i^* = \text{--------}$$

Is it interpolation or extrapolation?

(j) Find and interpret the value of the coefficient of determination.

LUDI 10.9. (KIDS R US) In a random sample of five kids, the age and height have averages of 24 months and 80 cms and standard deviations of 4.85 months and 2.92 cms respectively. Their Z-scores based on the individual's age and height are given below:

Kids					
Age	-1.3	-0.4	+0.8	+0.4	+0.5
Height	-1.1	-0.2	+0.8	+0.2	+0.3

Fig. 10.37. Growing kids.

(a) Find the correlation coefficient between age and height.
(b) Predict the height of a 27 months old child.
(c) Predict the height of a child after 100 years and comment.

LUDI 10.10. (GOLF LOVERS) The following are the golf scores of 10 members of a men's golf team in a tournament:

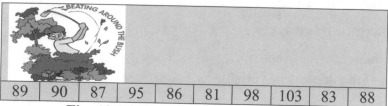

| 89 | 90 | 87 | 95 | 86 | 81 | 98 | 103 | 83 | 88 |

Fig. 10.38. Standardizing performance.

Complete the following table:

	x_i	$(x_i - \bar{x})$	$(x_i - \bar{x})^2$	$Z_x = \dfrac{(x_i - \bar{x})}{s_x}$
Sum				

Hint: Use the following in the above table: $n = ----$, $\displaystyle\sum_{i=1}^{n} x_i = ----$,

$\bar{x} = \dfrac{1}{n}\displaystyle\sum_{i=1}^{n} x_i = ----$, $s_x^2 = \dfrac{1}{n-1}\displaystyle\sum_{i=1}^{n}(x_i - \bar{x})^2 = ----$, $s_x = \sqrt{s_x^2} = ----$

LUDI 10.11. (UNEXPLAINED VARIATION) It is a well-known fact that the height, y, of children increases as they age, x. Assume that we selected a random sample of $n = 12$ children between the ages of 18 months and 29 months and found descriptive statistics as:

$\bar{y} = 80.87$ cms, $\bar{x} = 23.50$ months, $s_y^2 = 24.39$ cms^2, $s_x^2 = 13.00$ months^2 and $r = 0.804$.

(a) Fit the line:

$$\text{Height} = a + b(\text{Age})$$

(b) Predict the height of a child if age = 30 months. Is it interpolation or extrapolation?

(c) Find $\sum_{i=1}^{n} e_i^2 =$ --------, and interpret its meaning.

(d) How much variation in the heights of the children remains unexplained by their age? Why?

LUDI 10.12. (HOW FAST DO PLANTS GROW?) In a park, a visitor threw a sunflower seed and it grew. Every Sunday, a child visited this park and recorded its height in centimeters as:

Visit No. x_i	Height y_i
1	56
2	62
3	68
4	73
5	74
6	75

Fig. 10.39. Story of a child and a sunflower.

(a) Construct a scatter plot for the height versus visit number.
(b) What are the signs of the relationship and slope?
(c) Complete the following table:

x_i	$(x_i - \bar{x})$	$(x_i - \bar{x})^2$	Z_{x_i}	y_i	$(y_i - \bar{y})$	$(y_i - \bar{y})^2$	Z_{y_i}	$Z_{x_i} Z_{y_i}$
Sum								

From the above table, find:

$$n = ----, \quad \sum_{i=1}^{n} x_i = ----, \quad \bar{x} = \frac{\sum_{i=1}^{n} x_i}{n} = ----, \quad s_x^2 = \frac{\sum_{i=1}^{n}(x_i - \bar{x})^2}{n-1} = ----,$$

$$s_x = \sqrt{s_x^2} = ---, \quad \sum_{i=1}^{n} y_i = ---, \quad \bar{y} = \frac{\sum\limits_{i=1}^{n} y_i}{n} = ---, \quad s_y^2 = \frac{\sum\limits_{i=1}^{n}(y_i - \bar{y})^2}{n-1} = ---,$$

and $s_y = \sqrt{s_y^2} = ---$

(d) Find the value of the correlation coefficient:

$$r = \frac{1}{n-1} \sum_{i=1}^{n} Z_{x_i} Z_{y_i} = -------$$

(e) Find the best fit line:

$$\hat{y}_i = a + bx_i = --------$$

(f) Interpret the value of regression coefficient b and constant a.

(g) Now the child wonders if he makes 1,000 such visits, will the sunflower plant touch the sky? Predict the height of the plant on the $1,000^{th}$ visit. Is it interpolation or extrapolation?

(h) For $x_{min} = ---$, find $\hat{y}_{min} = ---$

For $x_{max} = ---$, find $\hat{y}_{max} = ---$

Join $P_1 \leftrightarrow (x_{min}, \hat{y}_{min})$ and $P_2 \leftrightarrow (x_{max}, \hat{y}_{max})$ to draw the predicted line.

(i) Find and interpret the value of coefficient of determination.

LUDI 10.13. (BABIES R US) The Z-scores of 5 babies based on age (x_i) between 18 months and 29 months and height (y_i) between 761 mm and 850 mm are given below:

Baby					
Z_{x_i}	-1.44	-0.51	0.19	0.65	1.11
Z_{y_i}	-1.06	-0.86	-0.03	0.64	1.31

Fig. 10.40. Z-scores maintain privacy of true scores.

(a) Find the value of the correlation coefficient, r.

(b) Given $\bar{x} = 24.20$ months , $\bar{y} = 801.00$ mm, $s_x = 4.32$ months, and $s_y = 37.42$ mm. Find the values of a and b in the best fit line given by:

Height $= a + b(\text{Age})$

(c) Interpret the values of a and b.

(d) Predict the height of a child if the age is 60 months. Is it interpolation or extrapolation?

LUDI 10.14. (SAVE FUEL) Bob bought an old car and recorded its odometer reading as 115,000 km and then used different amounts of fuel (in liters) at the appearance of the fuel indicator as follows:

Odometer (km)	115,250	115,580	116,280	116,360	116,530
Fuel (liter)	25	35	30	10	20

Fig. 10.41. Save fuel for the future.

Due to Bob's credit card limit, he can only buy 15 liters of fuel. He wonders how far he can go with it. Fit a regression line to predict his next destination. Is it interpolation or extrapolation?

LUDI 10.15. (NEVER MISS YOUR CLASS, BECAUSE YOU PAY FOR IT) At the end of a semester, an instructor took a random sample of six students and counted the number of classes missed and their scores on the final exam as follows:

Students			
Missed Classes	0	3	10
Scores	97	75	51
Students			
Missed Classes	1	5	2
Scores	93	70	88

Fig. 10.42. Scores and number of missed classes.

(a) Construct a scatter plot for score versus number of missed classes. (Be careful while choosing x and y)

(b) What is the sign of the linear relationship? What is the sign of the regression coefficient?

(c) Complete the following table:

	x_i	$(x_i - \bar{x})$	$(x_i - \bar{x})^2$	Z_{x_i}	y_i	$(y_i - \bar{y})$	$(y_i - \bar{y})^2$	Z_{y_i}	$Z_{x_i} Z_{y_i}$
Sum									

From the above table, find:

$$n = ---, \quad \sum_{i=1}^{n} x_i = ---, \quad \bar{x} = \frac{\sum_{i=1}^{n} x_i}{n} = ---, \quad s_x^2 = \frac{\sum_{i=1}^{n}(x_i - \bar{x})^2}{n-1} = ---,$$

$$s_x = \sqrt{s_x^2} = ---, \quad \sum_{i=1}^{n} y_i = ---, \quad \bar{y} = \frac{\sum_{i=1}^{n} y_i}{n} = ---, \quad s_y^2 = \frac{\sum_{i=1}^{n}(y_i - \bar{y})^2}{n-1} = ---,$$

and $s_y = \sqrt{s_y^2} = ---$

(d) Find the value of the correlation coefficient:

$$r = \frac{1}{n-1} \sum_{i=1}^{n} Z_{x_i} Z_{y_i} = --------$$

(e) Find the best fit line:

$$\hat{y}_i = a + bx_i = --------$$

(f) Interpret the value of b.

(g) Interpret the value of a.

(h) Predict the score if a student misses 7 classes. Is this interpolation or extrapolation?

(i) Predict the score if a student misses 20 classes. Is this interpolation or extrapolation?

(j) For $x_{min} = ---$, find $\hat{y}_{max} = ---$. For $x_{max} = ---$, find $\hat{y}_{min} = ---$. Join $P_1 \leftrightarrow (x_{min}, \hat{y}_{max})$ and $P_2 \leftrightarrow (x_{max}, \hat{y}_{min})$ to plot the predicted line on your scatter plot.

(k) Find and interpret the value of the coefficient of determination.

LUDI 10.16. (BEAUTIFUL DREAMS ARE SIGNS OF GOOD HEALTH) A child, Bob, was playing with his friends and listening and reading stories about evolutionary chains (Bob was a very intelligent, but peculiar child). That night, Bob had a bad dream that he saw a monkey whose height (in cm) was increasing after every two minutes as shown below:

Height	20	30	40	50	80	100	120	180
Time	0	2	4	6	8	10	12	14

Fig. 10.43. Always read good stories for good dreams.

Bob was afraid of such a big monkey and he woke up to tell his mom about the dream. Bob wonders how tall the monkey would have grown if he had slept another 10 minutes. Fit a regression line to predict the monkey's height and assure Bob that no one can touch the sky, which will at least relieve Bob's anxiety about giant monkeys.

LUDI 10.17. (GOOD SCORES ARE AN ASSET) A university uses GRE and TOEFL scores as one criterion for admission. Experience has shown that the distribution of GRE and TOEFL scores among the university's entire population of applicants is such that:

GRE score X	$\mu_X = 620$	$\sigma_X^2 = 2500$
TOEFL score Y	$\mu_Y = 580$	$\sigma_Y^2 = 1600$

Fig. 10.44. Good score needs hard work.

(I) Mean: (a) What is the mean of the total score $X + Y$ among students applying to this university?
(b) What is the mean of the difference between score $X - Y$ among students applying to this university?

(II)Variance: GRE and TOEFL scores are not independent, because students who score high on one exam tend to score high on the other also. The correlation between the GRE and TOEFL score is: $\rho = 0.8$.

(a) What is the variance of the total score $X + Y$ among students applying to this university?

(b) What is the variance of the difference between scores $X - Y$ among students applying to this university?

(III) Standard deviation: (a) What is the standard deviation of the total score $X + Y$ among students applying to this university?

(b) What is the standard deviation of the difference between scores $X - Y$ among students applying to this university?

LUDI 10.18. (ALWAYS ATTEND YOUR CLASS, BECAUSE YOU PAY FOR IT) In a semester, generally there are approximately 45 classes to attend. At the end of a semester, an instructor took a random sample of six students and counted the number of classes attended and their scores on the final exam as:

Students			
Classes Attended	42	39	29
Scores	97	75	51
Students			
Classes Attended	41	37	40
Scores	93	70	88

Fig. 10.45. Class attendance is must.

(a) Construct a scatter plot for score versus number of classes attended. (Be careful while choosing x and y)

(b) What is the sign of the relationship and slope?

(c) Complete the following table:

x_i	$(x_i - \bar{x})$	$(x_i - \bar{x})^2$	Z_{x_i}	y_i	$(y_i - \bar{y})$	$(y_i - \bar{y})^2$	Z_{y_i}	$Z_{x_i} Z_{y_i}$
Sum								

From the above table, find:

$$n = ---, \quad \bar{x} = \frac{1}{n}\sum_{i=1}^{n} x_i = ---, \quad s_x^2 = \frac{1}{n-1}\sum_{i=1}^{n}(x_i - \bar{x})^2 = ---, \quad s_x = \sqrt{s_x^2} = ---,$$

$$\sum_{i=1}^{n} y_i = ---, \quad \bar{y} = \frac{1}{n}\sum_{i=1}^{n} y_i = ---, \quad s_y^2 - \frac{1}{n-1}\sum_{i=1}^{n}(y_i - \bar{y})^2 = ----, \text{ and}$$

$$s_y = \sqrt{s_y^2} = ---$$

(d) Find the value of the correlation coefficient.

(e) Find the best fit line: $\hat{y}_i = a + bx_i = --------$

(f) Interpret the values of the slope b and intercept a.

(g) Predict the score if a student attends 15 classes.
Is this interpolation or extrapolation?

(h) Predict the score if a student attends 44 classes.
Is this interpolation or extrapolation?

(i) Find and interpret the value of the coefficient of determination.

LUDI 10.19. (THE JOB OF A DETECTOR IS DIFFICULT)

Bob's study group consists of 5 students. To compare himself with others, one day he computed the following Z-scores of all 5 students in verbal and math skills as follows:

TODAY'S DON'TS	Students	Amy	Bob	Cara	Don	Eric
	Verbal (Z_x)	-1.6	1.8	-1.1	0.4	0.5
	Math (Z_y)	-1.4	1.6	-1.1	0.6	0.3

Fig. 10.46. Bob on the board.

Bob claims that his Z-scores are positive and higher, thus he is better than others. Cara claims that Bob's calculations are wrong. Justify Cara's argument by verifying the properties of the Z-score.

LUDI 10.20. (GOLF LOVERS) Here are the golf scores of 5 members of a college men's golf team in two rounds of tournament play.

Player	1	2	3	4	5
Round-1	88	90	87	95	90
Round-2	97	85	89	89	90

Fig. 10.47. Practice again and again for good scores.

(a) Make a scatter plot of the data, taking the first round score as the explanatory variable. Is there any association between the two scores? If so, is it positive or negative?
(b) Find and interpret the value of the correlation coefficient r.
(c) Derive and interpret the value of the coefficient of determination.

LUDI 10.21. (BYE-BYE ELEPHANT) Bob bought 5 elephants, and their weights (in kg) and diets (in kg) are given below:

Elephants					
Weight	5,000	4,000	3,000	2,000	6,000
Diet	360	325	150	100	400

Fig. 10.48. Weighing elephants.

(a) Find the value of the correlation coefficient between the weight and diet of the elephants.
(b) Predict the weight of an elephant if its diet is 200 kg. Is this interpolation or extrapolation?
(c) Predict the weight of an elephant if its diet is 50 kg. Is this interpolation or extrapolation?
(d) Predict the weight of an elephant if its diet is 450 kg. Is this interpolation or extrapolation?
(e) Bob bought a new elephant whose weight is 4,500 kg. What is the predicted diet of this new elephant? Is this extrapolation or interpolation? Can we use the same model as used in (b) to (d)?

Table I. Pseudo Random Numbers

Random Numbers Table

Column Number

Row	1	2	3	4	5	6	7	8	9	10	11	12	13	14	15	16	17	18	19	20
1	3	7	5	6	9	6	1	6	0	2	0	1	0	3	3	0	4	3	0	6
2	1	9	8	8	7	2	2	7	9	6	3	6	2	0	5	8	4	4	6	9
3	5	2	8	4	0	8	0	1	8	9	6	8	4	5	0	6	0	0	5	0
4	9	9	6	2	6	7	5	8	3	3	0	6	0	2	3	1	3	2	6	2
5	3	1	0	2	0	4	7	3	1	9	5	6	8	7	8	0	0	6	6	3
6	3	6	7	0	0	4	0	6	1	0	6	2	7	7	0	1	4	0	3	4
7	1	4	8	3	9	3	1	4	3	9	8	1	9	9	7	1	0	4	1	1
8	0	8	9	4	3	9	2	0	5	0	8	3	4	1	1	0	3	9	3	0
9	0	5	0	4	0	6	2	0	6	4	2	4	7	6	7	5	6	5	4	5
10	0	0	6	5	8	9	6	1	8	3	3	8	9	7	8	4	0	4	3	8
11	0	8	5	3	5	3	8	8	3	2	7	7	7	3	1	2	0	8	9	9
12	0	1	0	1	2	0	7	9	7	2	4	4	9	3	2	1	9	4	3	8
13	0	5	5	7	9	4	3	0	7	0	1	0	7	0	2	2	3	9	5	5
14	0	0	9	3	5	6	4	5	8	7	5	3	9	8	2	9	0	5	2	0
15	0	8	4	3	1	4	2	8	6	2	6	5	4	9	3	8	3	6	8	8
16	9	1	6	5	5	2	7	1	8	5	9	6	6	6	3	9	7	3	2	0
17	6	9	6	7	8	8	0	6	7	4	7	2	0	6	9	2	4	5	3	5
18	1	9	6	6	6	7	6	7	2	5	4	1	0	7	7	0	2	9	7	7
19	6	0	4	7	1	4	0	3	8	4	6	1	7	6	3	9	1	5	4	0
20	5	1	2	4	3	5	8	1	8	9	3	4	9	1	1	3	0	6	1	8
21	0	1	0	4	6	0	8	9	7	9	7	2	9	1	4	1	3	0	5	3
22	6	3	0	5	3	1	9	4	0	2	1	9	4	3	3	0	4	7	6	3
23	9	1	6	5	3	2	4	0	5	0	6	5	6	6	4	9	6	6	1	3
24	5	8	7	4	0	4	5	6	3	8	9	5	6	6	3	7	3	3	1	8
25	2	0	4	4	1	0	5	0	3	1	2	0	4	4	8	1	1	8	6	3

Source: Generated in Excel using Randbetween (0,9)

Table II. Area under the standard normal curve

					Area to the left side of Z value				

Area to the left side of Z value

In this table:
  Minimum value of Z = -3.49
  Maximum value of Z = +3.49

Z	0.00	0.01	0.02	0.03	0.04	0.05	0.06	0.07	0.08	0.09
-3.4	0.0003	0.0003	0.0003	0.0003	0.0003	0.0003	0.0003	0.0003	0.0003	0.0002
-3.3	0.0005	0.0005	0.0005	0.0004	0.0004	0.0004	0.0004	0.0004	0.0004	0.0003
-3.2	0.0007	0.0007	0.0006	0.0006	0.0006	0.0006	0.0006	0.0005	0.0005	0.0005
-3.1	0.0010	0.0009	0.0009	0.0009	0.0008	0.0008	0.0008	0.0008	0.0007	0.0007
-3.0	0.0013	0.0013	0.0013	0.0012	0.0012	0.0011	0.0011	0.0011	0.0010	0.0010
-2.9	0.0019	0.0018	0.0018	0.0017	0.0016	0.0016	0.0015	0.0015	0.0014	0.0014
-2.8	0.0026	0.0025	0.0024	0.0023	0.0023	0.0022	0.0021	0.0021	0.0020	0.0019
-2.7	0.0035	0.0034	0.0033	0.0032	0.0031	0.0030	0.0029	0.0028	0.0027	0.0026
-2.6	0.0047	0.0045	0.0044	0.0043	0.0041	0.0040	0.0039	0.0038	0.0037	0.0036
-2.5	0.0062	0.0060	0.0059	0.0057	0.0055	0.0054	0.0052	0.0051	0.0049	0.0048
-2.4	0.0082	0.0080	0.0078	0.0075	0.0073	0.0071	0.0069	0.0068	0.0066	0.0064
-2.3	0.0107	0.0104	0.0102	0.0099	0.0096	0.0094	0.0091	0.0089	0.0087	0.0084
-2.2	0.0139	0.0136	0.0132	0.0129	0.0125	0.0122	0.0119	0.0116	0.0113	0.0110
-2.1	0.0179	0.0174	0.0170	0.0166	0.0162	0.0158	0.0154	0.0150	0.0146	0.0143
-2.0	0.0228	0.0222	0.0217	0.0212	0.0207	0.0202	0.0197	0.0192	0.0188	0.0183
-1.9	0.0287	0.0281	0.0274	0.0268	0.0262	0.0256	0.0250	0.0244	0.0239	0.0233
-1.8	0.0359	0.0351	0.0344	0.0336	0.0329	0.0322	0.0314	0.0307	0.0301	0.0294
-1.7	0.0446	0.0436	0.0427	0.0418	0.0409	0.0401	0.0392	0.0384	0.0375	0.0367
-1.6	0.0548	0.0537	0.0526	0.0516	0.0505	0.0495	0.0485	0.0475	0.0465	0.0455
-1.5	0.0668	0.0655	0.0643	0.0630	0.0618	0.0606	0.0594	0.0582	0.0571	0.0559
-1.4	0.0808	0.0793	0.0778	0.0764	0.0749	0.0735	0.0721	0.0708	0.0694	0.0681
-1.3	0.0968	0.0951	0.0934	0.0918	0.0901	0.0885	0.0869	0.0853	0.0838	0.0823
-1.2	0.1151	0.1131	0.1112	0.1093	0.1075	0.1056	0.1038	0.1020	0.1003	0.0985
-1.1	0.1357	0.1335	0.1314	0.1292	0.1271	0.1251	0.1230	0.1210	0.1190	0.1170
-1.0	0.1587	0.1562	0.1539	0.1515	0.1492	0.1469	0.1446	0.1423	0.1401	0.1379
-0.9	0.1841	0.1814	0.1788	0.1762	0.1736	0.1711	0.1685	0.1660	0.1635	0.1611
-0.8	0.2119	0.2090	0.2061	0.2033	0.2005	0.1977	0.1949	0.1922	0.1894	0.1867
-0.7	0.2420	0.2389	0.2358	0.2327	0.2296	0.2266	0.2236	0.2206	0.2177	0.2148
-0.6	0.2743	0.2709	0.2676	0.2643	0.2611	0.2578	0.2546	0.2514	0.2483	0.2451
-0.5	0.3085	0.3050	0.3015	0.2981	0.2946	0.2912	0.2877	0.2843	0.2810	0.2776
-0.4	0.3446	0.3409	0.3372	0.3336	0.3300	0.3264	0.3228	0.3192	0.3156	0.3121
-0.3	0.3821	0.3783	0.3745	0.3707	0.3669	0.3632	0.3594	0.3557	0.3520	0.3483
-0.2	0.4207	0.4168	0.4129	0.4090	0.4052	0.4013	0.3974	0.3936	0.3897	0.3859
-0.1	0.4602	0.4562	0.4522	0.4483	0.4443	0.4404	0.4364	0.4325	0.4286	0.4247
-0.0	0.5000	0.4960	0.4920	0.4880	0.4840	0.4801	0.4761	0.4721	0.4681	0.4641

Area to the left side of Z value									

Area

0   Z

Z	0.00	0.01	0.02	0.03	0.04	0.05	0.06	0.07	0.08	0.09
+0.0	0.5000	0.5040	0.5080	0.5120	0.5160	0.5199	0.5239	0.5279	0.5319	0.5359
+0.1	0.5398	0.5438	0.5478	0.5517	0.5557	0.5596	0.5636	0.5675	0.5714	0.5753
+0.2	0.5793	0.5832	0.5871	0.5910	0.5948	0.5987	0.6026	0.6064	0.6103	0.6141
+0.3	0.6179	0.6217	0.6255	0.6293	0.6331	0.6368	0.6406	0.6443	0.6480	0.6517
+0.4	0.6554	0.6591	0.6628	0.6664	0.6700	0.6736	0.6772	0.6808	0.6844	0.6879
+0.5	0.6915	0.6950	0.6985	0.7019	0.7054	0.7088	0.7123	0.7157	0.7190	0.7224
+0.6	0.7257	0.7291	0.7324	0.7357	0.7389	0.7422	0.7454	0.7486	0.7517	0.7549
+0.7	0.7580	0.7611	0.7642	0.7673	0.7704	0.7734	0.7764	0.7794	0.7823	0.7852
+0.8	0.7881	0.7910	0.7939	0.7967	0.7995	0.8023	0.8051	0.8078	0.8106	0.8133
+0.9	0.8159	0.8186	0.8212	0.8238	0.8264	0.8289	0.8315	0.8340	0.8365	0.8389
+1.0	0.8413	0.8438	0.8461	0.8485	0.8508	0.8531	0.8554	0.8577	0.8599	0.8621
+1.1	0.8643	0.8665	0.8686	0.8708	0.8729	0.8749	0.8770	0.8790	0.8810	0.8830
+1.2	0.8849	0.8869	0.8888	0.8907	0.8925	0.8944	0.8962	0.8980	0.8997	0.9015
+1.3	0.9032	0.9049	0.9066	0.9082	0.9099	0.9115	0.9131	0.9147	0.9162	0.9177
+1.4	0.9192	0.9207	0.9222	0.9236	0.9251	0.9265	0.9279	0.9292	0.9306	0.9319
+1.5	0.9332	0.9345	0.9357	0.9370	0.9382	0.9394	0.9406	0.9418	0.9429	0.9441
+1.6	0.9452	0.9463	0.9474	0.9484	0.9495	0.9505	0.9515	0.9525	0.9535	0.9545
+1.7	0.9554	0.9564	0.9573	0.9582	0.9591	0.9599	0.9608	0.9616	0.9625	0.9633
+1.8	0.9641	0.9649	0.9656	0.9664	0.9671	0.9678	0.9686	0.9693	0.9699	0.9706
+1.9	0.9713	0.9719	0.9726	0.9732	0.9738	0.9744	0.9750	0.9756	0.9761	0.9767
+2.0	0.9772	0.9778	0.9783	0.9788	0.9793	0.9798	0.9803	0.9808	0.9812	0.9817
+2.1	0.9821	0.9826	0.9830	0.9834	0.9838	0.9842	0.9846	0.9850	0.9854	0.9857
+2.2	0.9861	0.9864	0.9868	0.9871	0.9875	0.9878	0.9881	0.9884	0.9887	0.9890
+2.3	0.9893	0.9896	0.9898	0.9901	0.9904	0.9906	0.9909	0.9911	0.9913	0.9916
+2.4	0.9918	0.9920	0.9922	0.9925	0.9927	0.9929	0.9931	0.9932	0.9934	0.9936
+2.5	0.9938	0.9940	0.9941	0.9943	0.9945	0.9946	0.9948	0.9949	0.9951	0.9952
+2.6	0.9953	0.9955	0.9956	0.9957	0.9959	0.9960	0.9961	0.9962	0.9963	0.9964
+2.7	0.9965	0.9966	0.9967	0.9968	0.9969	0.9970	0.9971	0.9972	0.9973	0.9974
+2.8	0.9974	0.9975	0.9976	0.9977	0.9977	0.9978	0.9979	0.9979	0.9980	0.9981
+2.9	0.9981	0.9982	0.9982	0.9983	0.9984	0.9984	0.9985	0.9985	0.9986	0.9986
+3.0	0.9987	0.9987	0.9987	0.9988	0.9988	0.9989	0.9989	0.9989	0.9990	0.9990
+3.1	0.9990	0.9991	0.9991	0.9991	0.9992	0.9992	0.9992	0.9992	0.9993	0.9993
+3.2	0.9993	0.9993	0.9994	0.9994	0.9994	0.9994	0.9994	0.9995	0.9995	0.9995
+3.3	0.9995	0.9995	0.9995	0.9996	0.9996	0.9996	0.9996	0.9996	0.9996	0.9997
+3.4	0.9997	0.9997	0.9997	0.9997	0.9997	0.9997	0.9997	0.9997	0.9997	0.9998

Source: Generated in Excel using the $\text{NORMSDIST}(x)$.

Table III. Critical values for t-score

df =degree of freedom		Area (c)								
		0.25	0.15	0.125	0.10	0.05	0.025	0.0125	0.01	0.005
		Values of t_c								
01		1.000	1.963	2.414	3.078	6.314	12.71	25.45	31.82	63.66
02		0.816	1.386	1.604	1.886	2.920	4.303	6.205	6.965	9.925
03		0.765	1.250	1.423	1.638	2.353	3.182	4.177	4.541	5.841
04		0.741	1.190	1.344	1.533	2.132	2.776	3.495	3.747	4.604
05		0.727	1.156	1.301	1.476	2.015	2.571	3.163	3.365	4.032
06		0.718	1.134	1.273	1.440	1.943	2.447	2.969	3.143	3.707
07		0.711	1.119	1.254	1.415	1.895	2.365	2.841	2.998	3.499
08		0.706	1.108	1.240	1.397	1.860	2.306	2.752	2.896	3.355
09		0.703	1.100	1.230	1.383	1.833	2.262	2.685	2.821	3.250
10		0.700	1.093	1.221	1.372	1.812	2.228	2.634	2.764	3.169
11		0.697	1.088	1.214	1.363	1.796	2.201	2.593	2.718	3.106
12		0.695	1.083	1.209	1.356	1.782	2.179	2.560	2.681	3.055
13		0.694	1.079	1.204	1.350	1.771	2.160	2.533	2.650	3.012
14		0.692	1.076	1.200	1.345	1.761	2.145	2.510	2.624	2.977
15		0.691	1.074	1.197	1.341	1.753	2.131	2.490	2.602	2.947
16		0.690	1.071	1.194	1.337	1.746	2.120	2.473	2.583	2.921
17		0.689	1.069	1.191	1.333	1.740	2.110	2.458	2.567	2.898
18		0.688	1.067	1.189	1.330	1.734	2.101	2.445	2.552	2.878
19		0.688	1.066	1.187	1.328	1.729	2.093	2.433	2.539	2.861
20		0.687	1.064	1.185	1.325	1.725	2.086	2.423	2.528	2.845
21		0.686	1.063	1.183	1.323	1.721	2.080	2.414	2.518	2.831
22		0.686	1.061	1.182	1.321	1.717	2.074	2.405	2.508	2.819
23		0.685	1.060	1.180	1.319	1.714	2.069	2.398	2.500	2.807
24		0.685	1.059	1.179	1.318	1.711	2.064	2.391	2.492	2.797
25		0.684	1.058	1.178	1.316	1.708	2.060	2.385	2.485	2.787
26		0.684	1.058	1.177	1.315	1.706	2.056	2.379	2.479	2.779
27		0.684	1.057	1.176	1.314	1.703	2.052	2.373	2.473	2.771
28		0.683	1.056	1.175	1.313	1.701	2.048	2.368	2.467	2.763
29		0.683	1.055	1.174	1.311	1.699	2.045	2.364	2.462	2.756
30		0.683	1.055	1.173	1.310	1.697	2.042	2.360	2.457	2.750
Infinity		**0.674**	**1.036**	**1.150**	**1.282**	**1.645**	**1.960**	**2.241**	**2.326**	**2.576**

Source: Generated with EXCEL using the TINV(α, df) function.

Table IV. Chi-square critical values, $X_c^2 = X_\alpha^2(df)$

df	0.10	0.05	0.025	0.01	0.005	0.90	0.95	0.975	0.99	0.995
1	2.71	3.84	5.02	6.64	7.88	0.02	0.00	0.00	0.00	0.00
2	4.61	5.99	7.38	9.21	10.60	0.21	0.10	0.05	0.02	0.01
3	6.25	7.81	9.35	11.35	12.84	0.58	0.35	0.22	0.11	0.07
4	7.78	9.49	11.14	13.28	14.86	1.06	0.71	0.48	0.30	0.21
5	9.24	11.07	12.83	15.09	16.75	1.61	1.15	0.83	0.55	0.41
6	10.64	12.59	14.45	16.81	18.55	2.20	1.64	1.24	0.87	0.68
7	12.02	14.07	16.01	18.48	20.28	2.83	2.17	1.69	1.24	0.99
8	13.36	15.51	17.53	20.09	21.96	3.49	2.73	2.18	1.65	1.34
9	14.68	16.92	19.02	21.67	23.59	4.17	3.33	2.70	2.09	1.73
10	15.99	18.31	20.48	23.21	25.19	4.87	3.94	3.25	2.56	2.16
11	17.28	19.68	21.92	24.73	26.76	5.58	4.57	3.82	3.05	2.60
12	18.55	21.03	23.34	26.22	28.30	6.30	5.23	4.40	3.57	3.07
13	19.81	22.36	24.74	27.69	29.82	7.04	5.89	5.01	4.11	3.57
14	21.06	23.68	26.12	29.14	31.32	7.79	6.57	5.63	4.66	4.07
15	22.31	25.00	27.49	30.58	32.80	8.55	7.26	6.26	5.23	4.60
16	23.54	26.30	28.85	32.00	34.27	9.31	7.96	6.91	5.81	5.14
17	24.77	27.59	30.19	33.41	35.72	10.09	8.67	7.56	6.41	5.70
18	25.99	28.87	31.53	34.81	37.16	10.86	9.39	8.23	7.01	6.26
19	27.20	30.14	32.85	36.19	38.58	11.65	10.12	8.91	7.63	6.84
20	28.41	31.41	34.17	37.57	40.00	12.44	10.85	9.59	8.26	7.43
21	29.62	32.67	35.48	38.93	41.40	13.24	11.59	10.28	8.90	8.03
22	30.81	33.92	36.78	40.29	42.80	14.04	12.34	10.98	9.54	8.64
23	32.01	35.17	38.08	41.64	44.18	14.85	13.09	11.69	10.20	9.26
24	33.20	36.42	39.36	42.98	45.56	15.66	13.85	12.40	10.86	9.89
25	34.38	37.65	40.65	44.31	46.93	16.47	14.61	13.12	11.52	10.52
26	35.56	38.89	41.92	45.64	48.29	17.29	15.38	13.84	12.20	11.16
27	36.74	40.11	43.19	46.96	49.64	18.11	16.15	14.57	12.88	11.81
28	37.92	41.34	44.46	48.28	50.99	18.94	16.93	15.31	13.56	12.46
29	39.09	42.56	45.72	49.59	52.34	19.77	17.71	16.05	14.26	13.12
30	40.26	43.77	46.98	50.89	53.67	20.60	18.49	16.79	14.95	13.79

Source: Generated with EXCEL using the $\text{CHIINV}(\alpha, df)$ function.

Table V. F-ratios at 5% level of significance

$$F_c = F_{0.05}(v_1, v_2)$$

Area α

v_2	v_1									F_c
1	**1**	**2**	**3**	**4**	**5**	**6**	**7**	**8**	**9**	
1	161.45	199.50	215.71	224.58	230.16	233.99	236.77	238.88	240.54	241
2	18.51	19.00	19.16	19.25	19.30	19.33	19.35	19.37	19.38	19
3	10.13	9.55	9.28	9.12	9.01	8.94	8.89	8.85	8.81	8
4	7.71	6.94	6.59	6.39	6.26	6.16	6.09	6.04	6.00	5
5	6.61	5.79	5.41	5.19	5.05	4.95	4.88	4.82	4.77	4
6	5.99	5.14	4.76	4.53	4.39	4.28	4.21	4.15	4.10	4
7	5.59	4.74	4.35	4.12	3.97	3.87	3.79	3.73	3.68	3
8	5.32	4.46	4.07	3.84	3.69	3.58	3.50	3.44	3.39	3
9	5.12	4.26	3.86	3.63	3.48	3.37	3.29	3.23	3.18	3
10	4.96	4.10	3.71	3.48	3.33	3.22	3.14	3.07	3.02	2
11	4.84	3.98	3.59	3.36	3.20	3.09	3.01	2.95	2.90	2
12	4.75	3.89	3.49	3.26	3.11	3.00	2.91	2.85	2.80	2
13	4.67	3.81	3.41	3.18	3.03	2.92	2.83	2.77	2.71	2
14	4.60	3.74	3.34	3.11	2.96	2.85	2.76	2.70	2.65	2
15	4.54	3.68	3.29	3.06	2.90	2.79	2.71	2.64	2.59	2
16	4.49	3.63	3.24	3.01	2.85	2.74	2.66	2.59	2.54	2
17	4.45	3.59	3.20	2.96	2.81	2.70	2.61	2.55	2.49	2
18	4.41	3.55	3.16	2.93	2.77	2.66	2.58	2.51	2.46	2
19	4.38	3.52	3.13	2.90	2.74	2.63	2.54	2.48	2.42	2
20	4.35	3.49	3.10	2.87	2.71	2.60	2.51	2.45	2.39	2
21	4.32	3.47	3.07	2.84	2.68	2.57	2.49	2.42	2.37	2
22	4.30	3.44	3.05	2.82	2.66	2.55	2.46	2.40	2.34	2
23	4.28	3.42	3.03	2.80	2.64	2.53	2.44	2.37	2.32	2
24	4.26	3.40	3.01	2.78	2.62	2.51	2.42	2.36	2.30	2
25	4.24	3.39	2.99	2.76	2.60	2.49	2.40	2.34	2.28	2
26	4.23	3.37	2.98	2.74	2.59	2.47	2.39	2.32	2.27	2
27	4.21	3.35	2.96	2.73	2.57	2.46	2.37	2.31	2.25	2
28	4.20	3.34	2.95	2.71	2.56	2.45	2.36	2.29	2.24	2
29	4.18	3.33	2.93	2.70	2.55	2.43	2.35	2.28	2.22	2
30	4.17	3.32	2.92	2.69	2.53	2.42	2.33	2.27	2.21	2

Source: Generated with EXCEL using the $\text{FINV}(\alpha, v_1, v_2)$ function.

IMPORTANT FORMULAE

Sample mean

$$x = \frac{1}{n}\sum_{i=1}^{n} x_i$$

Population mean

$$\mu = \frac{1}{N}\sum_{I=1}^{N} X_I$$

Sample variance

By definition

$$s^2 = \frac{\sum_{i=1}^{n}(x_i - \bar{x})^2}{n-1}$$

Computing formula

$$s^2 = \frac{n\left(\sum_{i=1}^{n} x_i^2\right) - \left(\sum_{i=1}^{n} x_i\right)^2}{n(n-1)}$$

Sample standard deviation

$$s = \sqrt{s^2}$$

Population variance

$$\sigma^2 = \frac{\sum_{I=1}^{N}(X_I - \mu)^2}{N}$$

Population standard deviation

$$\sigma = \sqrt{\sigma^2}$$

Empirical rule

the interval $\bar{x} \mp s$ contains approximately 68% data values

the interval $\bar{x} \mp 2s$ contains approximately 95% data values

the interval $\bar{x} \mp 3s$ contains approximately 99% data values

Expected value of X

$$E(X) = \mu = \sum_{i} p_i x_i$$

Variance of X

By definition

$$\sigma^2 = \sum_{i} p_i(x_i - \mu)^2$$

Computing formula

$$\sigma^2 = \sum_{i} p_i x_i^2 - \mu^2$$

Binomial distribution

$$P(X = x) = \binom{n}{x} p^x q^{(n-x)}, \quad x = 0,\ 1,\ 2...,\ n$$

$$\text{Mean} = E(x) = \mu = np$$

$$\text{Variance} = \sigma^2 = npq$$

Poisson distribution

$$P(X = x) = \frac{e^{-\lambda}\lambda^x}{x!}, \quad x = 0, 1, 2,..., \infty$$

$$\text{Mean} = \lambda = np$$

Uniform distribution

$$X \sim U(a,\ b)$$

$$f(x) = \begin{cases} \dfrac{1}{(b-a)} & \text{if} \quad a \le x \le b \\ 0 & \text{otherwise} \end{cases}$$

$$\text{Mean} = \text{Medain} = \frac{a+b}{2}$$

$$\text{Variance} = V(X) = \sigma^2 = \frac{(b-a)^2}{12}$$

Normal distribution

$$X \sim N(\mu,\ \sigma)$$

$$f(X) = \frac{1}{\sqrt{2\pi}\sigma} e^{-\frac{1}{2}\left(\frac{X-\mu}{\sigma}\right)^2}, \quad -\infty < X < +\infty$$

$$Z = \frac{X - \mu}{\sigma}$$

$$f(Z) = \frac{1}{\sqrt{2\pi}} \exp\left\{ -\frac{Z^2}{2} \right\}, \quad -\infty < Z < +\infty$$

$$Z \sim N(0,\ 1)$$

Distribution of sample proportions

$$\mu_{\hat{p}} = P$$

$$\sigma_{\hat{p}} = \sqrt{\frac{P(1-P)}{n}}, \quad \hat{\sigma}_{\hat{p}} = \sqrt{\frac{\hat{p}(1-\hat{p})}{n}}$$

$$Z = \frac{\hat{p} - P}{\sqrt{\dfrac{P(1-P)}{n}}}$$

$$\hat{p} \sim N\left(P,\ \sqrt{\frac{P(1-P)}{n}}\right)$$

Distribution of sample means

$$\mu_{\bar{x}} = \mu$$

$$\sigma_{\bar{x}} = \frac{\sigma}{\sqrt{n}}, \quad \hat{\sigma}_{\bar{x}} = \frac{s}{\sqrt{n}}$$

$$Z = \frac{(\bar{x} - \mu)}{\sigma/\sqrt{n}}$$

$$\bar{x} \sim N\left(\mu, \ \sigma/\sqrt{n}\right)$$

Interval estimate

Single mean
(*large sample*)

$$\bar{x} \ \pm \ Z_{\frac{\alpha}{2}}\left(\sigma_{\bar{x}}\right)$$

Single proportion
(*large sample*)

$$\hat{p} \ \pm \ Z_{\frac{\alpha}{2}}\left(\hat{\sigma}_{\hat{p}}\right)$$

Single mean
(*small sample*)

$$\bar{x} \ \pm \ \left[t_{\alpha/2}(\mathrm{df} = n-1)\right](\hat{\sigma}_{\bar{x}})$$

Difference between
two means
(*large sample*)

$$(\bar{x}_1 - \bar{x}_2) \pm Z_{\alpha/2}\sqrt{\frac{\sigma_1^2}{n_1} + \frac{\sigma_2^2}{n_2}}$$

Pooled variance

$$s_p^2 = \frac{(n_1 - 1)s_1^2 + (n_2 - 1)s_2^2}{n_1 + n_2 - 2}$$

Difference between
two means
(*small sample*)

$$(\bar{x}_1 - \bar{x}_2) \pm \left[t_{\alpha/2}(df = n_1 + n_2 - 2)\right]\sqrt{s_p^2\left(\frac{1}{n_1} + \frac{1}{n_2}\right)}$$

Difference between
two proportions
(*large sample*)

$$(\hat{p}_1 - \hat{p}_2) \pm Z_{\alpha/2}\sqrt{\frac{p_1(1 - p_1)}{n_1} + \frac{p_2(1 - p_2)}{n_2}}$$

Testing hypotheses

Single proportion
(*large sample*)

$$Z_{\mathrm{cal}} = \frac{\hat{p} - P_0}{\sqrt{\frac{P_0(1 - P_0)}{n}}}$$

Single mean
(*large sample*)

$$Z_{cal} = \frac{(\bar{x} - \mu_0)}{\sigma/\sqrt{n}}$$

Single mean
(*small sample*)

$$t_{cal} = \frac{(\bar{x} - \mu_0)}{s/\sqrt{n}}$$

Difference between
two means
(*large sample*)

$$Z_{cal} = \frac{(\bar{x}_1 - \bar{x}_2) - 0}{\sqrt{\frac{\sigma_1^2}{n_1} + \frac{\sigma_2^2}{n_2}}}$$

Difference between two means *(small and independent samples)*	$t_{cal} = \dfrac{(\bar{x}_1 - \bar{x}_2) - d}{\sqrt{s_p^2\left(\dfrac{1}{n_1} + \dfrac{1}{n_2}\right)}}$
Difference between two means *(small and dependent samples)*	$t_{cal} = \dfrac{\bar{d} - 0}{s_d / \sqrt{n}}$
Difference between two proportions *(large and independent samples)*	$Z_{cal} = \dfrac{(\hat{p}_1 - \hat{p}_2) - 0}{\sqrt{\hat{p}\hat{q}\left(\dfrac{1}{n_1} + \dfrac{1}{n_2}\right)}}$
Pooled proportion	$\hat{p} = \dfrac{n_1 \hat{p}_1 + n_2 \hat{p}_2}{n_1 + n_2} = \dfrac{x_1 + x_2}{n_1 + n_2}, \quad \hat{q} = (1 - \hat{p})$
Pooled standard error for the difference between two proportions	$\hat{\sigma}_{(\hat{p}_1 - \hat{p}_2)(\text{pooled})} = \sqrt{\hat{p}\hat{q}\left(\dfrac{1}{n_1} + \dfrac{1}{n_2}\right)}$
Chi-square test for contingency table	$X_{cal}^2 = \sum\limits_{i=1}^{R} \sum\limits_{j=1}^{C} \dfrac{(O_{ij} - E_{ij})^2}{E_{ij}}$
Chi-square test for single variance	$X_{cal}^2 = (n-1)s^2 / \sigma_0^2$
Equality of two Variances	$F_{cal} = s_1^2 / s_2^2$

Correlation analysis

$$r = \frac{n(\sum x_i y_i) - (\sum x_i)(\sum y_i)}{\sqrt{n(\sum x_i^2) - (\sum x_i)^2}\sqrt{n(\sum y_i^2) - (\sum y_i)^2}}$$

$$r = \frac{1}{n-1} \sum_{i=1}^{n}\left(\frac{y_i - \bar{y}}{s_y}\right)\left(\frac{x_i - \bar{x}}{s_x}\right)$$

$$-1 \leq r \leq +1$$

Regression analysis

$$b = \frac{n(\sum x_i y_i) - (\sum x_i)(\sum y_i)}{n(\sum x_i^2) - (\sum x_i)^2}$$

$$b = r\frac{s_y}{s_x}$$

$$a = \bar{y} - b\bar{x}$$

$$\hat{y} = a + bx^*$$

$$e_i = (y_i - \hat{y}_i)$$

$$r^2 = 1 - \sum_{i=1}^{n} e_i^2 \bigg/ \sum_{i=1}^{n}(y_i - \bar{y})^2$$

BIBLIOGRAPHY

Let us go to the:

to find the following books:

Aliga, M. and Gunderson, B. (2003). *Interactive Statistics*, 2nd ed. Prentice Hall.

Best, J. (2001). *Damned lies and statistics*. University of California Press.

Bulmer, M.G. (1979). *Principles of statistics*. Dover Pubns.

Downing, D. and Clark, J. (1997). *Statistics: The easy way*. Barrons Educational Series.

Geis, I. and Huff, D. (1984). *How to lie with statistics*. W.W. Norton and Company.

Gonick, L. and Smith, W. (1994). *The cartoon guide to statistics*. HarperResource.

Graham, A. (2003). *Teach yourself statistics*. McGraw-Hill.

Jaisingh, L.R. (2000). *Statistics for the utterly confused*. McGraw-Hill.

*Kendall, M.G. (1963). Ronald Aylmer Fisher, 1890-1962. *Biometrika*, 50 (1-2), 1-15.

Langley, R.A. (1971). *Practical statistics simply explained.* Dover Publications.

Maxwell, N. (2004). *Data Matters: Conceptual statistics for a random world.* Key College Publishing, California.

Moore, D.S. and McCabe, G.P. (1999). *Introduction to the Practice of Statistics*, 3rd ed., DWH Freeman and Company.

Paulos, J.A. (1996). *A mathematician reads the newspaper.* Anchor.

*Pearson, E.S. (1936). Karl Pearson: An Appreciation of Some Aspects of His Life and Work. *Biometrika,* 28(3-4), 193-257.

Rumsey, D. (2003). *Statistics for dummies.* John Wiley and Sons (Series for Dummies).

Salkind, N.J. (2004). *Statistics for people who hate statistics.* SAGE Publications.

Slavin, S.L. and Slavin, S. (1998). *Chances Are: The only statistics book you'll ever need.* Medison Books.

Voelker, D.H. and Orton, P.Z. (2001). *Statistics.* Cliffs Notes.

Wild, C.J. and Pfannkuch, M. (1999). Statistical thinking in empirical enquiry. *International Statistical Review*, 67(3), 223-265.

*Pictures of Sir R.A. Fisher and Professor Karl Pearson in the text are printed with permission. For detail refer to the Preface.

HANDY SUBJECT INDEX

The following subject index may help you!

T

U

V

Z

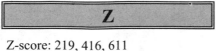